REPRODUCTIVE PHYSIOLOGY OF PLANTS

ADVANCES IN PLANT PHYSIOLOGY SERIES—10

Reproductive Physiology of Plants

D. K. Arora
Seema Gupta

Anmol Publications Pvt Ltd
NEW DELHI-110002 (INDIA)

Anmol Publications Pvt. Ltd.
4374/4B, Ansari Road, Daryaganj,
New Delhi-110002

Reproductive Physiology of Plants

First Edition 1996

ISBN 81-7488-244-8

PRINTED IN INDIA

Published by J. L. Kumar for Anmol Publications, Pvt. Ltd. New Delhi and printed at Chetna Offset Press, Delhi.

Preface

'Reproductive Physiology of Plants' offers a comprehensive account of the fundamental principles of plant physiology. It incorporates and organizes the information concerning Reproductive Physiology of Plants from relevant authentic sources. It is an humble attempt to present a correct and precise account of the subject matter of this important branch of Botany which forms an integral part of the studies undertaken by the undergraduate and post-graduate students of this subject. The principles have been outlined in considerable detail alongwith this fundamental facts and theories that explain the life processes of plants. The main object of the present book is to provide a comprehensive and well illustrated account of the prescribed subject matter in a very simple and lucid language and concise manner yet—still maintaining its desired academic standard.

Most of illustrations have been taken from professional papers, standard books and journals. Subject matter has also been obtained from a large number of standard books, reviews and research papers by internationally renowned workers and established authors. That is why there can be no claim of originality except in the manner of treatment.

Though the authors have taken special care to present a current account, yet they are fully aware of their limitations, and the readers may come across the mistakes of various types. For all types of mistakes they extend their due apology.

The authors express their thanks to their friends and colleagues whose continuous inspirations have invited them to bring out this text.

Contents include : Structure of Flower, Pollination, Fertilization, Production of Seeds, Germination of Seeds, Physiology of A sexual Propagation, Physiology of Grafting and Budding. and so on.

Authors

Contents

1

Structure of Flower

The flower may be regarded as a leafy shoot highly specialised for the performance of reproductive functions. The function of a flower is to produce seed and fruit, and the various parts (stem and leaf organs) are specially suited to the performance of that function. The essential structures in the flower, therefore, are the organs, which are more immediately concerned in the production of seed. We cannot appreciate the significance or morphological nature of the organs without a knowledge of the reproductive organs of the Vascular Cryptogams and Gymnosperms, and some of the terms used in this chapter will only be understood when these plants have been considered.

The axis (stem portion) of the flower usually shows two regions—the pedicel, and the receptacle. The pedicel is, popularly, the stalk of the flower. It may be present or absent. If present, the flower is *pedicellate* ; if absent, *sessile*. The receptacle is the portion of the axis to which the floral leaves are attached. In many flowers there are four sets or series of floral leaves. To the outside are the sepals ; collectively, they constitute the calyx. Internal to these are the petals, constituting the corolla. Then come the stamens forming the androecium ; and finally, in the centre of the flower, are the carpels, forming the gynaeceum or pistil.

The buttercup (*Ranunculus*) flower will serve as a convenient type for introduction to these structures (Fig. 1.1). In the

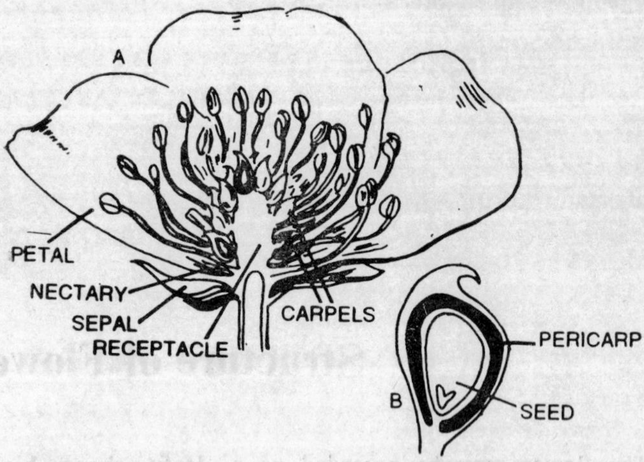

Fig. 1.1. Buttercup.
A. Longitudinal section, Flower ; B. Longitudinal section, Achene.

buttercup the carpels are separate from each other, and *each* shows a hollow basal portion called the ovary, above which are the parts known as style and stigma (Fig. 1.1B). In many flowers the carpels are united and form a single ovary.

The following facts support the above view of the morphological character of the flower : (*a*) the flower, like an ordinary foliage-shoot, arises as a bud, very often in the axil of a leaf (*bract*). (*b*) The receptacle has the general structure of a stem, and the sepals and petals in their structure and development resemble leaves. (*c*) While in most cases the stamens and carpels are quite unlike leaves, there are certain conditions in which they become distinctly leaf-like. Thus, in many cultivated flowers, *e.g.* rose, the stamens are transformed into petals ; in double cherry the gynaeceum is represented by a tuft of small green leaves ; in water lily there is a gradual transition between petals and stamens.

Ontogeny of the Flower. The flower develops from an apical meristem that differs in only minor ways from that of a leafy stem. The early stages of floral-appendage development are closely like those of leaves. All organs develop laterally on the apical meristem, initiated by periclinal divisions below the surface layer. The sequence of development of the whorls of organs varies but is commonly acropetal. All organs have, at least for a time, an apical meristem, and those that are laterally expanded have marginal initials. Procambium develops acropetally in all organs as does the phloem. The xylem matures either acropetally or both acropetally and basipetally from one or more points of beginning near the base of the organ. In carpels, especially in syncarpous ovaries, connection of the early-formed xylem with the xylem of the receptacle may be long delayed, even until the fruit is partly developed. The carpel remains an immature organ until the fruit is mature. In various parts it retains meristematic tissues of different types. The other organs are usually mature in the flower. Where fusion is present, the fusion may be ontogenetic, as it is frequently between carpel margins, or it may be phylogenetic, the fused organs arising from a common meristem, as in typical inferior ovaries. The latter type of union has been called "congenital fusion." In adjacent organs that arise as independent organs from independent apical meristems, for example the members of a gamopetalous corolla, the initiating regions may shift from the apex to the base and merge, and the organs then continue growth as a single structure with only the vascular tissue showing its compound nature. Adnate organs, in any number, may arise independently but later continue growth from a single meristematic region as a fused but superficially single structure. Such apparently simple structures may show evidence of their compound nature in the presence of independent vascular bundles for every organ within the common structure, or the vascular skeletons of the organs may also be so intimately fused as to show little or no evidence of their multiple morphological nature.

THE VASCULAR SKELETON OF THE FLOWER

The Floral Stele

In structure, the pedicel is a typical stem with a ring of vascular bundles or an unbroken cylinder of vascular tissue. As the stele enters the receptacle, it conforms in shape to the shape of that part of the flower, commonly expanding and then constricting in its upper part. From the receptacular stele depart the vascular traces to the various floral organs, traces which in origin, structure, and behaviour are similar to those of leaves. Gaps accompany the exit of these traces, and the gaps and the crowding of the organs break up the cylinder into a network of strands. Often the vascular cylinder is further dissected, as in the stems of herbaceous plants with reduced vascular cylinders. The receptacular stele may be simple but is usualiy highly complex.

The traces to sepals, petals, stamens, and carpels are given off successively from the stele in whorls or in spirals, according to the manner of arrangement of these organs in the flower. (Many organs that to external inspection are apparently whorled, such as the petals of some Ranunculaceae and the stamens of some Rosaceae, can be seen from anatomical evidence to be arranged in flat spirals.) The distal bundles of the receptacular stele commonly become the traces of the uppermost carpels, but in many floral types stelar bundles continue beyond the point where the last traces are given off and gradually fade out in the top of the receptacle. Such bundles are vestigial and usually consist of procambium or phloem only.

The Inflorescence

The floral, or reproductive region of the plant is usually distinctly marked off from the foliage, or vegetative region, and is known as the inflorescence. Sometimes the main vegetative-axis of the plant ends in a single terminal flower, *e.g.* tulip and wood anemone. Here the flower is said to be solitary and terminal. In other cases, the flowers are developed singly in the axils of ordinary foliage leaves, and are called solitary and axillary. These are simple types of inflorescence. Usually the flowers are aggregated on a more or less complex branch-system. According to the nature of the branching, and other features, many different kinds of such inflorescence are recognised.

The main or primary axis of the inflorescence, together with any secondary axes which may be developed (apart from the pedicels of the flowers), is called the peduncle. This term is applied instead of pedicel to the stalks of solitary terminal, and solitary axillary flowers. If thc peduncle is an unbranched leafless axis which arises from the midst of radical leaves and bears flowers at its apex, it is called a scape, *e.g.* cowslip and members of the Amaryllidaceae.

Bracts, etc.

When the flower arises as a lateral bud, the axis on which it is borne is called the mother-axis. This may or may not be the primary axis of the inflorescence. The side of the flower which is towards the mother-axis (or towards the growing point of the mother-axis) is said to be posterior ; the side away from the motheraxis is anterior. In a solitary terminal flower it is evident that these terms are not applicable.

If the flower arises in the axil of a leaf-structure, this leaf-structure is called a bract.

Bracts present great variety of form and colour. The bracts may be ordinary foliage leaves, as in solitary axillary flowers, or may more or less differ from them. Frequently they are small, green, and scale-like. In many plants they are reduced to small, tooth-like structures. When they are not green, but coloured like the petals of a flower, they are said to be *petaloid*. In many flowers the flower-stalk bears small out-growths of the nature of reduced leaves. These are called bracteolcs. When present, there are *usually* two in Dicotyledons, placed laterally, and one in Monocotyledons, situated on the posterior side.

Perianth

The outer series of floral leaves, distinct from stamens and carpels, constitute the perianth. Flowers without perianth are naked or achlamydeous. Flowers with a single perianth series are haplo- or monochlamydeous and the floral leaves (tepals) are either bracteoid or petaloid. Flowers with a double perianth (two series of floral leaves) are diplochlamydeous. When both series of floral leaves (tepals) are similar they are homoiochlamydeous, but fusion of the two series may render them apparently haplochlamydeous. When the two series of floral leaves are dissimilar, heterochlamydeous, the outer series is sepaloid (sepals, calyx) and the inner series petaloid (petals, corolla). When petals are absent as a result of abortion the flower is apopetalous.

The Essential Organs

The androecium and the gynaeceum, because they bear the reproductive bodies, microspores (pollen-grains) and ovules, necessary for the production of seed, are called the essential organs.

If both are present *in the same flower* (the rule in Angiosperms) the flower is hermaphrodite or *bisexual* (symbol ⚥).

When they are borne on different flowers, as is sometimes found, the flowers are imperfect or *unisexual.* The unisexual flowers bearing the stamens are male (♂) or *staminate* ; those bearing carpels, female (♀) or *pistillate.* If staminate and pistillate flowers are borne on the same plant (*e.g.* hazel), the plant is monoecious ; if on different plants (*e.g.* willow and some species of campion), dioecious. A plant is *polygamous* if it bears staminate, pistillate, and hermaphrodite flowers (*e.g. Fraxinus*, *Mangifera*, *Anacardium*, *Rhus*, etc.). Flowers in which both stamens and pistil have been lost are *neuter* (*e g.* ray florets of *Senecio* species, *Volutarella*, *Helianthus*).

Floral Phyllotaxis

In most flowers the series of floral leaves are arranged in whorls, and the phyllotaxis is *cyclic.* Sometimes, however, all the floral leaves are in a spiral (*e.g.* cactus) and the flower is said to be *acyclic.* If some of the series are cyclically arranged, others spirally, the flowers are *hemicyclic.* In buttercup, for example, the calyx and corolla are whorled, while the stamens and carpels are spiral.

Number of Parts

In general, then, flowers have four definite series of floral leaves—calyx, corolla, androecium and gynaeceum—often with the same number of parts in each series. This, however, is far from true for many flowers where the number of parts in a series may be increased, or much reduced by the fusion of parts, or the loss of one or more parts from any series.

The following examples will illustrate these points ; *Viola* has five sepals, five petals, five stamens, three carpels ; the pea has five sepals, five petals, ten stamens, one carpel ; mustard has four sepals in two whorls, four petals in one whorl, six stamens in an outer whorl of two and an inner whorl of four, two sterile and two fertile carpels ; many flowers have a large number of stamens in several whorls (*e.g. Rosa*). It should be noticed that owing to the abbreviation of the receptacle and other causes, it is often difficult to distinguish the separate whorls, *e.g.* the two whorls of sepals in mustard, the two whorls of stamens in pea.

Neglecting the reduction of parts met with in particular series, and more especially in the gynaeceum, we find that in Dicotyledons the series of floral leaves are, *as a rule*, arranged in twos, fours, or fives, or multiples of these numbers. In other words, the arrangement is *dimerous*, *tetramerous*, or *pentamerous*, rarely trimerous. The *trimerous* arrangement, *i.e.* in threes or multiples of three, is characteristic of Monocotyledons.

Alternation of Parts

The general rule is that the leaves of the different series alternate in position with each other—the petals alternate with the sepals, the stamens with the petals, etc. If there are several whorls of stamens, these whorls alternate with each other.

But there are exceptions. In spiral flowers, the parts are sometimes superposed. In cyclic flowers the departure from regular alternation arises from various causes. In *Primula*, for example, there are five sepals, five petals, five stamens, and the stamens are opposite to the petals (*antipetalous*). Sometimes, where there are two alternating whorls of stamens, the outer whorl is opposite the petals. This is known as the *obdiplostemonous* condition. The carpels, where fewer in number, clearly cannot alternate with the leaves of the outer series.

Regular and Irregular Flowers

In regular flowers the parts in each series have the same size and form, *i.e.* the sepals resemble each other, so also the petals, etc. Irregular flowers are those in which some of the floral leaves in any one series have a different shape or size from the others—for example, the petals of pea or violet.

Floral Symmetry

Flowers may be radially symmetrical (actinomorphic), isobilateral (zygomorphic), or asymmetrical. The planes of symmetry may be median or antero-posterior, diagonal, or lateral (Figs. 1.2 and 1.3). Zygomorphy is frequently due to

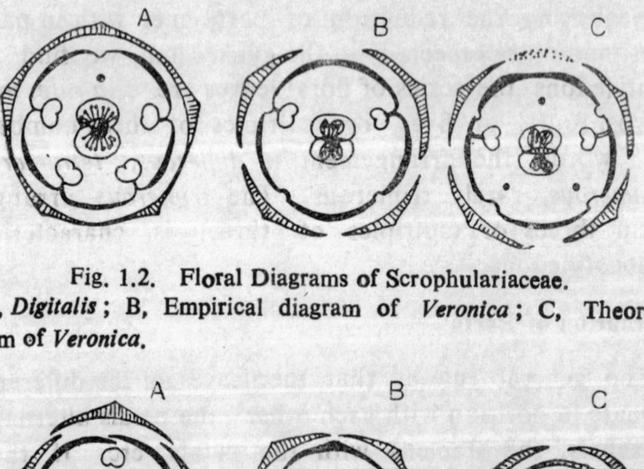

Fig. 1.2. Floral Diagrams of Scrophulariaceae.
A, *Digitalis* ; B, Empirical diagram of *Veronica* ; C, Theoretical diagram of *Veronica.*

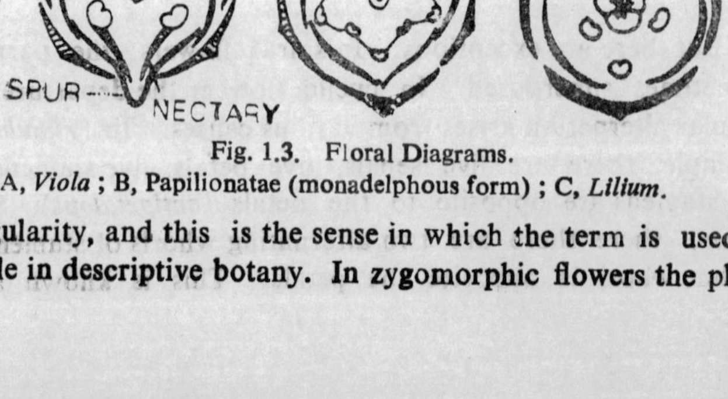

Fig. 1.3. Floral Diagrams.
A, *Viola* ; B, Papilionatae (monadelphous form) ; C, *Lilium.*

irregularity, and this is the sense in which the term is used as a rule in descriptive botany. In zygomorphic flowers the plane of symmetry is, in most cases, antero-posterior or median, *i.e.* it is the plane passing through the anterior and posterior sides of the flower, *e.g.* pea, violet (Fig. 1.3 A and B). Asymmetrical flowers are usually spiral, as for example in the Cactaceae.

The Receptacle : Insertion of Floral Leaves

The receptacle is nearly always short. Only occasionally is it elongated between the whorls of floral leaves, as in some species of campion (Fig. 1.4). The form of the receptacle varies considerably. It may be convex and more or less dilated, or flattened, or hollow and cup-shaped. The insertion of floral leaves varies according to the form of the receptacle.

In many flowers (*e.g. Ranunculus, Silene, Papaver*) the receptacle is more or less convex. The gynaeceum is developed at the apex of the receptacle ; the stamens, petals, and sepals, are inserted, *in order*, on the side of the receptacle below the gynaeceum. This is to hypogynous arrangement (Fig. 1.4 A).

Suppose now that the receptacle is not convex, but forms a flattened circular disc. The apex of the receptacle is, of course, in the middle of the disc, and the flattened form is due to the sides of the receptacle having grown up to the same level. The gynaeceum is developed in the middle of the disc, and the sepals, petals, and stamens round the rim or margin. They are not *underneath* the gynaeceum, but *round about* it. Hence this is called a perigynous arrangement (Fig. 1.4 C). Sometimes the carpels are borne on a conical protuberance in the middle of the disc ; this would represent a continued growth of the apex (*e.g. Fragaria, Rubus*, etc., Fig. 1.4 B).

It is with the perigynous condition that the student will experience most difficulty ; there are so many degrees of it. The receptacle may not be flat, but hollowed out, and more or less cup-like. This is due to the sides of the receptacle continuing to grow above the apex, which lies at the bottom of the cup (Fig. 1.4 D). The carpels (gynaeceum) are developed in the cup ; the sepals, petals, and stamens from the rim of the cup. This also is a perigynous condition. A still more

extreme form of perigyny is found in wild rose (Fig. 1.4 E). Here there is a very deep cup.

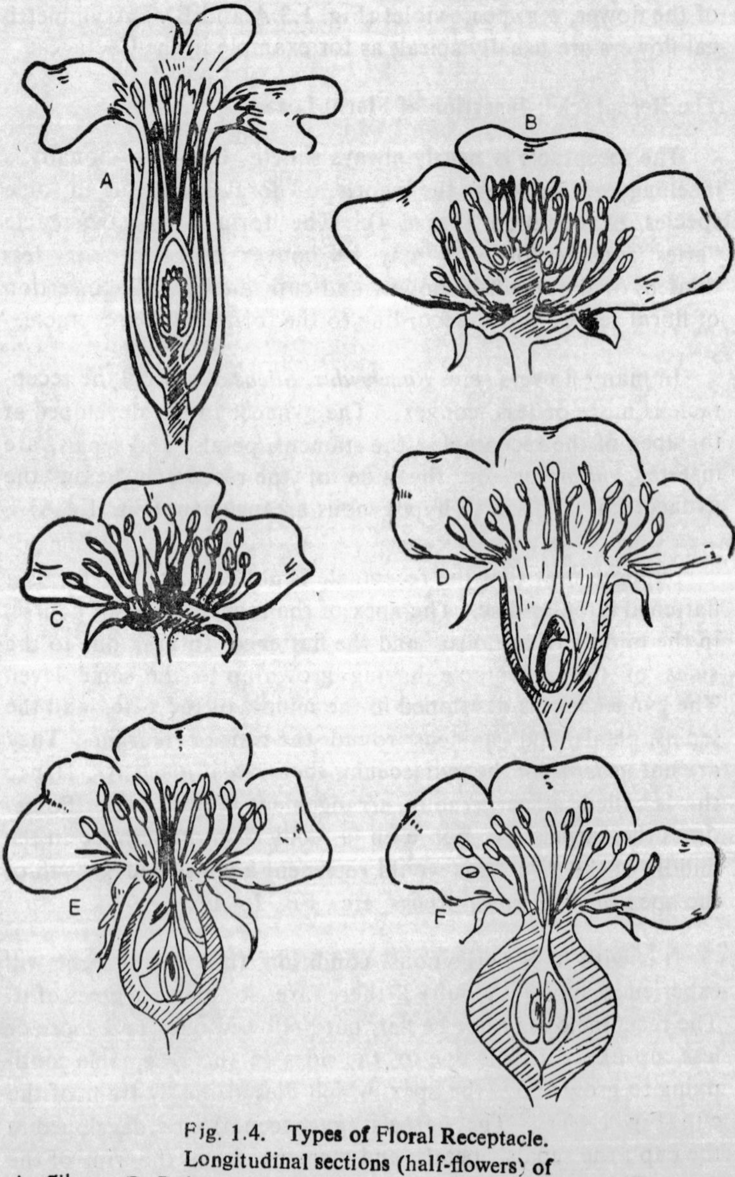

Fig. 1.4. **Types of Floral Receptacle.**
Longitudinal sections (half-flowers) of
A. Silene ; B. Rubus ; C. Filipendula ; D. Prunus ; E. Rosa ; F. Pyrus.

Finally, in the epigynous condition (Fig. 1.4 F) the receptacle forms a deep cup as in the extreme forms of perigyny ; but the carpels are from the first adherent to the receptacle, which is for this reason considered as part of the ovary. Thus in epigynous flowers the sepals, petals, and stamens are inserted *on* the gynaeceum. In the perigynous condition the receptacle wall remains distinct from the ovary.

Nectaries

The receptacle frequently bears a fleshy or glandular outgrowth, such as is found on the top of the inferior ovary in Umbelliferae and in the ivy (*Hedera*). This is termed the disc. In *Rubus* the disc lines the outer concave part of the receptacle. Usually the disc is lobed (vine), and frequently it secretes nectar. Nectaries may, however, develop from, or upon, any part of the flower. Thus in *Viola* the outgrowths borne by two stamens secrete nectar into a hollow spur borne by the anterior petal. In buttercup a small scale at the base of each of the petals covers a pocket-shaped nectary, while in christmas rose there are hollow tubular nectaries between the androecium and perianth. Nectaries occur upon the gynaeceum of *Gentiana*, and each sepal of *Althaea* bears a nectary on its inner surface.

The Calyx

The calyx may consist of numerous sepals showing a primitive spiral arrangement, as in cactus and water lily ; but usually it consists of from two to five sepals. If the sepals are free, the calyx is polysepalous. When they are united laterally, however slightly, the calyx is gamosepalous. The gamosepalous condition is due, not to the actual fusion of originally separate sepals, but to common basal growth during development. In all hypogynous and perigynous arrangements the calyx is described as *inferior* ; in the epigynous flower the calyx is described as *superior*.

In some flowers, *e.g. Fragaria* (strawberry), the sepals are stipulate. The stipules fuse in pairs between the sepals and produce an outer series of small sepal-like structures, forming what appears like an outer calyx. This is known as the epicalyx (Fig. 1.5).

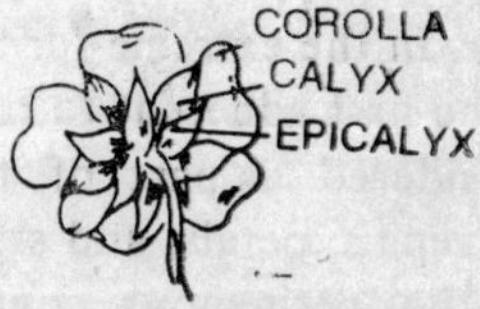

Fig. 1.5. Epicalyx, Calyx and Corolla of Strawberry.

An epicalyx may also be produced by the aggregation of bracts or bracteoles beneath the calyx, *e.g.* Malvaceae, shoe flower.

The calyx usually has a protective function. It commonly serves to protect the parts of the young flower in the bud. When the flower opens the calyx may fall off, *e.g.* in poppy, in which case it is said to be *caducous*, or the sepals fold back as in the wild rose. The calyx is *deciduous* if it falls off when the flower withers. But frequently it *persists* till fruiting takes place and protects the young fruit, which is developed from the ovary of the flower (*e.g.* strawberry and Labiatae). A gamosepalous calyx not only affords a more efficient protection to the flower-bud than a polysepalous one, but also gives support and protection to the base of the adult flower and to the developing fruit. Hence a gamosepalous calyx is never caducous.

In Umbelliferae, where the flowers are closely aggregated, and in many Compositae, where in addition they are surrounded by a ring of bracts, a protective calyx is not required and is either very small or quite absent.

The calyx, however, may perform other functions. Thus, in many Compositae (*e.g. Taraxacum*, *Senecio*, and *Cnicus*) there is a rudimentary calyx represented by hairs, forming a silky pappus, which undergoes further development after flowering, and serves to disperse the fruit. In some flowers also the sepals instead of being green, as is usually the case, are brightly coloured, and attractive. In this case the calyx is described as petaloid (Fig. 1.6).

In a polysepalous calyx the outline of the individual sepals is described in the same terms as are used for ordinary foliage

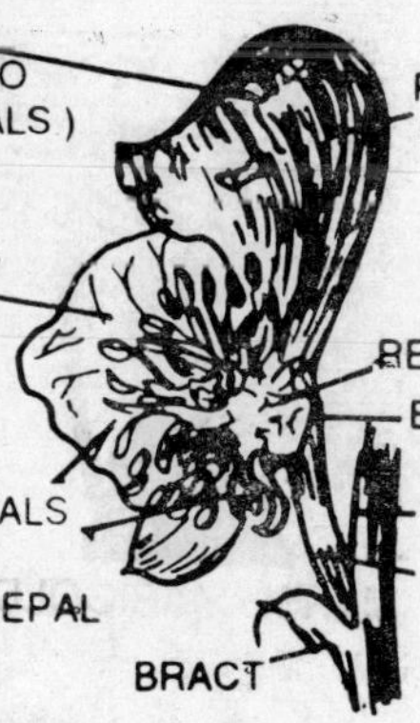

Fig. 1.6. Longitudinal Section, Flower of *Aconitum napellus*.

leaves. The number of sepals in a gamosepalous calyx is usually indicated by divisions or teeth. If the divisions pass almost to the base of the calyx, it may be described according to their number as 3-, 4-, 5-*partite* ; if about half-way down, 3-, 4-, 5-*fid* ; if the divisions are small, 3-, 4-, 5-*toothed.*

COROLLA

It is in the corolla where the greatest diversity of shape, colour and arrangement occurs.

In some flowers the petals are entirely absent, for example, the willow (*Salix* species). In many cases the petals join to form a tube, as in the primrose (Fig. 1.7). The number of petals, too, varies considerably, even more than in the case of the sepals. For example, the wallflower has four, the pink (*Dianthus* species) and the buttercup, five, and some, for example, the white water-lily (*Nymphaea alba*), have an indefinite number.

Some flowers are regular in the arrangement of their petals. For example, in the case of the butter cup or the wallflower, the petals and, indeed, all the floral organs are symmetrical about any axis. That is, it does not matter through what vertical plane the flower is cut, the two halves produced are the mirror images of each other. Such regular flowers are said to be *actinomorphic* (Fig. 1.7).

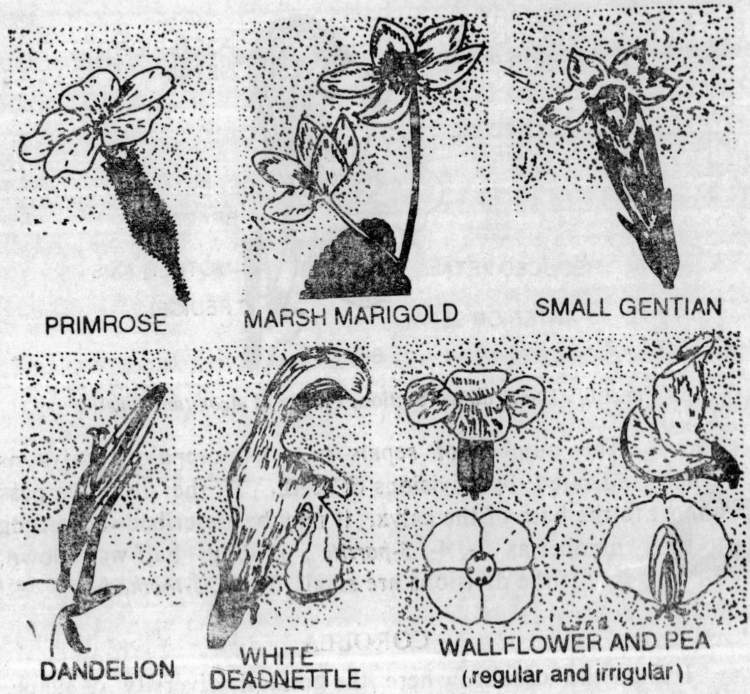

Fig. 1.7. **Types of Floral Form.**

On the other hand, many flowers are irregular. In the sweet pea (*Lathyrus odoratus*), for example, there are five petals, but they are not all of the same shape. Looking straight towards the inside of the flower, there is one large petal standing up at the back. It is larger and more spreading than any of the others, and is called the standard. Then, there are two wing-like petals, one on each side of the standard. Each is called a wing. At the bottom are two still smaller petals, facing each other and appearing similar to a ship's keel. The two together are therefore called the keel. In this flower it is quite obvious that there is only one vertical plane through which the flower could be cut in order to produce two symmetrical halves. The plane would pass down the middle of the standard and between the two wings and the two petals forming the keel. Such irregular flowers as these are said to be *zygomorphic*

(Fig. 1.7). Violets (*Viola* species) and the white deadnettle (*Lamium album*) are also examples of zygomorphic flowers.

In a large number of flowers, it is impossible to distinguish a separate calyx and corolla. Good examples of this are the tulip, bluebell, crocus, lily, etc. There are, however, two whorls of brightly coloured segments, an outer one of three and an inner one of three, which alternate with the three outer ones. In such examples, where there is no differentiation into sepals and petals, the two whorls are known collectively as the *perianth.*

The corolla may consist of spirally arranged, free, non-coherent petals, *e.g.* Cactaceae ; or of a single whorl (*Digitalis, Malva*) or, more rarely, a double whorl of petals (*Papaver*). In water lily the petals are arranged in a close spiral.

The corolla may be *polypetalous* or *gamopetalous* (cf. calyx), *regular* or *irregular* and, as it to a large extent determines the symmetry of the flower, the terms *zygomorphic actinomorphic,* are applied to it. According to the insertion of petals the corolla is described as hypogynous, perigynous, or epigynous.

The corolla serves, in most cases to attract insects to the flower in connexion with the process of pollination. It also protects the stamens and carpels, especially when the petals are united in the form of a tube enclosing these essential organs. The tube also serves as a receptacle for nectar. After fertilisation the seeds begin to develop, and the corolla is usually shed. The withered corolla, however, may persist in a few cases.

The petals are usually brightly coloured, but sometimes green (sepaloid). They may be absent, *e.g. Alchemilla* and some Ranunculaceae (*Clematis, Anemone*) or represented by nectar-secreting structures, *e.g. Aconitum* and *Helleborus.*

In a polypetalous corolla the outlines of the individual petals are described in the same terms as are used for the foliage leaf, and, as in the calyx, the gamopetalous corolla may be described as 3, 4, 5-partite,-fid, or toothed.

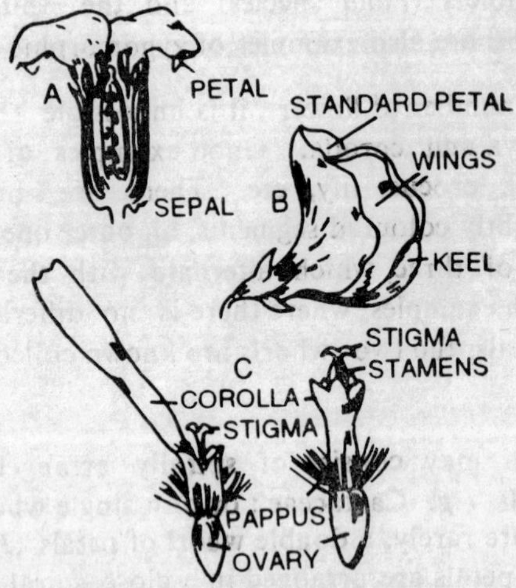

Fig. 1.8. Forms of Flowers. (*a*) *Cheiranthus* (Cruciferae), longitudinal section ; (*b*) *Lathyrus* (Papilionatae), external ; (*c*) Ligulate and tubular florets of *Aster* (Compositae).

The following special terms are applied to polypetalous corollas ; ***Cruciform*** where the corolla consists of four clawed petals arranged crosswise, *i.e.* in the diagonal planes of the flower (*e.g.* mustard and Cruciferae generally, see Fig. 1.8 *a*) ; *rosaceous* (Fig. *b-e*), if it consists of five spreading petals, not clawed, and attached perigynously (Rosaceae) ; *caryophyllaceous* (Fig. 1.4 *a*), if it consists of five clawed petals, with spreading limbs attached hypogynously to the receptacle inside a slender tubular calyx (pinks and many Caryophyllaceae) ; *papilionaceous* (from the supposed resemblance to a butterfly), if it consists of five petals, one large—the standard, two lateral—wings, and two fused to form a boat-shaped structure —the keel, *e.g.* pea and other members of the Papilionatae (Leguminosae, Fig. 1.8 *b*).

The Corona

This is the term applied to the whole series of outgrowths developed on the corolla or perianth of certain flowers. In *Narcissus*, where the perianth is gamophyllous, the outgrowths are coherent, and the corona is cup-shaped. Ak or akanda *Calotropis gigantea* furnishes a good example of a corona.

Aestivation

This has already been referred to. The aestivation of the perianth (or calyx and corolla) only can be studied. The folding of the individual floral leaves is described in the same terms as are used for foliage leaves. The aestivation of calyx or corolla may be *valvate*, *imbricate*, or *contorted (twisted)*. Aestivation may be recognised either by taking transverse sections of young flower-buds, or carefully removing the young floral leaves one after the other.

The Androecium

A stamen (Fig. 1.9) consists of three parts—filament, anther, and connective. The anther is usually two *lobed*, and

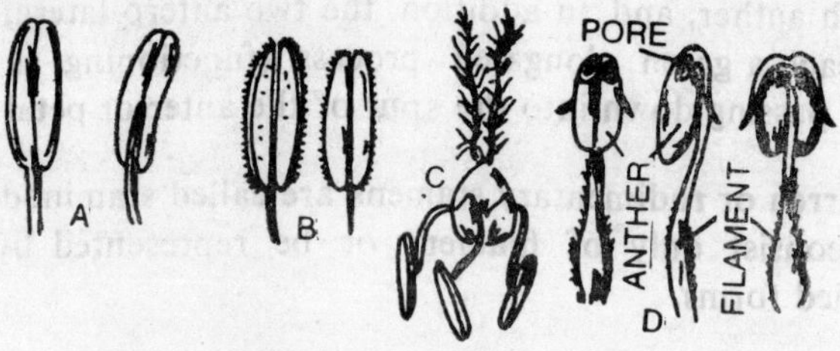

Fig. 1.9. **Stamens, External View. (*a*) Adnate, dorsifixed (*Nicotiana*); (*b*) Innate, basifixed (*Datura*); (*c*) Versatile, dorsifixed (*Triticum*); (*d*) Poricidal dehiscence (*Arctostaphylos*).**

contains microspores or pollen-grains. These lie in four cavities, the microsporangia or pollen-sacs (Fig. 1.10), of which there are two in each anther-lobe.

The anther-lobes are connected by a strip of tissue containing a vascular bundle. This is the *connective*. It is usually narrow, and the anther-lobes lie close together, but may be elongated so that the lobes are widely separated, as in some Labiatae.

In a few cases (*e.g. Malva, Corylus, Carpinus*) the stamens when quite young undergo division or segmentation, and thus

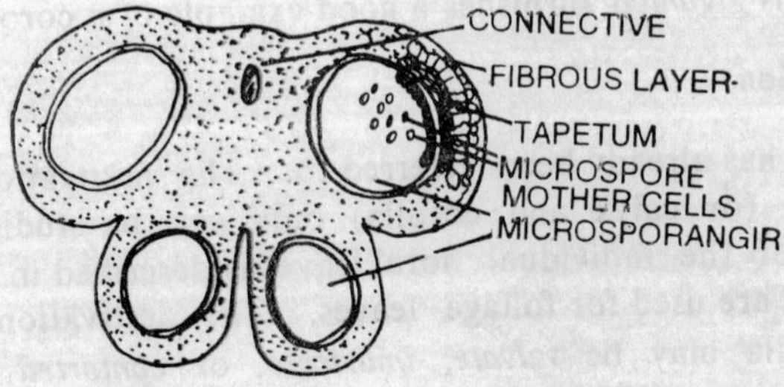

Fig. 1.10. **Transverse section, young undehisced anther of *Lilium*.**

in the fully developed flower the anthers have only one anther-lobe with two pollen-sacs.

Sometimes special appendages are developed on stamens. These generally arise as outgrowths of the connective. In *Viola* there is a membranous orange-coloured outgrowth on top of each anther, and, in addition, the two antero-lateral stamens have each a green elongated process (functioning as a nectar gland) passing down into the spur of the anterior petal.

Barren or rudimentary stamens are called staminodes. They may consist only of filament or be represented by various modified forms.

Some genera such as *Rosa* and *Prunus* which have "single" flowers in the wild state, possess "double" flowers in cultivated forms. If such "double" flowers are dissected it is frequently possible to trace a transition from stamens, and often from carpels also, to petals. Where doubling is complete the flower is sterile and such horticultural varieties are propagated vegetatively.

The stamens may be hypogynous, perigynous, or epigynous; but sometimes, owing to common basal growth, they adhere to the corolla (or perianth). They appear then to be developed on the petals, and are said to be epipetalous (*epiphyllous*, if on a perianth). This is found in many gamopetalous, or gamophyllous, orders of Angiosperms, *e.g.* Compositae, Labiatae, Convolvulaceae. Sometimes the stamens are adhe-

rent to the gynaeceum, *e.g.* in orchids ; this is the *gynandrous* condition.

If the stamens are free from each other, *i.e.* not coherent, the androecium is polyandrous (diandrous, triandrous, pentandrous, etc., according to the number). If united, the union may be of two kinds. (*a*) The stamens cohere by their filaments : this is the adelphous condition—monadelphous if all are united to form a tube round the pistil, diadelphous if united in two groups, *polyadelphous* if in several groups. The monadelphous condition is found, for example, in *Malva* and some Papilionatae ; the diadelphous, in other Papilionatae (*e.g.* pea) where, of the ten stamens, nine are fused and the tenth is free ; the polyadelphous, in Anonaceae and Rutaceae. (*b*) The stamens cohere by their anthers, the filaments being free. This is characteristic of Compositae, some Solanaceae (*e.g.* *Solanum* species), etc. It is the syngenesious condition.

Where the stamens in a flower have different lengths, special terms are sometimes applied to the androecium. Thus, in the Family Cruciferae, there are frequently four long and two short stamens (Fig. 1.8 *a*), and the androecium is said to be tetradynamous. In Labiatae and Scrophulariaceae where there are two long and two short stamens, it is didynamous. These are the only common orders in which these terms are used.

Insertion of the Anthers

The attachment of the anther to the filament should be noticed (Fig. 1.9). It is *innate* or *basifixed* if the anther is fixed directly on top of the filament ; *adnate* if the connective is well marked, and there is no articulation of the filament to the base of the anther, so that the filament seems to run up the back of the anther ; *dorsifixed* if the filament is attached to the back of the anther and the anther is immovable ; *versatile* if the attachment is similar, but the anther swings on the filament.

Development of Stamen

The stamen arises as a protuberance on the receptacle, and soon shows an external differentiation into anther and filament.

At an early stage in the development of the anther, the two anther lobes appear. Further development can be followed by means of sections of progressively older material, which has been carefully fixed and stained, viewed under a microscope.

Development of the Anther (Fig. 1.11)

In each anther-lobe two groups of hypodermal cells divide by periclinal walls, separating an outer layer of cells from an inner. The latter is the primary archesporium. The cells of the outer layer further divide by two successive periclinal walls to form three layers of cells. The outermost of these layers becomes the *fibrous layer* of the anther wall, and the innermost becomes the *tapetum.* The primary archesporium divides repeatedly to form ultimately the spore-mother cells. Each spore-mother cell, by tetrad division, gives rise to four *microspores* or pollen-grains. The process of nuclear division of the spore-mother cell to form four daughter cells is known as reduction division or meiosis.

We have already seen that mitosis involves the resolving of the cell nucleus into a number of chromosomes. Each chromosome splits longitudinally and the halves separate and move to opposite poles, where they associate together as two daughter nuclei. Thus in mitosis the daughter nucleus has the same number of chromosomes as the parent nucleus.

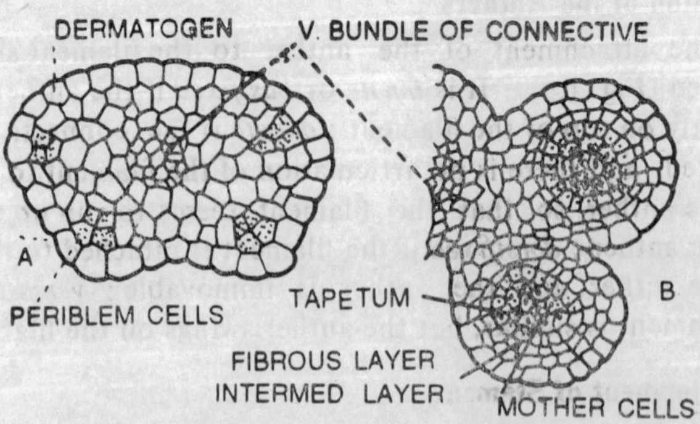

Fig. 1.11. Development of Anther.

In meiosis, on the other hand, each daughter nucleus has only half the number of chromosomes of the parent nucleus. The reason for this is that at the first division of the spore-mother-cell nucleus, whole chromosomes move to opposite poles. This division is quickly followed by a second division corresponding to ordinary mitosis. These two divisions constitute the process of meiosis.

The behaviour of the nucleus during meiosis is as follows. The spore-mother-cell nucleus resolves itself into distinct slender threads, the chromosomes. These associate in pairs, but not at random, because it has been shown that one chromosome of each pair comes from the male, the other from the female parent of the preceding generation. The partners of each pair of chromosomes are thus said to be homologous. Such a pair of chromosomes (*a* and *b*) is represented diagrammatically in Fig. 1.12 (I), and their behaviour subsequently traced. They tend to shorten, thicken, and inter-twine (II), and at this stage, if not before, it is quite clear that each partner of the pair is split longitudinally into two except at the centromere. The two halves of each chromosome are called chromatids (numbered 1, 2, 3. and 4 in the diagram). The paired chromosomes lie at the nuclear (equatorial) plate of the mother cell (Prophase). The partners of each pair tend to fall apart somewhat, but still maintain intimate contact in one or more regions along their length (Metaphase). One of these points of contact is the centromere. Corresponding with other points of contact we may get a transverse division of two chromatids, one of each pair of chromosomes, as in II, 1 and 3. These transverse divisions divide the chromatids into segments of corresponding length, and the points at which they occur are the chiasmata. From the centromere, forces operate which bring about the complete separation of the two chromosomes, except that in the process of being forced or pulled apart similar segments of two chromatids of the homologous pair of chromosomes may change places. Thus, in II, III, and IV, the segment at the upper end of chromatid 3 changes place with the corresponding segment from chromatid I. This process is called crossing over. It involves an interchange of nuclear material between two chromatids of an homologous pair of chromosomes.

The separated chromosomes (*a* and *b*) continue to move along a spindle which has meanwhile been forming, to opposite

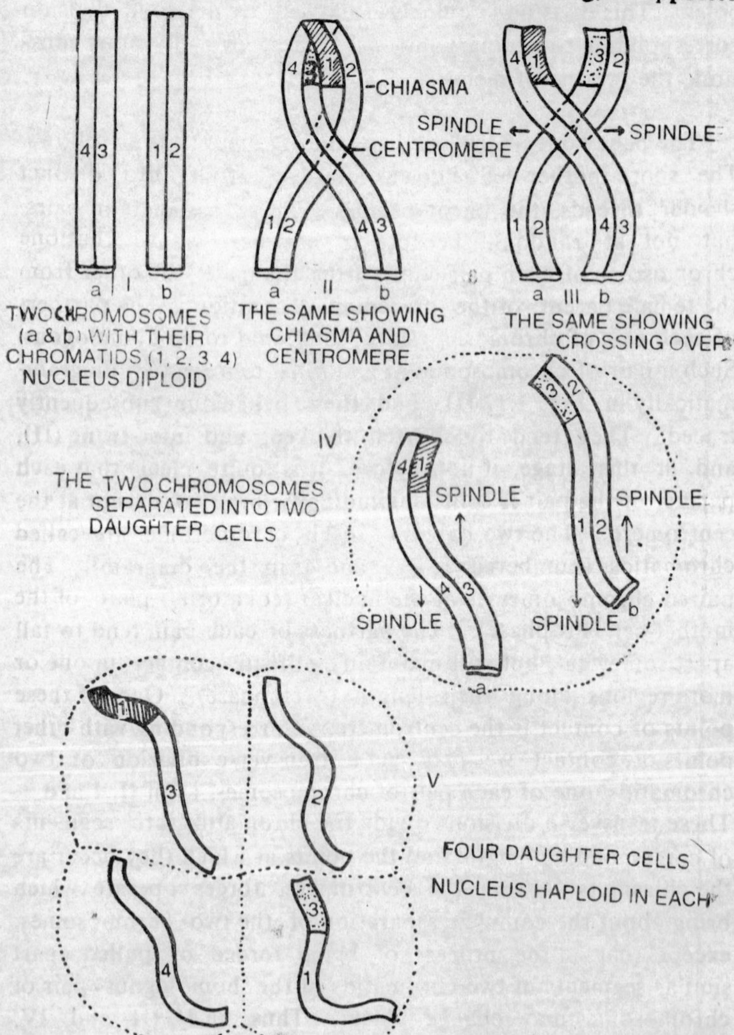

Fig. 1.12. Diagram illustrating meiosis and crossing over.

poles of the cell (Anaphase). If crossing over has taken place, these chromosomes are not exactly identical with those which entered into association at the beginning (compare Fig. 1.12, I, *a* and *b*, with IV, *a* and *b*). This has an important bearing on genetics.

Usually no cell-wall is formed at this stage between the two sets of chromosomes thus separated (Telophase), and the second division follows immediately. In this division the chromatids separate, 1 from 2, 3 from 4. The result is four daughter nuclei, each of which has only half the chromosome complement of the original mother nucleus of the spore-mother-cell. Each of these four nuclei is said to be haploid. This condition is conveniently represented by the symbol n, and the diploid condition by $2n$.

The spore-mother-cell nucleus has now given rise to four daughter nuclei. Each of these becomes associated with cytoplasm, and around each cytoplasmic mass is secreted a cell-wall. The nutritive material for this development, which also involves expenditure of energy, is supplied by the tapetum. The four cells thus formed may remain associated together in tetrads, frequently forming regular geometrical figures such as triangular pyramids, but more usually they separate as individual microspores. Meanwhile the anther has been growing, and the microspores occupy the four cavities in most mature anthers. Simultaneously the anther wall develops and some differentiation takes place in it in relation to a dehiscence mechanism. This may take the form of a fibrous layer, immediately below the epidermis, the cells of which have characteristically thickened walls (Fig. 1.13).

Dehiscence of the Anther

In order that the microspores may be liberated to bring about pollination, the anther dehisces. When this takes place in many stamens a longitudinal split appears between each pair of pollensacs in each anther lobe. This split gradually enlarges by the separation and curling back of the anther wall on either side of it (Fig. 1.13). The wall separating the two pollen-sacs also breaks down, and thus the pollen-sacs are exposed as one cavity. Two such cavities will therefore be formed in each anther, one on either side of the connective. If these cavities face towards the centre of the flower, the

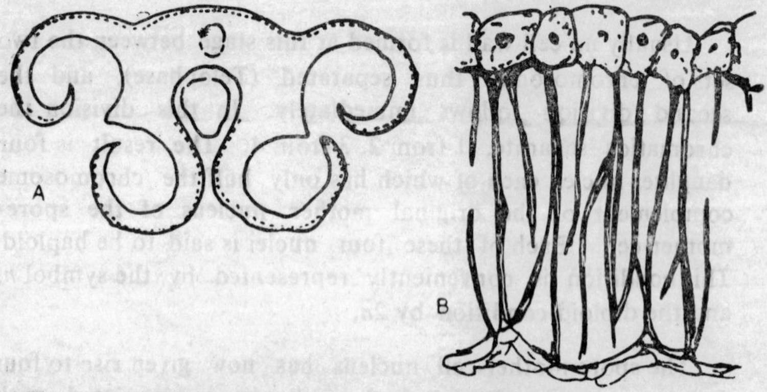

Fig. 1.13. (*a*) Dehisced anther, transverse section (diagrammatic)
(*b*) One cell of fibrous layer much enlarged to show detail of thickening of anticlinal walls.

dehiscence is said to be *introrse* ; if towards the outside, *extrorse* ; if sideways, *lateral.*

The splitting and curling of the anther walls is brought about by the action of the fibrous layer. As these cells lose water by evaporation they contract, but the arrangement of the thickening of the cell-walls permits of contraction only in the plane parallel to the surface. The strain thus set up causes a break in the anther wall along a prepared line between the pair of pollen-sacs in each anther lobe. The outer cell-walls of the cells of the fibrous layer contract at a much greater rate than the inner walls on account of the differential thickening. Hence the curling outwards and backwards of the anther walls.

The microspores vary much in size, form and colour, in different plants. At first they are unicellular and the wall consists of two membranes. The outer coat, the *exine*, frequently has protuberances, spines, etc,, and also one or more areas where the wall is thin and forms a pore. The inner coat, the *intine*, is thin and consists of cellulose. In some plants, *e.g.*, orchids, the microspores are not loose, but those of each anther lobe are aggregated into a single mass called a pollinium.

Before seed can be produced the microspores must be transferred to the stigma, either of the same flower, or of another flower of the same species ; a process called pollination.

The Gynaeceum or Pistil

The gynoecium shows many interesting forms. One of the simplest is that of the buttercup, where all the carpels are separate. The number, too, is high and indefinite. In many flowers, on the other hand, the carpels are few and definite in number, and are very often joined to each other.

The methods of joining of the carpels are interesting and varied. The best way to examine the various types would be first of all to imagine the carpel as an open leaf, bearing its ovules on the margins. It must be remembered, however, that ovules never are borne naked and exposed in this way in Angiosperms ; but, starting with this *hypothetical* case, it is easy to see the various ways in which the ovules could conceivable become enclosed by the carpel.

The simplest method for the carpel to enclose its ovules would be to fold in half, thus bringing together its two margins bearing the ovules. Then, imagine these margins to become fused. Thus one carpel enclosing a vertical line of ovules would be produced. In the bean, for example, there is a series of about eight ovules arranged as one would expect them in such a case.

The next method of enclosure would be for two carpels, facing each other, to join at their margins. This would give two carpels forming an ovary with one common cavity into which the ovules would project in two longitudinal rows. This is the case in the gooseberry (*Ribes uva-crispa*). A variation of this is seen in the tomato. Here, the margins, four in all, two from each carpel, meet at a common centre. Thus there are two carpels, with two cavities, and a row of ovules projecting into each cavity. In many other cases the same method as that of the gooseberry applies, except that there are three carpels, with a common cavity into which three longitudinal rows of ovules project. This is seen in the violet.

Then a type similar to that of the tomato (*Solanum lycopersicum*) except that it has three carpels with three cavities, into each of which one row of ovules projects, is also possible; indeed, it is very common among flowers. It is seen, for example, in the tulip. A completely different type is seen in the primrose. Here, there are five carpels joined together; but the ovules, instead of forming rows at the fused carpellary margins, are borne on a projection from the base of the ovary cavity.

The gynaeceum or pistil consisting of megasporophylls or carpels forms the inner essential organ of the flower. It may consist of one or of several carpels. In the latter case, according to the number, it is bicarpellary, tricarpellary, etc.

The old classical view represents it as a single carpellary leaf folded with the lower side outermost and its margins coherent along a line (*the ventral suture*). and, further, with the apex of the leaf elongated and slightly swollen at the tip.

In such a monocarpellary pistil (Fig. 1.14) the hollow basal portion of the folded carpel is the ovary, from which the fruit is afterwards developed; it contains a varying number of ovoid or rounded bodies, the ovules or megasporangia which after fertilisation develop into seeds. The slender prolongation, of varying length, on top of the ovary in the style, which usually contains a central cavity communicating with the cavity of the ovary, but may be composed of loose tissue throughout. The apical portion of the style, called the stigma, is usually swollen and covered with hairs or glandular papillae; as we shall see later, it forms the receptive surface for the microspores. When there is no style, the stigma is said to be sessile on the ovary.

If we examine the ovary, we find that the ovules are developed in two rows, each row on a longitudinal ridge or cushion

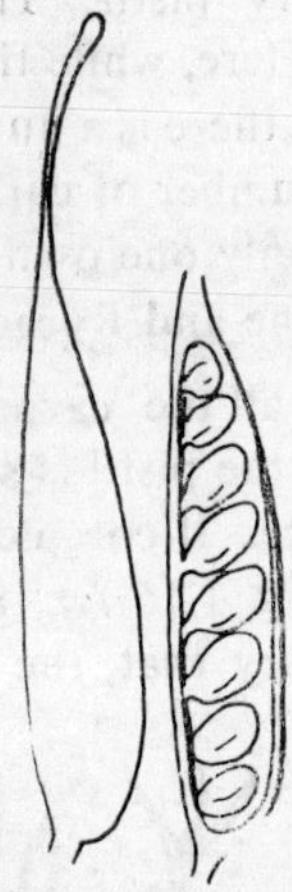

Fig. 1.14. Monocarpellary Gynaeceum, Carpel of Leguminous Flower.

of tissue called the placenta, along the ventral suture, on the inner surface of the ovary wall. The dorsal suture is opposite to it. The placentation (*i.e.*, the position or arrangement of placentas in an ovary) in the *simple* ovary is described as marginal. The pistil of the Papilionatae (pea, bean, etc.) is a good example of this.

Although morphologists have compared the formation of the monocarpellary pistil to the folding of a carpellary leaf, we must not assume that this process can be observed during the development of the flower. But in certain Gymnosperms the ovules are not enclosed in an ovary, but are borne on the open megasporophylls. In *Cycas* they are borne on the margins of the megasporophylls. They imagined such a megasporophyll folded so as to bring the margins together and form a cavity enclosing the ovules, thus giving the condition found in such Angiosperms as the pea.

Polycarpellary Gynaeceum

Of this there are two conditions, according as the carpels are or are not united with each other. If the carpels are free, *each* forms a simple ovary, style and stigma, like the single carpel of the monocarpellary pistil. This is the apocarpous condition. (Fig. 1.4, B.C.) Here, while there is a single gynaeceum or pistil in the flower, there is a number of simple ovaries. The number indicates the number of carpels. The placentation is marginal. Frequently, only one ovule is developed in each loculus (many Ranunculaceae and Rosaceae).

In the second condition all the carpels are united to form a single *compound* ovary, and the pistil is syncarpous. The union may or may not be complete. If complete, the ovary bears a single style and stigma (Fig. 1.15 A), and it is only by the internal structure of the ovary that the number of carpels can

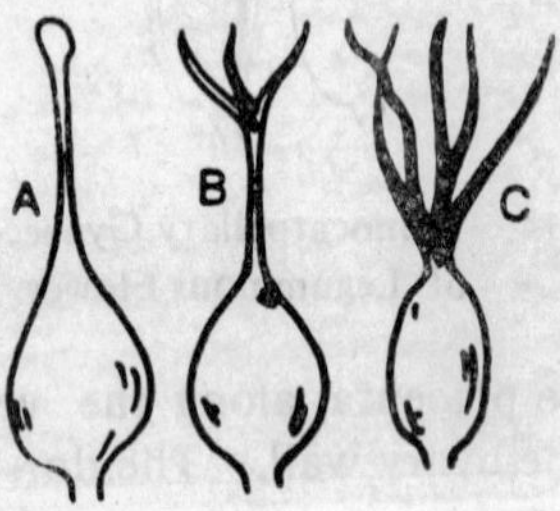

Fig. 1.15. Degrees of Fusion of Style.
A *Primula* : B. *Luzula* ; C, *Stellaria.*

be determined. If incomplete, a number of styles or stigmas are borne on the single ovary (Fig. 1.15 B, C), owing to the apices of the carpels remainin free. With few exceptions the number of styles or stigmas is the same as the number of carpels. Hence, *e.g.*, in Compositae (Fig. 1.8 C), where the style is single, but there are two stigmas, we assume that the pistil is bicarpellary.

The structure of the ovary and the placentation in the syncarpous pistil differ in different cases. The following conditions should be noticed :

(*a*) If the carpels are fused by their adjacent margins (Fig. 1.16) they form a uniocular ovary. The fused margins are the placentas bearing ovules, and the placentation is marginal and parietal. The number of parietal placentas indicates the number of carpels.

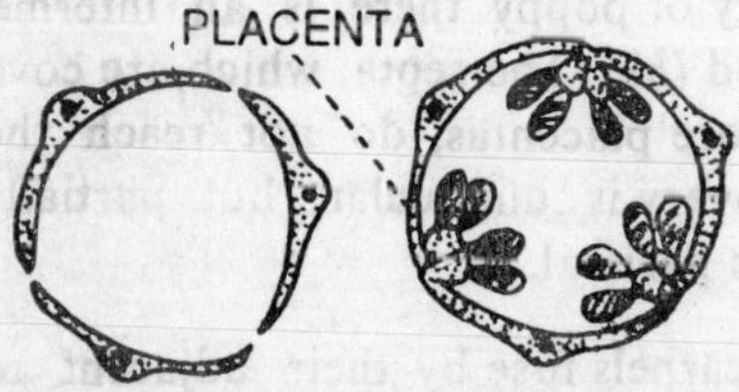

Fig. 1.16. Diagrammatic Representation of the Union of Three Carpels to form a unilocular ovary with parietal placentation, *e.g., Viola*, seen in transverse section.

(*b*) If the carpels are folded on themselves before fusing, that is, if the fused margins run in to the middle of the ovary (Fig. 1.17), a multilocular ovary results and the marginal

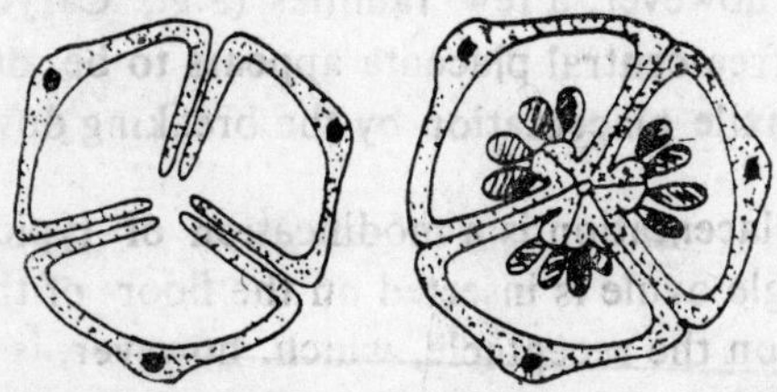

Fig. 1.17. **Diagrammatic representation of the union of three carpels to form a trilocular ovary with axile placentation. *e.g., Lilium*, seen in transverse section.**

plancentas of all the carpels fuse in the centre to form a central or axile column. The placentation is marginal and axile. The number of loculi, or the number of septa by which the ovary is divided, indicates the number of carpels (except where *false* septa are formed—see below). Sometimes only one ovule is developed in each loculus (Fig. 1.18).

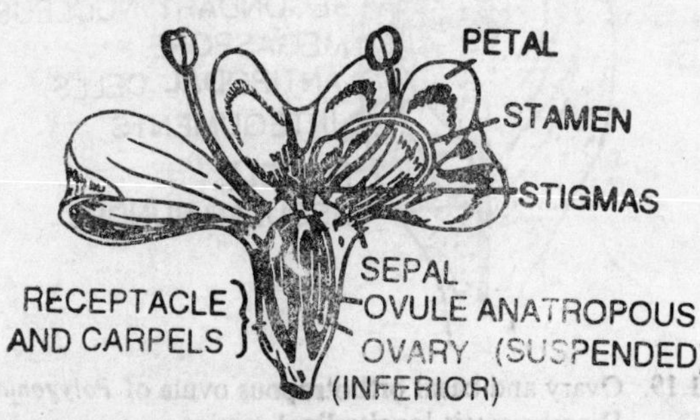

Fig. 1.18. **Flower or Umbelliferae. (Half-flower in vertical section.)**

In the ovary of poppy there is an intermediate condition between (*a*) and (*b*). The septa, which are covered with ovules and are therefore placentas, do not reach the middle of the ovary. The ovary is unilocular, but partially divided. The placentation is parietal.

(*c*) If the carpels fuse by their adjacent margins and the ovary is unilocular as in (*a*), and the ovules are not developed on the carpellary margins, but are borne on a central axis running through the middle of the ovary, the placentation is free-central. In typical cases (Myrsinaceae) the central axis is a prolongation of the receptacle into the ovary. The ovules are developed on the axis of the flower, not on the carpels. There are, however, a few families (*e.g.*, Caryophyllaceae) in which the free-central placenta appears to be derived from an originally axile placentation by the breaking down of the septa.

Basal placentation is a modification of typical free-central. Here a single ovule is inserted on the floor of the ovary. It is developed on the receptacle, which, however, is not prolonged as an axis into the ovary (*e.g.* Polygonaceae, Fig. 1.19, and Compositae).

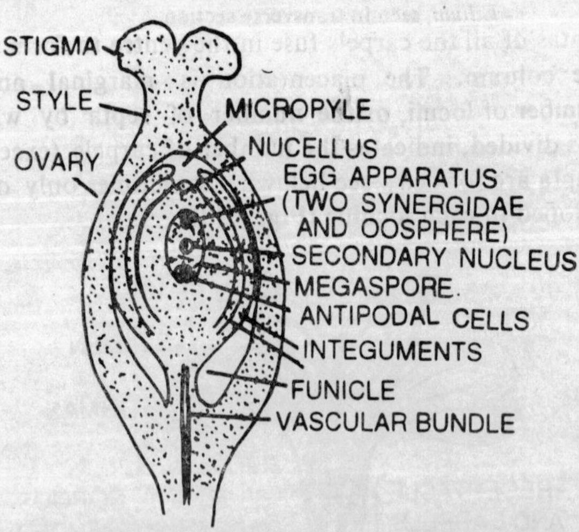

Fig. 1.19. Ovary and basal orthotropous ovule of *Polygonum*. Diagrammatic longitudinal section.

The ovules are, in rare cases, developed, not on the margins of the carpels, but over the whole inner surface, *e.g.*, the flowering rush (apocarpous) and white water lily (syncarpous). This is called superficial placentation.

True and False Septa

True septa, or dissepiments, are those which represent the inturned margins of carpellary leaves. Septa formed in any other way, *e.g.*, as ingrowths from the surfaces of the carpels, are *false*. In the ovary of Cruciferae (Fig. 1.20), for example, the false septum is formed by two membranes, which grow in from the two parietal placentas, and meet and fuse in the centre.

Superior and Inferior Ovaries

In all hypogynous and perigynous conditions, the ovary is described as *superior*; in the epigynous condition, as inferior. It might seem out of place to describe the ovary as superior, and the calyx inferior, in such perigynous conditions as are shown in Fig. 1.4, D and E. But it must be realised that the ovary here is developed at the morphological apex of the receptacle and is free from the calyx.

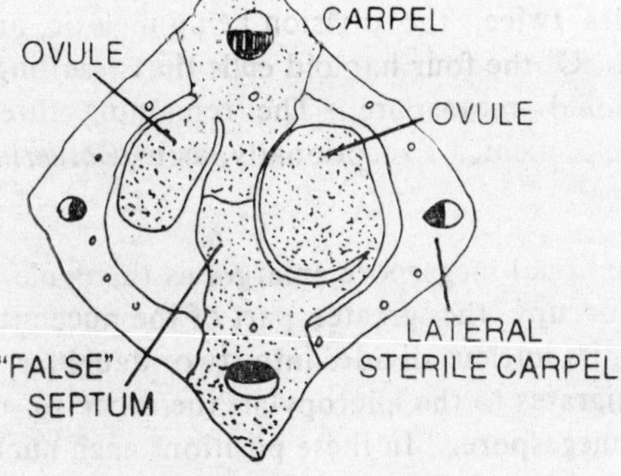

Fig. 1.20. Ovary of Cruciferae According to Saunders' Theory. Four carpels : 2 median, solid, fertile ; 2 lateral, valve, sterile.

Development of the Ovule

The development of the ovule is followed by examining series of longitudinal sections of ovules at progressively older stages of development. The material must be specially fixed and stained for this purpose. Such ovaries as those of pea, lily and marsh-marigold are useful material for examination, and if cut transversely at different stages of development the sections will show ovules cut more or less longitudinally.

The young ovule arises as a small protuberance from the placenta and gradually increases in size and complexity. The body of the ovule is called the nucellus. Quite early in its development, one or two ridges of cell tissue appear at its base and in their further growth encroach over the still growing nucellus so as ultimately to enclose it, except for a small pore left at the apex which persists as the micropyle. The one or two coverings thus formed are the integuments. Meanwhile, from the basal region of the ovule, or chalaza, below the integuments, an elongation forms the funicle, or "stalk" of the ovule (Fig. 1.19). This joins the chalaza at the hilum.

At an early stage in the development of the nucellus a hypodermal cell at the apex enlarges (Fig. 1.21A). It usually divides into two, an upper (outer) tapetal cell, and a lower (inner) sporemother cell (Fig. 1.21B). The nucleus of the latter divides twice, the division being meiotic, and cell-walls are formed. Of the four haploid cells thus resulting, only one is a functional megaspore. The remaining three show no further development. Exceptionally, as in *Casuarina*, all four are functional.

The functional megaspore enlarges as the ovule grows, and comes to occupy the greater part of the nucellus (Fig. 1.22). Meanwhile, its nucleus divides into two daughter nuclei, one of which migrates to the micropylar, the other to the chalazal end of the megaspore. In these positions each nucleus divides further, giving rise to four nuclei at each pole. Three of the nuclei at the micropylar pole become surrounded by cytoplasm and form the two synergidae and the oosphere or egg-cell

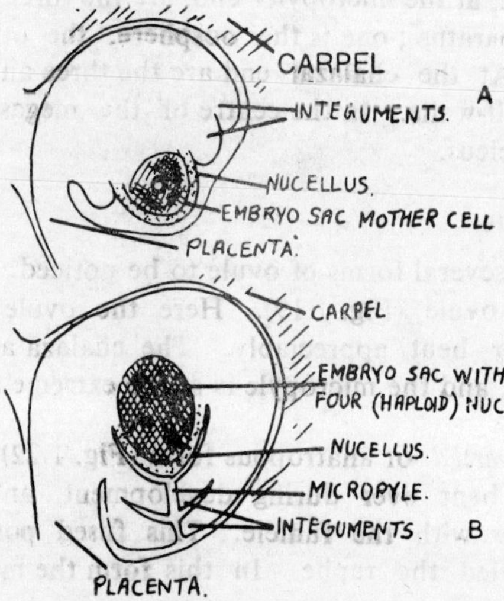

Fig. 1.21. Development of ovule of *Lilium*.
A, Hypodermal cell at apex of young nucellus :
B, Result of meiotic division giving 4 haploid nuclei.

(female gamete). Three of the nuclei at the chalazal pole become surrounded by cytoplasm and cell-walls, and form the antipodal cells. The remaining two polar nuclei, one at each pole, migrate towards the centre of the megaspore, where they fuse to form the secondary nucleus (Fig. 1.19), often referred to as the primary endosperm nucleus.

Before cytological work revealed the true nature of the megaspore, it was known by the name embryo-sac, and it is still frequently referred to by this name.

Structure of the Mature Ovule

Fig. 1.22 illustrates the structure of the ovule of *Caltha* at the time when it is ready for fertilisation. The slender funicle attaches the ovule to the basal placenta of the ovary and carries a vascular bundle to the chalaza. The nucellus is invested by the two integuments, and the micropyle is seen as a narrow passage or pore at the apex. Almost filling the nucellus is the enlarged megaspore (or embryo-sac). Within the megaspore, at the micropyler end, are the three protoplasts of the egg-apparatus ; one is the oosphere, the other two the synergidae. At the chalazal end are the three antipodal cells enclosed in cell-walls. In the centre of the megaspore is the secondary nucleus.

Forms of Ovule

There are several forms of ovule to be noticed. One is the orthotropous ovule (Fig. 1.19). Here the ovule is straight, not curved or bent appreciably. The chalaza and hilum lie close together, and the micropyle is at the extreme apex.

In the *inverted* or anatropous form (Fig. 1.22) the body of the ovule has bent over during development, and fused for some distance with the funicle. This fused portion of the funicle is called the raphe. In this form the micropyle and

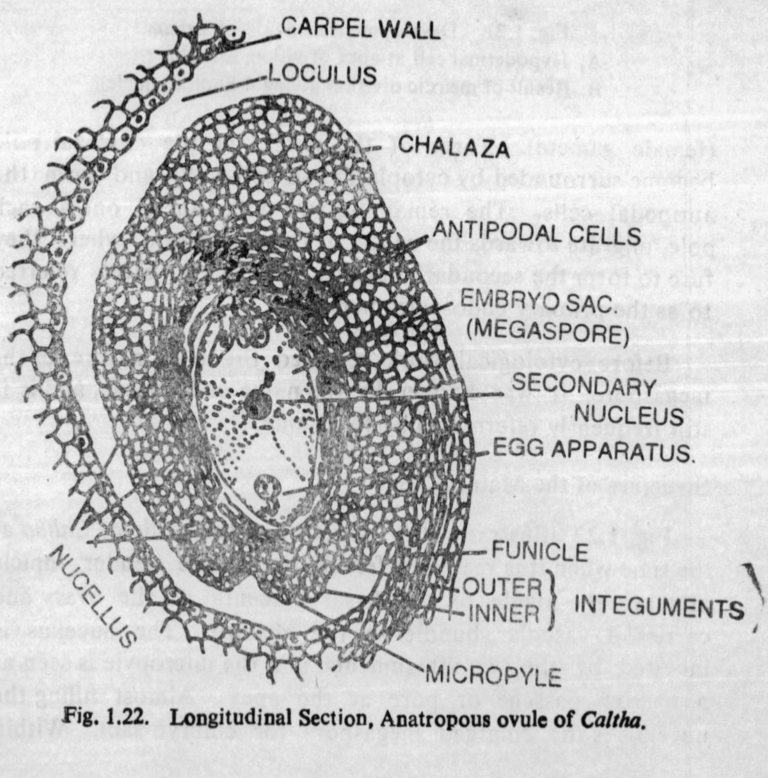

Fig. 1.22. Longitudinal Section, Anatropous ovule of *Caltha*.

hilum lie close together, and the chalaza is towards the other end.

In the curved or campylotropous form (Fig. 1.23A) the body is curved and bent round, so that the micropyle lies near the funicle ; but there is no fusion with the funicle. Hilum, chalaza, and micropyle all lie close together. The amphitropous ovule is an intermediate form in which the body of the ovule is straight, but has been twisted round, so that its long axis is at right angles to the funicle (Fig. 1.23B).

Of these forms the anatropous is most frequently met with. Examples of the campylotropous ovule are found in many Cruciferae and Papilionatae (pea, bean, etc.). The orthotropous ovule is less frequently found, *e.g. Polygonum* (Fig. 1.19). The Loganiaceae and some Cruciferae give examples of the amphitropous ovule.

Cohesion and Adhesion

The student must be clear as to the meaning of these terms. *Cohesion* is union between members of the same series of floral leaves. Thus gamosepalous, polysepalous, polyandrous, syngenesious, apocarpous, syncarpous are terms signifying cohesion or want of cohesion. *Adhesion* means union between members of different series, as when the stamens are epipetalous. We have already explained that cohesion or adhesion of parts in the flower is not due to the actual fusion of parts originally separate, but to common basal growth during development.

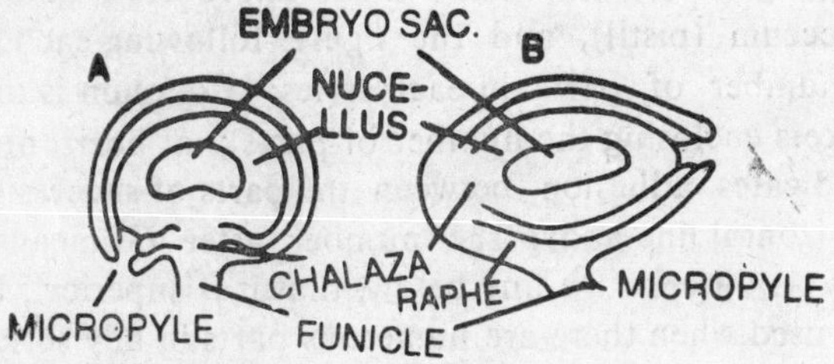

Fig. 1.23. Forms of ovule.
A, Campylotropous ovule, *e.g.* Cruciferae ;
B, Amphitropous ovule, *e.g. Strychnos Nux-vomic.*

Vertical Sections and Floral Diagrams

The general structure and arrangement of parts in a flower may be shown in drawings of longitudinal or vertical sections (Figs. 1.4, 1.6) (half-flower), in floral diagrams and floral formulae.

The floral diagram may be described as a ground-plan of the flower showing the relation of the parts to each other and to the mother-axis. In making a floral diagram cohesion of parts may be indicated by connecting lines (Fig. 1.2 A), and this may also be done in the floral formula which should accompany the diagram. The aestivation may be indicated as in Fig. 1.3A and B. An *empirical* diagram (Fig. 1.2B) is one showing only the relative positions of the parts actually present. A *theoretical* one indicates, as well, by conventional means, the relative positions of parts which we assume were originally present, but are now lost (Fig. 1.2C).

The floral formula, together with the diagram and longitudinal section, enables us to represent the essential *morphological* features of the flower without a word of description being necessary.

The symbols ⊕ and † respectively denote radially and bilaterally symmetrical (zygomorphic) flowers, the direction of the arrow indicating the plane of symmetry along which the flower can be divided into equal halves. The signs ♂, ♀, ☿ respectively denote staminate, carpellary, and hermaphrodite ("perfect") flowers. The letters K, C, and P represent calyx, corolla, and perianth, A and G the androecium (stamens) and gynaeceum (pistil), and the figure following each letter gives the number of parts in each series. Cohesion is indicated by brackets enclosing the number of parts ; a horizontal bracket ⌒ indicates adhesion between the parts of successive whorls ; a horizontal line above the number after G means that the ovary is inferior, a line below, that it is superior ; the symbol ∞ is used when there are numerous parts in any series.

Thus the typical floral formula of Myrsinaceae

☿ ⊕K(5) $\widehat{C(5)\ A5}$ G(5)

reads—hermaphrodite, radially symmetrical flower, gamosepalous calyx of five sepals, gamopetalous corolla of five petals, androecium of five free epipetalous stamens, syncarpous pistil of five carpels with a superior ovary.

MODIFICATIONS OF THE FLOWER

Although all flowers follow the same structural plan, they exhibit numerous modifications in design. Certain of these modifications increase the chances of pollination. Furthermore, they often reveal evolutionary relationships among plants, thus permiting botanists to place related forms together and to classify flowering plants according to descent, from the most simple to the most advanced and specialized forms.

Number and arrangement of floral parts. The parts of the flower are usually arranged in whorls, or circles. These whorls are either four or five in number : one of sepals, one of petals, one or two of stamens and one of carpels, united into a compound pistil. The number of parts in each whorl varies with the species, but it is usually constant and small.

The two large subclasses of angiosperms (flowering plants) are distinguished by the number of floral organs in each whorl. In the dicotyledons, the number is four or five or multiples of these numbers. A flower may have, for example, five sepals, five petals, ten stamens, and five carpels. However, the number of carpels in a compound pistil may be only two. In the monocotyledons, the number of organs in a whorl is three or multiples of three. In the tulip, for example, there are six perianth parts, six stamens, and three carpels.

In both subclasses, however, certain families have numerous stamens and carpels—generally more than ten. Among the families in which this condition occurs are the magnolia (Fig. 1.24), buttercup rose, and waterlily of the dicotyledons, and the water plantain family of the monocotyledons. In some of these families the numerous stamens and carpels are attached to the receptacle spirally rather than in whorls. This combination of spiral arrangement and numerous stamens and carpels is considered indicative of a more primitive stage

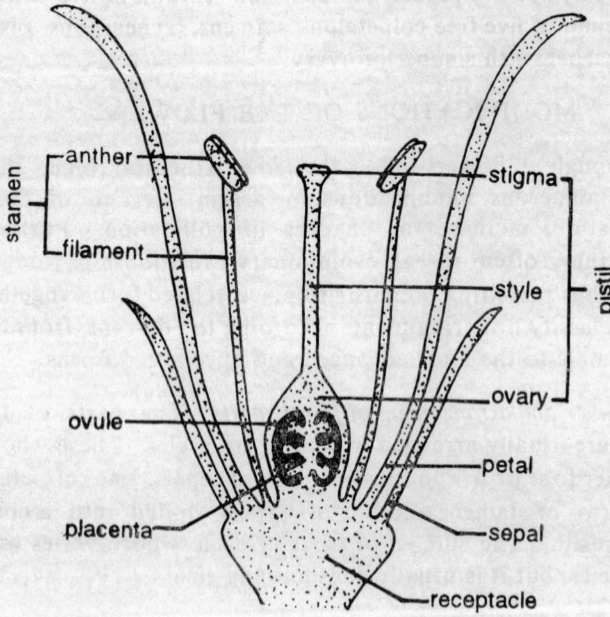

Fig. 1.24. Diagram showing the floral organs.

in evolutionary development than the whorl arrangement with the parts reduced in number.

Fusion of floral organs. Among the important modifications found in the flower are fusions of the organs of the same whorl and of different whorls. The dicotyledons, for example, are divided into two large groups, based upon the fusion of the petals. In one group the petals are separate. In the other group, the petals are joined to form a *corolla tube*, which is lobed at its margin (Fig. 1.25). The lobes usually correspond to the number of petals. The sepals are usually distinct, but the basal edges are often joined into a shallow tube. The stamens are usually free from one another but may be joined by their filaments or by their anthers into one or several groups. Sometimes free stamens are fused to the corolla tube at the base. The carpels, as we have seen, are commonly united into a compound pistil.

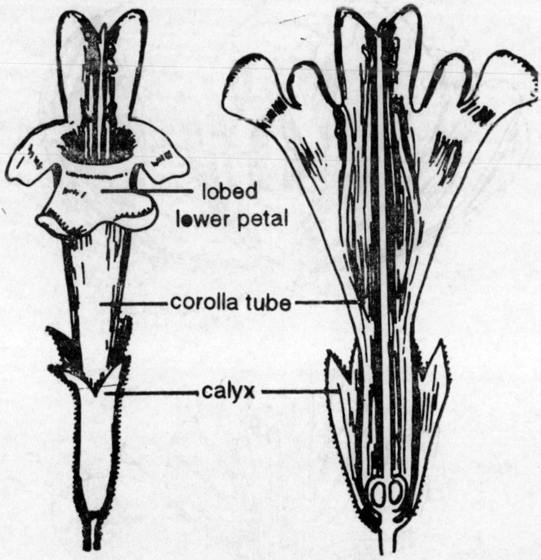

Fig. 1.25. Flower of a mint, ground ivy (*Glecoma bederacea*), illustrating fusion of floral parts and irregular symmetry of the flower.

In a flower with free parts, the sepals, petals, and stamens are attached to the receptacle beneath the ovary. Since the ovary is situated above the zone of attachment, it is said to be *superior* (Fig. 1.26 A). This is the condition in a very large number of flowering plants. In plum, cherry, blackberry, and other species, the basal portions of the sepals, petals, and stamens are united to form a cup-shaped *floral tube* (also called calyx tube). The lobes of the sepals and petals extend from the margins of this tube together with the unfused portions of the stamens. The ovary is still superior in this type of flower (Fig. 1.26 B).

One other type of fusion is that of the floral tube with the ovary. Then the sepals, petals, and stamens apparently grow from the top of the ovary, and the ovary is said to be *inferior* (Fig. 1.26 C). Actually the pistil is still the uppermost structure of the flower, just as in flowers with superior ovaries. The ovary appears to be inferior because of its union with the floral

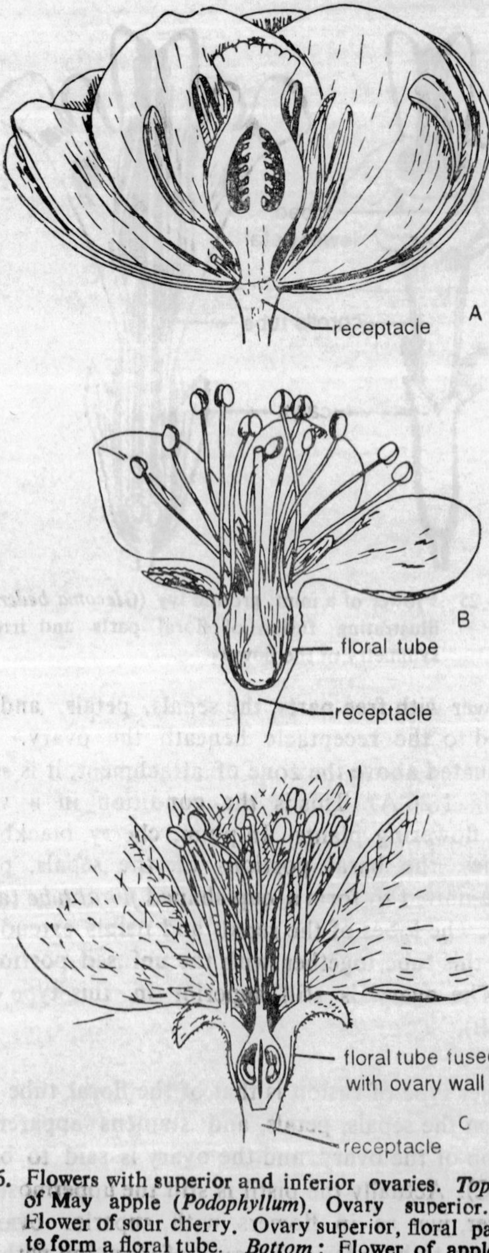

Fig. 1.26. Flowers with superior and inferior ovaries. *Top* : Flower of May apple (*Podophyllum*). Ovary superior. *Center* : Flower of sour cherry. Ovary superior, floral parts fused to form a floral tube. *Bottom* : Flower of apple. Ovary inferior, the result of the fusion of the floral tube to the ovary wall.

tube. Inferior ovaries are found in such plant families as the carrot, bluebell, sunflower, evening primrose, iris, cactus, and orchid. The position of the ovary, whether inferior or superior, is significant in plant classification and also helps in interpreting the nature of many kinds of fruits.

The fusion of floral parts—of petals to form a tube, of carpels to form a compound pistil, and of ovary wall and floral tube—is an evolutionary development. The various organs do not become fused during the development of the flower but are united from the very beginning of growth in the embryonic cells at the apex of the floral axis.

Symmetry of the flower. In many flowers, all the parts of each whorl are alike in size and shape, and the flower may be divided into two similar parts in more than one longitudinal plane (Fig. 1.27). Examples are tomato, apple, petunia,

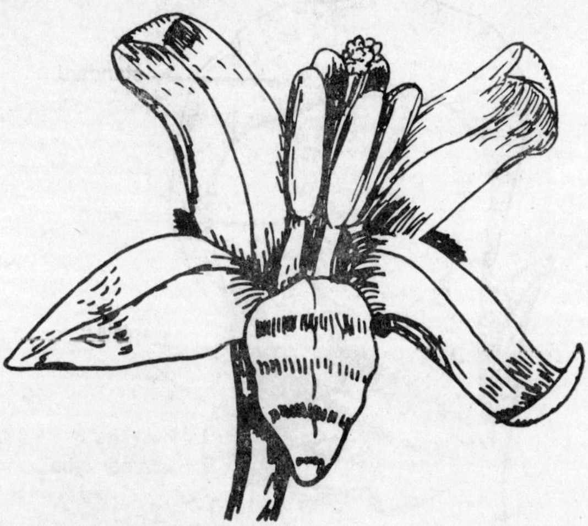

Fig. 1.27. Radially symmetrical flower of common nightshade (*Solanum nigrum*).

morning-glory, and many others. Such flowers are said to be *regular* or *radially* symmetrical. The flowers of other species are *irregular*, or bilaterally symmetrical, for they are capable of division into two equal parts along only one longitudinal plane. In such flowers, the corolla is commonly more modified than other floral parts. In the garden pea (Fig. 1.28) the one top petal, the *standard*, is much enlarged. Two lateral ones,

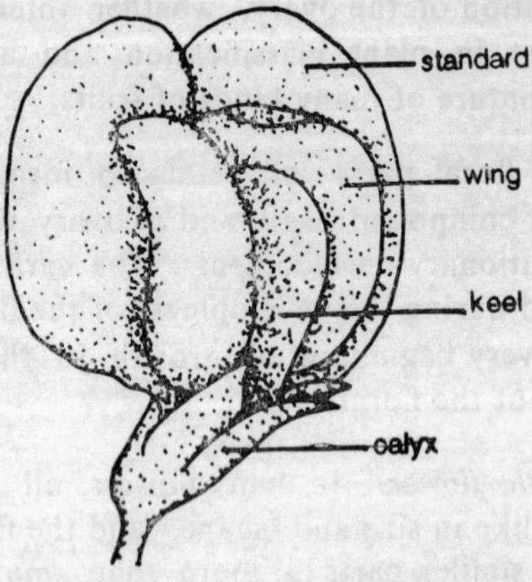

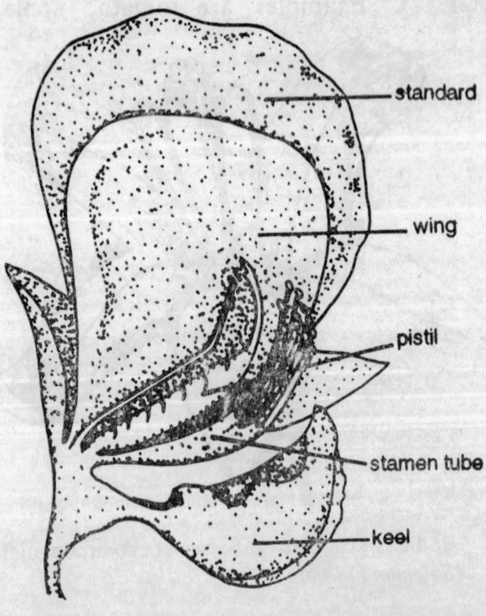

Fig. 1.28. Irregular flower of the garden pea. *Top* : Exterior view of the flower. *Bottom* : Section view of the flower.

the *wings*, are much smaller, and the two remaining lower petals are united along one edge to form the *keel*, which encloses the stamens and pistil. Nine of the ten stamens are united into a tube which surrounds the pistil. The tenth stamen is free. Other examples of irregular flowers are the snapdragons, mints (Fig. 1.25), violets, and orchids. Irregular flowers are usually pollinated by insects, commonly bees.

Reduction in the flower. The flower in which four readily distinguishable kinds of floral organs occur—sepals, petals, stamens, and pistils—is not the only kind. The flower may become reduced so that some organs are no longer evident, although they may be present in a vestigial form which can be recognized only after careful investigation. Plants in which the perianth parts have become minute or inconspicuous, or have disappeared, include the grasses and many trees such as willows and poplars, maples (Fig. 1.29), oaks, hickories, and

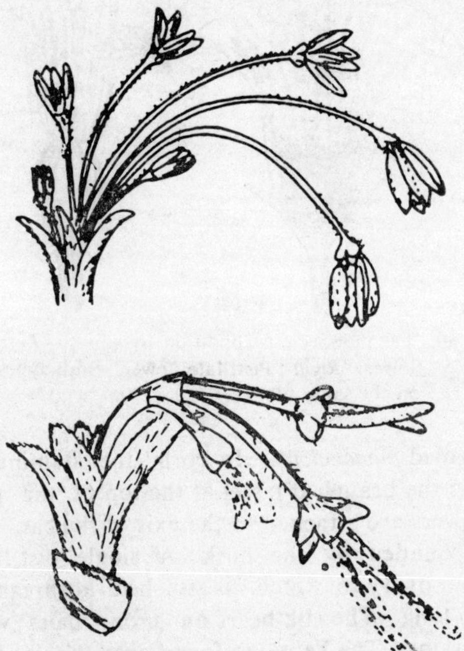

Fig. 1.29. Reduced unisexual flowers of box elder (*Acer Negundo*). *Top* : Staiminate flowers. *Bottom* : Pistillate flowers. The perianth is greatly reduced. This species is dioecious, for the two kinds of flowers are borne on different trees.

elms. Contrary to popular belief, a conspicuous perianth is not an essential feature of the flower.

Reduction in the number of parts may also extend to the stamens and pistils. A flower in which both of these organs are present is said to be *perfect*, or *bisexual.* If one of these organs is missing, the flower is *imperfect* or *unisexual.* Unisexual flowers are termed *staminate* if they lack a pistil or *pistillate* if they lack stamens.

Both kinds of unisexual flowers may be present on the same plant, as in corn, most begonias, squash (Fig. 1.30), pumpkin and cucumber, cattail, calla lilies, and many woody plants, such as oaks, beeches, birches, walnuts, and hickories. Such

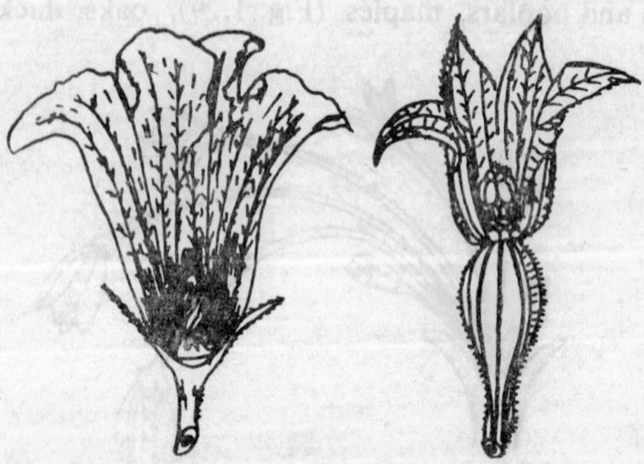

Fig. 1.30. The monoecious condition in squash. *Left* : Staminate flower. *Right* : Pistillate flower. Both types are borne on the same plant.

plants are called *monoecious.* In corn, the staminate flowers are borne on the branched tassel at the top of the plant. The pistillate flowers are attached to the axis of the ear, or "cob", and are surrounded by the husk. A single pistillate flower consists of an ovary to which is attached a thread-like silk 4 to 12 inches long. The silk bears numerous hairs which catch the pollen grains. The hairs are found even within the portion of the silk covered by the husk but are most numerous near the tip. Each thread of silk thus serves as a combined style and stigma (Fig. 1.31).

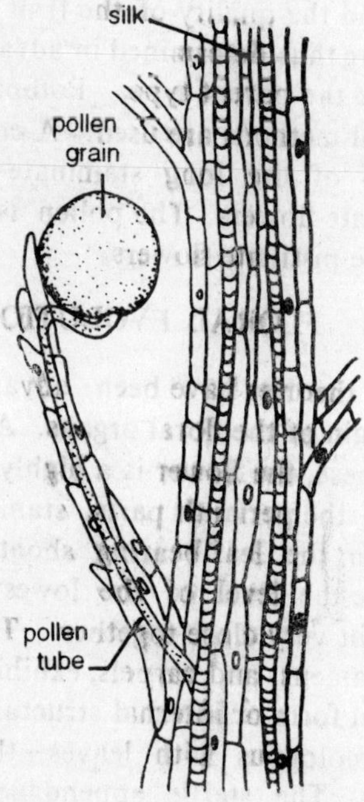

Fig. 1.31. Germination of pollen grain of corn on a hair of the silk.

In *dioecious* plants the staminate and pistillate flowers occur on different plants of the some species. Examples of this condition are found in ashes and maples, in spinack, tree of heaven (*Ailanthus*), asparagus, Indian hemp, the date palm, hops, willows, holly, and poplars.

The data palm probably represents the earliest recognized example of the dioecious condition, for records show that the tree was cultivated in Mesopotamia as early as 3500 B.C. The farmers of this region recognized that there were male and female trees and practiced artificial pollination by bringing clusters of staminate flowers in contact with the pistillate flowers. The methods of bringing about fruit production in the date palm today differ but little from those of ancient times. Staminate trees to provide pollen are grown with the pistillate trees, which bear the fruit. Propagation is achieved by planting offshoots from the base of the parent tree. The sex of the tree and the quality of the fruit which the pistillate plant will bear are thus determined in advance, for the offshoots always reproduce the parent type. Pollination is still done by hand, and several methods are used. A common procedure is to invert several of the long staminate clusters among the clusters of pistillate flowers. The pollen is shed and drifts to the stigmas of the pistillate flowers.

FLORAL EVOLUTION

A number of theories have been advanced to explain the evolutionary origin of the floral organs. According to the most widely held of these, the flower is a highly modified axis, bearing appendages—the perianth parts, stamens, and carpels. It is comparable to the leaf-bearing shoot. The internodes are suppressed above the level of the lowest sepals, and so the nodes are brought very close together. The floral organs, the sepals, petals, stamens, and carpels, exhibit many similarities to leaves in external form or internal structure and are generally regarded as homologous with leaves—that is, having had a common origin. The sterile appendages of the floral axis, constituting the perianth, show a greater similarity to leaves than do the uppermost, fertile appendages, the stamens and carpels. Nevertheless, none of the floral parts is believed to be descended directly from photosynthetic organs; they are thought to have merely evolved along similar lines, undergoing many transformations during their evolution. The carpel is thus interpreted as a modified leaflike organ folded inward along the midrib. Its margins are considered to have grown together, enclosing the ovules. The apical region of the appendage has become modified into the stigma, which receives the pollen.

This interpretation of the nature of the flower has led to the application of the term *sporophyll* (literally, spore leaf) to the stamen and carpel, since certain structures essential to seed formation, known at *spores*, are located within the ovule and anther. The flower is therefore defined as a group of a sporophylls, usually accompanied by a perianth.

The view that the parts of the flower are transformed leaf-like organs has met with considerable criticism in recent years, and the problem of the essential nature of the flower, from an evolutionary viewpoint, must be regarded as yet unsolved. Nevertheless, there is sound evidence that the flowers of certain plants, such as the buttercups and magnobas, are relatively primative is structure and represent types from which other forms may have arisen.

ANATOMY OF SIMPLE FLOWERS

Examples of anatomically (and primitively) simple flowers provide a basis for the understanding of more complex types.

Aquilegia. Figure 1.32 show the vascular structure of the flower. The stele continues in vestigial form beyond the level of departure of the uppermost appendages. Figure 1.32 shows the structure of the flower as shown in a series of diagrams of cross sections at successive levels from the pedicel to the top of the flower.

The pedicel has five bundles (Figs. 1.32*a*) which unite at the base of the flower, forming an unbroken cylinder. The sepal traces are then given off (Fig. 1.32*b*), three to each sepal, the three together leaving one gap. The traces to the petals, one to each, are next given off (Fig. 1.32*c*) just below the level at which the sepal gaps close. Then the stamen traces pass off, one to each organ, in several whorls (Fig. 1.32*d—e*). Above the uppermost whorl of stamens the vascular cylinder becomes complete again. Then the dorsal traces of the carpels are given off (Fig. 1.32*f*), followed by the pairs of ventral traces to each carpel. These ventral traces are derived from the sides of the gaps formed by the dorsal traces (Fig. 1.32*g*), and their gaps therefore merge with those of the dorsal traces. The ventral traces immediately twist, passing out inverted into the carpel. Figure 1.32*h* shows the condition in the receptacle just above the outward passage of the ventral carpel traces. The base of the loculus of each carpel is seen with the dorsal trace at the outside and the pair of ventrals on the inside. Internal to the ventrals the remainder of the vascular cylinder is seen as five masses of less strongly developed tissue (shown by dotted lines), Fig. 1.32*h*, chiefly phtoem, with some weak xylem. This tissue (Fig. 1.32, center), becomes smaller and weaker toward the rounded top of the receptacle and fades out. (This fading out of vascular tissue is similar to that often seen in the vascular cylinder of determinate leafy stems.) Figure 1.32*i* shows the carpels nearly free from the receptacle, with their three-bundle supply. Figure 1.32*j* (above the receptacle) shows cross sections of the typical follicles with their dorsal bundles, and their pairs of inverted ventral bundles.

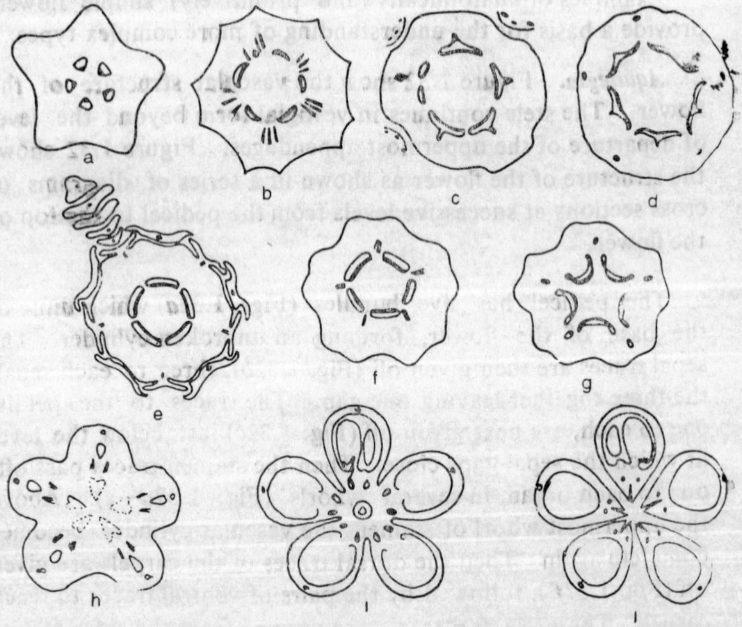

Fig. 1.32. Diagrams representing vascular structure of flower of *Aquilegia* as seen in series of transverse sections from pedicel to level above top of receptacle. Detailed description in text.

ANATOMY OF MORE COMPLEX FLOWERS

Fusion in the Floral Skeleton. In the specialization of the flower, fusion and reduction of appendages have been the outstanding changes. Stages of *cohesion*—where members of the same whorl are fused to one another—and of adnation—where members of a whorl are fused to members of a whorl above or below—are found in many families. Evidence of reduction is to be seen in vestigial structures and through comparative study of related forms. In the evolutionary history of fusion, organs are at first united

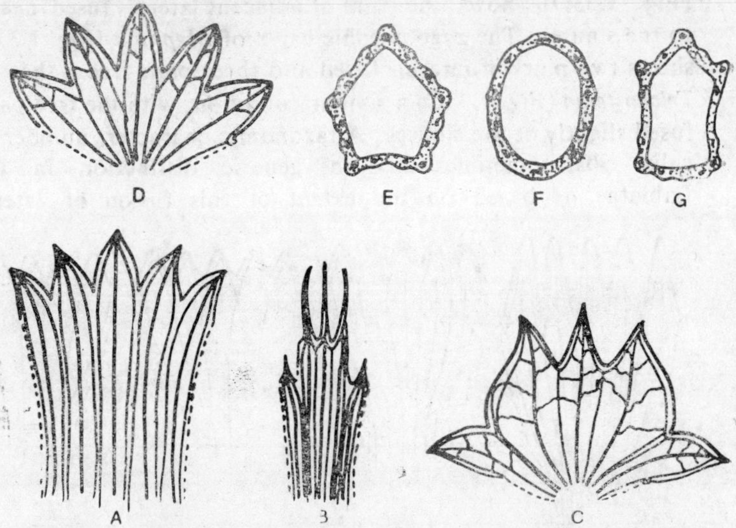

Fig. 1.33. Anatomy of calyces. showing stages in fusion cf lateral bundles under cohesion, *A. Nepeta veronica* ; *B. Blephilia hirsuta* ; *C. Trichostema dichotomum* ; *D. Ajuga reptans. E, F, G* (cross sections) : *E. Monarda didyma* ; *F. Physostegia virginiana* ; *G. Salvia patens.* Detailed description in text.

externally, and internal structure is not modified. The line of fusion is histologically obscure or absent ; the vascular tissues are unaffected. Ultimately the fusion becomes more intimate and, wherever the vascular bundles lie close to one another, fusion may involve them also. Fusion may be between the traces alone or between the traces and the bundles within the appendages, but it seldom extends throughout the length of the bundles. Often there is ontogenetic and histological evidence of the fusion.

Fusion under Cohesion. The effect of cohesion on the vascular skeleton is chiefly upon lateral traces and bundles that are brought close together by gamosepaly, gamopetaly, syncarpy, and the marginal fusion of the upfolded margins of the carpel. It can be simply show by examples. The gamosepalous calyces of the mints shown all stages of fusion of the vascular tissue. In *Nepeta* (Fig. 1.33A) each sepal has a midrib and two lateral bundles ; there is no fusion. The calyx of *Ajuga* (Fig. 1.33D) shows the pairs of adjacent laterals fused nearly to the sinuses. The zygomorphic calyx of *Blephilia* (Fig. 1.33B) shows two pairs of laterals fused and three pairs free ; that of *Trichostema* (Fig. 1.33C) a similar condition, with the free pairs fused slightly at the sinuses. A taxonomic character, number of calyx ribs, sometimes used in generic distinction in the Labiatae, is based on the extent of this fusion of lateral

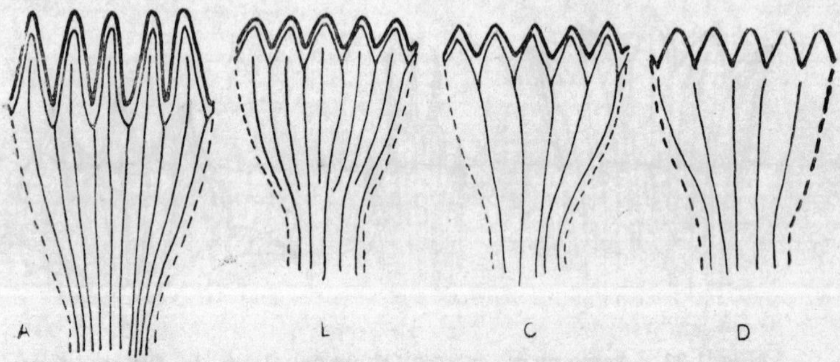

Fig. 1.34. Vascular system of gamopetalous corollas showing fusion of lateral petal-bundles and loss of bundles in reduction. *A. Helianthus divaricatus* ; *B. Senecio Fremontii* ; *C. Chrysanthemum Leucanthemum* ; *D. Xanthium orientale*. In all examples fusion of lateral bundles is complete nearly to the sinuses ; in *Chrysanthemum*, the free ends of the lateral bundles have been lost. Stages in loss of median petal bundles are shown : *A.* bundles unreduced ; *B.* basal parts lost ; *C, D,* bundles absent.

bundles. Examples are seen in *Monarda* (Fig. 1.33E), with 15 bundles, no fusion ; *Physostegia* (Fig. 1.33F), with 10 bundles, all pairs of laterals fused; and *Salvia* (Fig. 1.33G), with 13 bundles, two pairs of laterals fused. In petals (Fig. 1.33) and carpels (Fig. 1.34) similar fusion is common.

The folding of he carpel presents the simplest type of fusion under cohesion in the gynoecium (Fig. 1.35). The two ventrals are brought close together, forming the pair of bundles characteristic of the ventral side of the primitive carpel, the follicle (Fig. 1.35*a*,*b*). In earliest phylogenetic stages the carpel margins merely touch or are lightly fused (Fig. 1.35*a*). (The carpel is even not completely closed in some genera). With closer fusion, the epidermal layers on the surface of contact are lost and the two bundles form a double bundle (Fig. 1.35*c*) ; with complete fusion they become a single bundle that shows no histological or ontogenetic evidence of its doubleness (Fig. 1.35*d*).

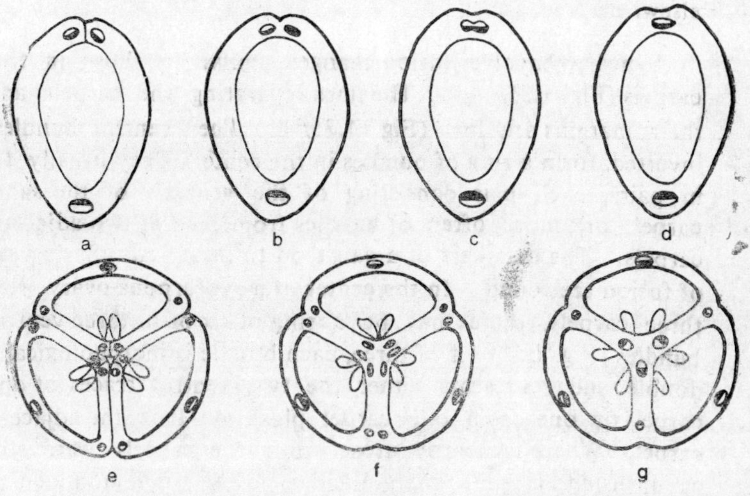

Fig. 1.35. Diagrammatic cross sections of typical ovaries showing structural effects of cohesion, especially upon the vascular bundles *a-d*, stages in the fusion of the ventral bundles of follicles ; *e—g*, stages in the development of syncarpy, the fusion of carpels and of their ventral and lateral bundles, *a*, margins of carpel appressed, but unfused, epidermal layers persisting ; *b*, margins fused, epidermal layers lost ; *c*, the two ventral bundles slightly fused ; *d*, the ventral bundles completely fused, without histological evidence of the fusion ; *e*, the carples cohering, but without structural fusion ; *f*, the lateral carpel-walls fused, the ventral and lateral bundles of adjacent carpels approximated in pairs but not fused ; *g*, the pairs of lateral and ventral bundles of adjacent carpels intimately fused.

Fusion in the vascular tissue of a carpel may be present in the ventral bundles from an origin as one trace throughout their length, or may exist only in part of the carpel (Fig. 1.36) ; where the ventral bundles arise as separate traces, they may unite at any point in their course. A carpel with two bundles, a dorsal and a ventral that consists of two bundles, may have one or two traces. Further, in much reduced carpels, especially in achenes, all three carpel traces may arise as one and separate only at the base of the loculus (Fig. 1.41*j*). In any carpel, fusion may also occur among the bundles in the distal part of the ovary or in the style, when the traces and bundles are elsewhere free.

Syncarpy involves fusion changes similar to those in free carpels (Fig. 1.35*e*—*g*). The lines separating the carpels and their margins are lost (Fig. 1.35*f*). The ventral bundles, inverted, form a ring of bundles in the center. They usually lie in pairs, each pair consisting of the ventrals of the same carpel, or, more often, of bundles from each of two adjacent carpels. The members of a pair tend to fuse and all degrees of fusion are found. In the center of a syncarpous ovary with three carpels, there may be a ring of six or of three ventral bundles (Fig. 1.35)—if of three, each bundle is morphologically double and represents either the two ventral traces of one carpel or one from one carpel plus one from the adjacent carpel. Where syncarpy arose while the carpels were still open, fusing margin to margin and enclosing a common loculus (Fig. 1.37)—or where the septa have been reduced or lost, and placentation is parietal—the ventrals of adjacent carpels lie in pairs, or fused, in the outer wall of the ovary (Fig. 1.37*A*). Carpel limits in such ovaries run through the placentas, which are morphologically double structures, each half belonging to a different carpel.

Fusion under Adnation. Whenever members of one whorl become fused to those of other whorls—whether the whorls consist of like or of different organs—their vascular bundles tend to fuse, as under cohesion. Fusion takes place between bundles that are near together either radially or tangentially and may involve any number of bundles belonging to two or

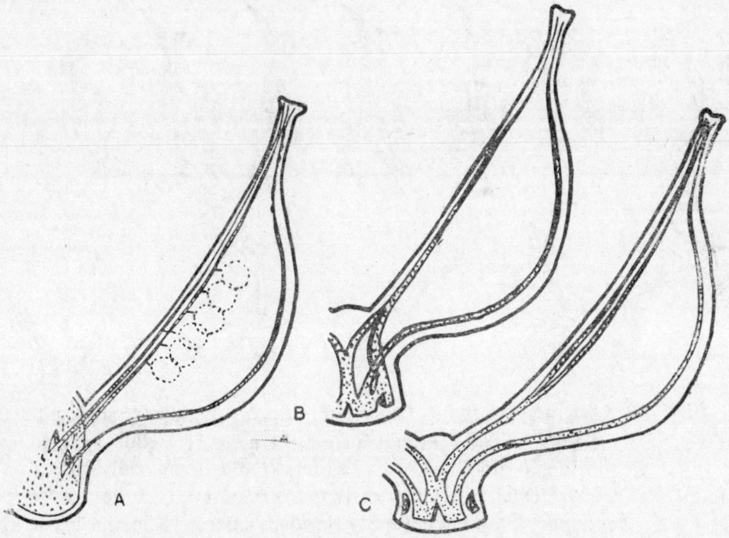

Fig. 1.36. Vascular supply of the carpel. *A*, three traces from three gaps, the bundles within the carpels unfused to the stigma ; *B*, three traces from a dissected stele, the ventrals soon fused, only two bundles in the carpel ; *C*. the ventral traces arising fused but becoming free in base of ovary. In *B* and *C*, the top of the floral stele is used up in the formation of the ventral traces.

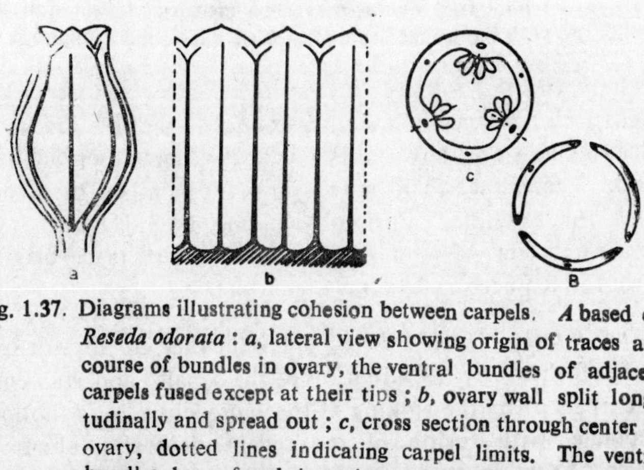

Fig. 1.37. Diagrams illustrating cohesion between carpels. *A* based on *Reseda odorata* : *a*, lateral view showing origin of traces and course of bundles in ovary, the ventral bundles of adjacent carpels fused except at their tips ; *b*, ovary wall split longitudinally and spread out ; *c*, cross section through center of ovary, dotted lines indicating carpel limits. The ventral bundles have fused in pairs and the fused bundles and placentae lie on the line of carpellary fusion. *B*, illustrating theoretical position of carpels before fusion to form syncarpous ovary of the type shown in *A*.

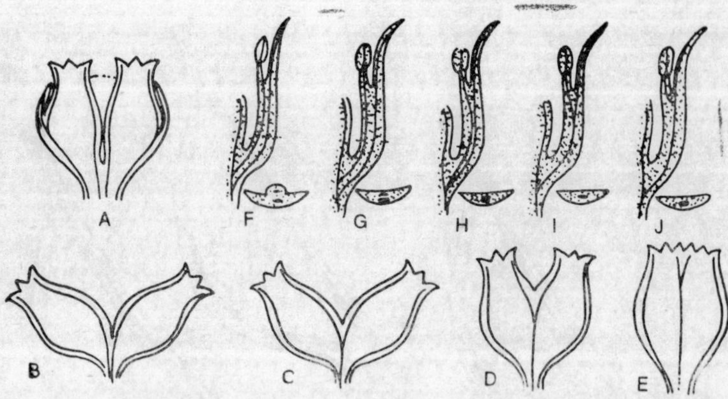

Fig. 1.38. Diagrams showing fusion of vascular bundles under adnation. *F—J*, stages in fusion of stamen to petal, as seen in longitudinal and cross sections: *F*, the organs only lightly fused, externallv still clearly two structures ; *G*, fusion more complete, double nature not evident externally in the lower part, but in vascular structure clear throughout ; *I.J.* the vascular bundles fused for various distances from the base, without histological evidence of double nature. *A*, adnation of bracts to flowers, *Lonicera caerulea*, the bracts adnate to the lower part of the inferior ovaries and the median bundle of the bract fused to one of the bundles in the ovary wall. *B—E*, adnation of flower to flower : *B*, *Lonicera canadensis* : *C.L. tatarica* ; *D.L. fragrantissima* : *E, L. oblongifolia*. The four species show increasing fusion of ovaries, with similar increasing fusion of the vascular supply of the two flowers.

more whorls. Stages in a simple example—that of the epipetalous stamen—are shown in Fig. 1.38 *F—J* and described in the legend. These stages can be seen in many families, the Crassulaceae, for example. Fusion involving several whorls, with union extending varying distances from the trace origin, is shown for Rosaceous flowers (Fig. 1.‹0*M—P*).

The Inferior Ovary. It is apparent that an inferior ovary can be formed by the adnation of the sepals, and stamens to the carpels or by the sinking of the gynoecium in a hollowed receptacle, with fusion of the receptacle walls about the carpels. From anatomical evidence, it is clear that the inferior ovary in nearly all families is made up of appendages fused to the carpels. In the ovary walls of some plants, *Alstroemeria*, *Hedera*, for example, the traces of all adnate organs run free to the base of the ovary in the positions they would have if there

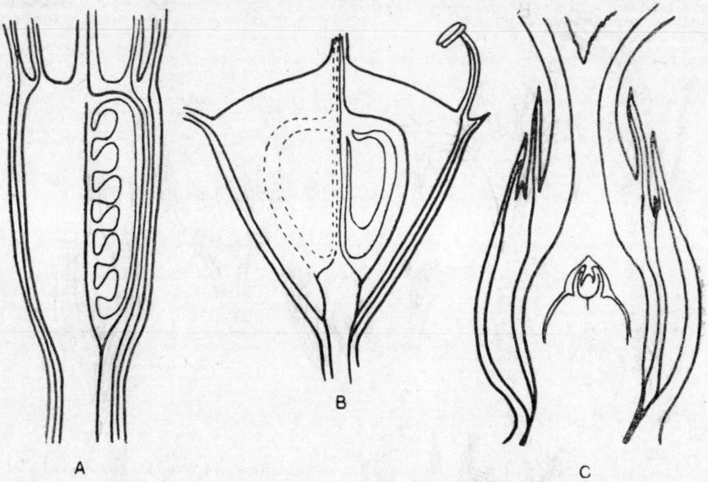

Fig. 1.39. Diagrams of vascular structure in inferior ovaries. *A*, *Alstroemeria*, and *B*. *Hedera helix* with bundles of the calyx, corolla, and stamens with no fusion with each other or with carpellary bundles in their course through the ovary wall : *C*, *Juglans nigra* (flower with adnate bracts and bractlets as well as sepals), bundles of the sepals, bractlets, and bracts fused with carpellary bundles for various distances from their origin. (Ventral carpellary bundles not in plane of section).

were no adnation (Fig. 1.39 *A*, *B*). There is no fusion. In most species, however, with inferior ovaries, there is fusion within the ovary wall for various distances, as in *Juglans* (Fig. 1.39*C*) and some of the Ericaceae (Fig. 1.40*A*—*F*). The "ovary wall" in such forms, as is shown by the position and course of the bundles, clearly consists of the true ovary wall plus the tissues of the adnate organs.

The sinking of the gynoecium in the receptacle necessarily involves invagination of the top of the stele. An "ovary wall" so formed contains two cylinders of stelar bundles, the inner inverted. Within the cavity, the distal part of the receptacular

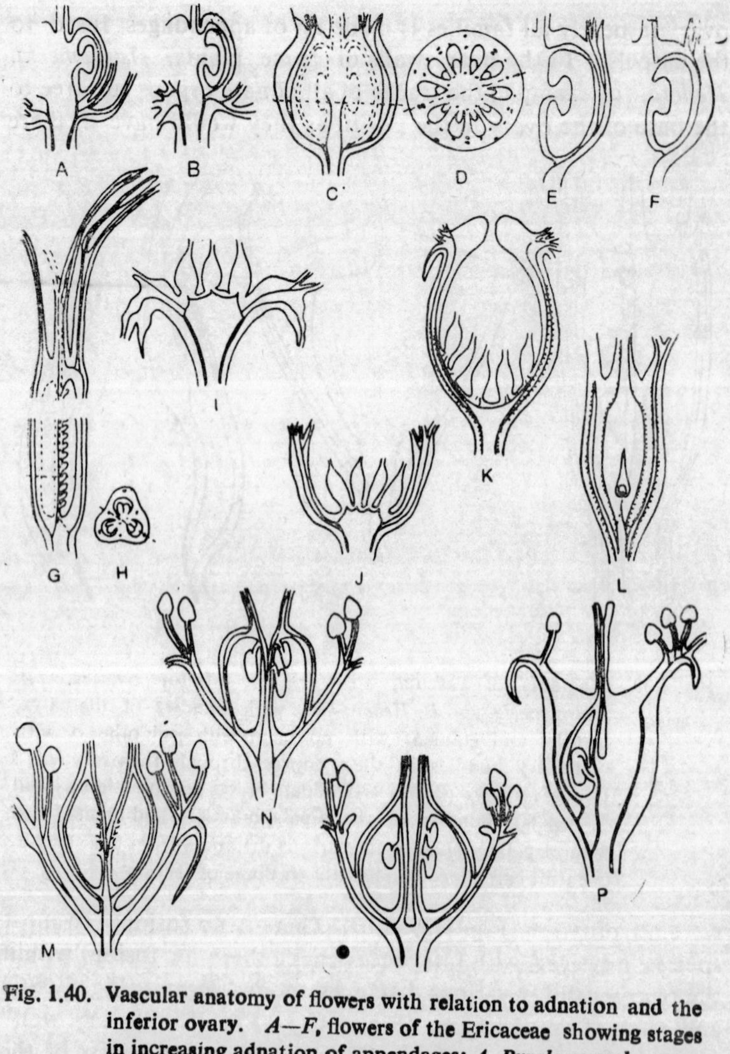

Fig. 1.40. Vascular anatomy of flowers with relation to adnation and the inferior ovary. *A—F*, flowers of the Ericaceae showing stages in increasing adnation of appendages: *A*, *Pyrola secunda*, ovary superior, no adnation ; *B*, *Andromeda glaucophylla*, ovary snperior, stamens fused to petals at base, with fusion of the vascular bundles of the two organs nearly to point of separation ; *C,D*, *Gaytuscacia frondosa*, ovary inferior, calyx, corolla, and stamens adnate to carpels but vascular supply of these organs free in their course through region of adnation, except close to base, where the bundles that lie in some radii—those

stele, from which the carpel traces are derived, turns downward and the morphologically uppermost carpels are the lowest. *Darbya* (Fig. 140*L*) shows this well for the Santalaceae, the only family in which the inferior ovary has been shown to be of receptacular nature. In a few genera the tip of the receptacle is sunken and the carpels line the cavity, as in *Rosa* (Fig. 1.40*K*) and *Calycanthus*, but the sides of the receptacular cup are not fused to the carpels and the ovaries remain superior. The fleshy fruit of the rose is, in its lower part, receptacle; in its upper part, it consists of fused appendages, as clearly shown by the course of the vascular bundles (Fig. 1.40*K*). Stages in the development of the peculiar rose-type of fruit are shown by related genera (Fig. 1.40*I*—*K*). It is commonly believed that the rose hip and the apple represent fleshy

of sepals and outer stamens and those that lie in alternate radii (those of petals and inner stamens)—fused ; *D.* cross section through center of ovary showing position and freedom of bundles of organs adnate to carpels (dorsal carpellary bundles not fused with other bundles because lying in different radii ; E.F. *Vaccinium corymbosum* and *V. macrocarpon*, showing fusion, progressively greater than in *C*, of petal and stamen bundles with dorsal carpellary bundles. *G.H. Iris versicolor*, typical inferior ovary, the bundles of the adnate organs in each radius fused throughout their length as far as the top of the ovary. *I, J, K, Dalibarda repens, Rubus triflorus, Rosa setigera*, respectively : *I, J*, showing flattening of receptacular tip ; *K*, showing inversion of tip. *K*, with stelar bundles inverted at about one-third distance to top ; the lower third, with its inverted stelar bundles is receptacular ; the upper two-thirds, "calyx tube"—sepals united with petals and stamens—similar to and homologous with that in *I* and *J* but in-arched, and covering the sunken carpels. *L. Darbya*, ovary inferior, formed by inversion of lop of receptacle as evidenced by inverted vascular bundles. *M—P*, series of related Rosaceous genera, showing stages in adnation of "calyx tube" to carpels : *M, Physocarpus opulifolius*, the carpels superior, the sepals, petals, and stamens, fused, but the "calyx tube" free from the carpels ; *N, Sorbus sorbifolia*, the "calyx tube" fused to the base of the carpels ; *O. Spiraea van Houtei*, the "calyx tube" adnate to carpels halfway to the top of the ovary ; *P. Malus pumila*, the "calyx tube" adnate to the ovary to the top, bundles in some of its radii fused with the dorsal carpellary bundles.

receptacles of similar nature. But the rose and the apple tree belong to widely separated groups within the rose family, as shown by taxonomic and cytological, as well as anatomical evidence, and their fruits are not homologous. The ovary and fruit of the apple tree and closely related genera are entirely appendicular—the traces to all the floral parts arise from a receptacle of normal form and there are no inverted stelar bundles. Adnation is extensive (Fig. 1.40*P*) : in one set of radii, sepal median trace, stamen trace, and dorsal-carpellary trace are united ; in radii alternate with these, the lateral traces of adjacent sepals, petal trace, and three stamen traces are united. Fusion of the traces of the various organs extends for different distances, that of the dorsal carpellary with other traces only a short distance. The ventral carpellary traces alone show no fusion with other traces. Histologically the limits of the carpels can be seen in some varieties of apple. Morphologically, the flesh of the apple consists of tissues of all parts of the flower, with the receptacle forming only an insignificant bit at the base. Genera related to the apple show stages in the evolutionary development of the ovary (Fig. 1.40 *M—P*).

The inferior ovary in most families—orchids, irises, evening primroses, for example—shows extreme fusion of the bundles of all appendages so that a cross section shows only a few bundles in the outer ovary wall (Fig. 1.40*G*,*H*).

Adnation of Flowers with Other Flowers and Other Organs. Two or more flowers borne close together on a peduncle may become anatomically fused, as in *Mitchella*, *Maclura*, some species of *Lonicera* and *Cornus*. Such fusion usually involves only the basal parts, especially the ovaries. Various species of *Lonicera* show all stages in the fusion of the ovaries of two flowers (Fig. 1.38 *B—E*); *Mitchella* shows complete fusion of the ovaries of two, rarely three and four flowers. In these genera, fusion is so complete in the most intimately fused species that some of the bundles of one flower are fused to those of the other. One bundle within the double ovary may supply perianth parts of two flowers.

Bracts may be fused to the flowers that they subtend, as in *Juglans*, and *Lonicera* (Figs. 1.38A, 1.39C). Fusion of all the parts of an inflorescence—flowers, bracts, and axis—is illustrated by the pineapple (*Ananas comosus*). The many flowers, each subtended by a bract, are closely packed in spirals. The inferior ovary of each flower is completely sheathed by bracts, the anterior half by its own bract, the remainder by parts of three other bracts. Fusion between the ovary and the four surrounding bracts is histologically complete to the top of the ovary. This fusion extends throughout the inflorescence, uniting all ovaries, all bracts, and the axis into one structure. The bundles of the flower are wholly free from each other and from those of the bracts, but bundles of adjacent bracts are often fused and supply two bracts.

Where vegetative parts are similarly fused, for example, bract to peduncle (*Tilia*); axillary peduncles to the adjacent stem (*Sparganium*, *Streptopus*)—giving supra-axillary inflorescences—the vascular tissue may also be fused and structurally complex.

Placental Vascular Supply. Because in many flowers the ventral bundles give rise directly to ovular traces, these bundles have been, incorrectly, called the "placental supply." In follicles, achenes, and some other types of ovaries, the placenta is merely a position, and there is no placental supply as such. Where the placenta is a definite, often fleshy enlargement of an area along the margin of the carpel, it is supplied by branches from the ventrals, which branching further, give rise to the ovule traces, as in the Ericaceae and Cucurbitaceae. The ovules on a placenta in a syncarpous ovary obviously may receive their vascular supply from different carpels. Where the number of ovules has been reduced from an earlier larger number, the surviving ovules may have strong vascular traces; in extreme reduction—some types of basal and free-central placentation—one or two ovules may have the vascular supply of more than one carpel (Juglandaceae; Polygonaceae). (Anatomical structure demonstrates that no angiosperm ovules are cauline.)

Vestigial Vascular Tissue. Evidence of the structure of ancestral floral types is frequently found in the vascular skeleton of modern flowers. The presence of vestiges of the vascular stele at the tip of the receptacle has been discussed. Remnants of the vascular supply of lost organs may be present in the tissues of the receptacle when all external evidence of the organs has disappeared. Only rarely does the vascular supply of a vestigial organ disappear while external remnants are still present. The vascular vestiges of lost appendages are usually stubs of traces buried in the cortex of the receptacle. The lost organs so represented may be entire whorls, especially of petals and stamens, or individual members of persisting whorls. Traces to lost petals are common in apetalous flowers : *Aristolochia* ; *Rhamnus* (apetalous forms) ; *Salix* ; *Quercus* (some species) ; to lost stamens, for example, in Scrophulariaceae, Labiatae, Cucurbitaceae, *Lysimachia*, *Chionanthus*, *Lychnis*, to lost carpels in Caprifoliaceae, Ericaceae, Rutaceae, Valerianaceae. In unisexual flowers there are commonly external remnants of lost organs ; in those in which such vestiges have disappeared, vascular traces may be present within the receptacle : Fagaceae ; Caryophyllaceae ; Urticaceae.

Traces to lost ovules are among the commonest vestigial structures in the flower. They are frequent in some families, especially the Ranunculaceae, Umbelliferae, and Rosaceae (Fig. 1.41 B,E,N).

Reduction of Vascular Supply within an Organ. As reduction occurs in an organ in evolutionary specialization—for example, as a follicle with several ovules becomes an achene with one ovule ; as a corolla becomes greatly reduced in size in the crowding together of small flowers in a head, as in the Compositae—the vascular supply of the organ is greatly reduced. In carpels either the dorsal or the ventral bundles may be shortened, the latter extending only to the ovules. In extreme reduction the entire carpellary supply consists of a small bundle running directly from the receptacular stele to the ovule. Stages and variety in this loss are shown in achenes (Fig. 1.41). In the stamen, the bundle, which commonly extends to the anther, may enter only the base of the filament. In the petal, either median or lateral bundles may be shortened or lost. In the gamopetalous corolla, the median bundle is more often lost (Fig. 1.34B,C,D) and the fused laterals persist ; the latter may be reduced in length (Fig. 1.34D).

The Sepal and the Petal. Both the sepal and the petal are commonly leaf-like in form and general appearance. Structurally also the former is like the leaf except when it is petaloid. The petal is leaf-like in grosser structure but histologically differs in many ways from the typical leaf (Fig. 1.42). The vascular system is commonly reduced in amount and in supporting tissues. The mesophyll is simple in structure : there is usually no palisade layer and the spongy tissue consists of few cell layers. Chromoplasts or colored cell sap, or both, are present. The epidermis is less simple than that of leaves. Its cells are commonly weak-walled and of complex form, often undulate, stellate, or irregularly lobed in outline and dovetailed with one another. With this shape and arrangement they doubtless form a layer mechanically stronger than one consisting of cells of simpler form. The cells of one or both surfaces are usually papillose. Stomata are fewer than those on leaves and frequently nonfunctional or absent. The guard cells lack chloroplasts except when the mesophyll also has these plastids. Secretory areas and glandular hairs are common. Intercellular spaces, usually lacking in the epidermis of leaves, are frequent. They lie in the loops or lobes of the cell wall (Fig. 1.43C, D,E) but are always covered by cuticle. The cuticle varies greatly in thickness being very thin on delicate, ephemeral petals.

The extreme types of petals show many variations from, and exceptions to, the typical structure described above. The less highly specialized type of petal has a strong vascular supply, a weak palisade layer, chloroplasts, a nonpapillose epidermis with many stomata and often a hypodermis, and supporting tissue about the larger veins. The most highly specialized type has the vascular tissue weak, with veinlets and even most of the main bundles lost, and wholly without accompanying supporting tissue. The mesophyll consists of one to three poorly defined layers of widely spaced cells which are lacking along the margins and in distal parts, so that part

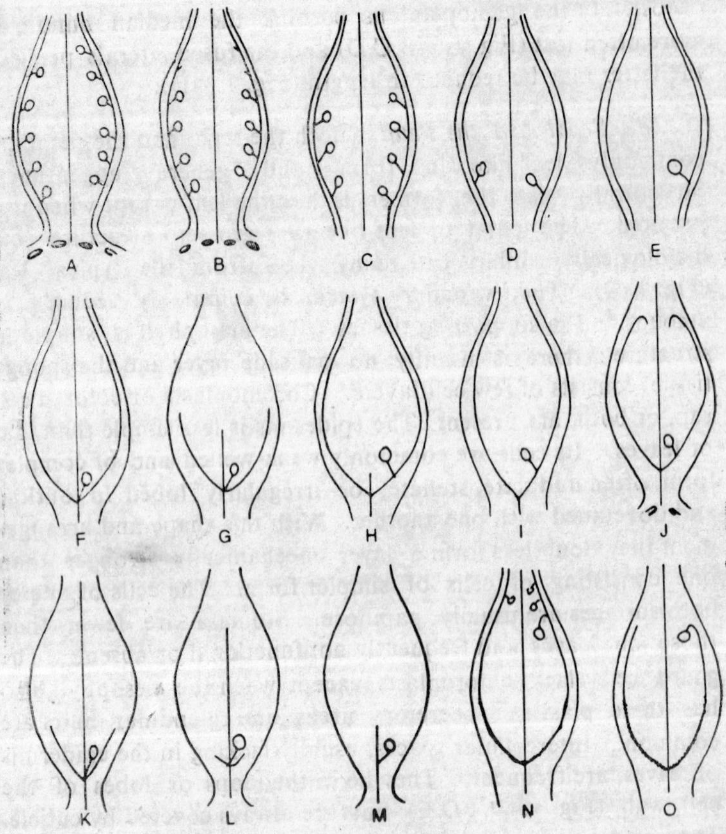

Fig. 1.41. Diagrams of carpel structure showing reduction of ovules and fusion and reduction of vascular supply. *A—E*, follicles : *A*, *Helleborus viridis*, typical follicle, with many ovules and three traces ; *B*, *Trollius laxus*, ovules reduced in number, traces to lost ovules persisting ; *C*, *Aquilegia canadensis*, upper ovules and their traces lost ; *D*, *Hydrastis canadensis*, all ovules but lower two lost, one of these abortive ; *E*, *Waldsteinia fragarioides*, all ovules but one lost, trace to one other persisting. *F—O*, achenes : *F—M* with basal ovule surviving ; *N*, *O*, with an upper ovule surviving. All with dorsal and ventral traces united at the base, in some as far as the ovule, which then appears to be attached to the dorsal (*F*, *G*, *K*, *L*, *M*). *F*, *Geum rivale*, with distal parts of all bundles present ; *G*, *Duchesnea indica*, dorsal and ventral bundles greatly shortened ; *H*, *Fragaria vesca*, with dorsal

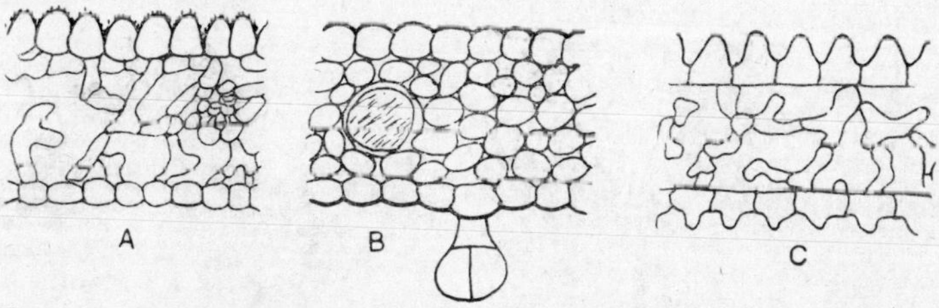

Fig. 1.42. Cross sections of petals. *A, Amelanchier laevis* ; *B. Lysimachia Nummularia*, showing glandular hair and secretory chamber ; *C, Pinguicula vulgaris.*

of the petal consists of epidermal layers only. The epidermis is papillose, with chromoplasts or coloured cell sap, and without stomata or glands.

The Stamen and the Carpel. Stamens and carpels consist chiefly of unspecialized parenchyma. Their traces continue in the organs as weak or strong, often amphicribral bundles, for varying distances, and with varying amounts of branching. In the stamen, the trace or traces commonly run unbranched to the anther, but in delicate stamens may fade out anywhere in the filament. In some stout stamens, as in those of *Magnolia*, small lateral branches are present. The traces of the carpel may pass as unbranched bundles to the style or stigma, or may branch in varying degrees, some showing leaf-like venation in the branching. Within the stigma the bundles frequently branch. In reduced carpels, all branching is lost, and the main bundles are shortened by loss of the distal portions. For example, the ventral bundles may not extend beyond the

bundle greatly shortened ; *I, Agrimonia striata*, with dorsal bundle lost beyond its union with the ventral bundles *J, Ranunculus Ficaria*, the ventral bundles recurving, not entering the style ; *K.R. Flammula*, the ventral bundles greatly shortened ; *L, R. Cymbalaria*, only vestiges of the ventral bundles persisting, the dorsal bundle greatly shortened ; *M, R. aquatilis*, the ventral bundles lost beyond the ovule, the dorsal bundle continuing hardly beyond it ; *N, Potentilla recta*, one ovule surviving, others vestigial, the ventral bundles unreduced ; *O, P. canadensus*, one ovule surviving, the ventrals lost beyond the ovule.

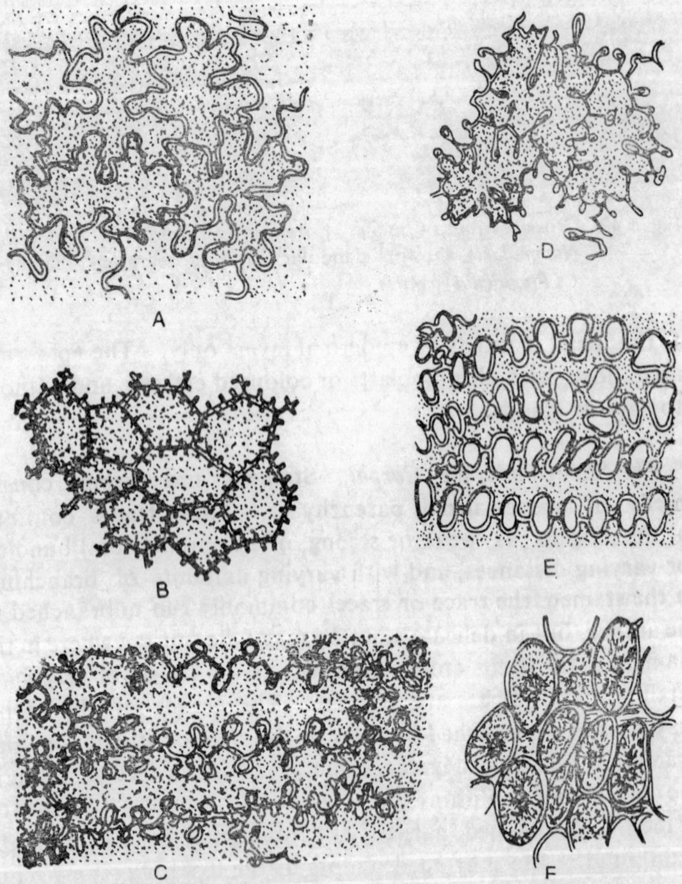

Fig. 1.43. Epidermis of corolla. *A, Calceolaria* ; *B, Pelargonium* ; *C, Clarkia* ; *D, Anchusa* ; *E, Linum* ; *F, Erythrina*.

placentae and all bundles may fail to enter the style. In extreme reduction, as in small achenes, the vascular supply may be reduced to little more than a basal remnant (Fig. 1.41M). Vestigial bundles are frequently found. These are the weak or abortive remnants of traces or bundles representing the supply to organs lost in evolutionary modification. The most common vestigial bundles are the traces to lost petals, stamens, and carpels, represented by stubs arising from the receptacular stele as normal traces but ending blindly in the cortex of the receptacle, and ovule traces that end blindly in the placenta or the margin of the carpel (Fig. 1.41B,E). The epidermis is of typical structure, the cells usually with straight walls and thin cuticle. Stomata occur on stamens only when these structures are expanded and leaf-like. In carpels they occur freely on the outer surface and are even found in the epidermis lining the ovarian cavity.

THE INFLORESCENCE

Introductory

The inflorescence is the part of a plant which is concerned with the production of flowers. It may or may not be distinct from the vegetative part. Thus, in some annual plants the entire branch-system of the shoot may be regarded as an inflorescence, whilst in biennials the second year's growth may be so regarded. Inflorescences are classified according to the type of branching into (*a*) indefinite or racemose and (*b*) definite or cymose. The leaves, whether ordinary vegetative leaves or modified ones which subtend the flowering branches are termed bracts. In racemose inflorescences the growing point of the primary axis does not terminate in a flower, but, throughout its growing period, it is continually developing the axis and new lateral members, *i.e.* it is a monopodium, whereas in cymose inflorescences (sympodia) the primary axis and the successive daughter axes terminate in flowers. It is, of course, possible for the primary axis to be monopodial, and the lateral branch-systems to be sympodial, thus forming a compound racemo-cymose inflorescence.

Simple Racemose Inflorescences

Of these we recognise four chief types :—

(*a*) The typical raceme (Fig. 1.44), in which the mother-axis (peduncle) is elongated, and the flowers are stalked (pedicellate). Examples are found in *Crotalaria* and many orchids.

Similar to this in essential characters is the corymb, which may be regarded as a modification of the typical raceme. The mother-axis is relatively shorter, and, owing to the elongation

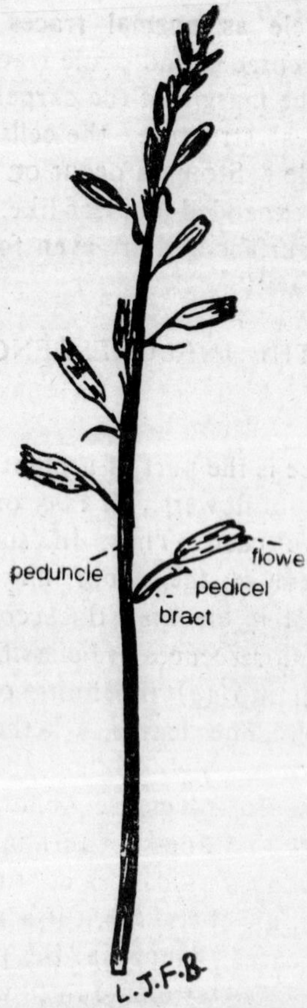

Fig. 1.44. Inflorescence of Bluebell (Raceme).

of the lower pedicels, all the flowers come to one level (Fig 1.45). This renders the whole inflorescence more con-

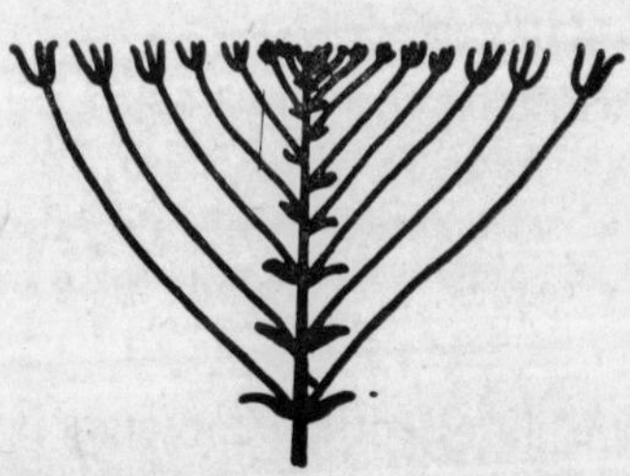

Fig. 1.45. Diagram of a Corymb.

spicuous to insects, and the individual flowers tend to remain small. Good examples are found in many Cruciferae (*e.g.* candytuft).

Inflorescences intermediate in character between the corymb and typical raceme are described as corymbose racemes, *e.g.* the wallflower, in which the inflorescence is corymbose when young, but lengthens out when fruiting.

(*b*) The spike is a racemose inflorescence in which the mother-axis is elongated, and the flowers are *sessile* (Fig. 1.46), *e.g.* spotted orchid and *Plantago*. By this arrangement small flowers may be aggregated in a cylindrical mass.

There are one or two special forms of the spike. The spadix is a massive fleshy spike, bearing small, usually unisexual flowers occurring especially in the Araceae. It is protected by a large enveloping bract, sometimes green, more usually petaloid, known as a *spathe*. The spathe and upper part of the spadix serve to attract insects, and sometimes, as in *Arum maculatum* (Fig. 1.47), there is a fly-trap mechanism in connexion with the pollination of the flowers. The catkin (Fig. 1.48) is generally a long, more or less pendulous, deciduous

Fig. 1.46. Diagram of a Spike.

compound spike, bearing unisexual flowers. It is found in many nut-bearing and other trees, *e.g.* birch, hazel, popular. The male catkin as a rule hangs loosely in the air so that the microspores (pollen), protected from rain by the catkin scales, are readily blown out by the wind when dry.

(*c*) The umbel (Fig. 1.49) is a racemose inflorescence in which the flowers are stalked, but, owing to the abbreviation

Fig. 1.47. Spadix of *Arum*. Par t of lower end of spathe removed to expose the flowers.

of the mother-axis, are given off at one level. The indefinite growing point develops a large number of lateral flowers, but does not give rise to an elongated mother-axis. Each flower of the umbel may be subtended by a bract, the collection of bracts forming the involucre of the umbel.

(*d*) The capitulum (Fig. 1.50) is a racemose inflorescence in which the flowers are sessile, and crowded together on a reduced or abbreviated mother-axis which is called the *disc*. It is sometimes flat, more frequently dilated and convex or conical. Examples are found chiefly in the Compositae (Fig. 1.50). The student must clearly recognise that the heads of the daisy, dandelion, etc., are not single flowers, but inflorescences bearing a large number of sessile flowers or florets. The capitulum is invested by a number of small scaly, overlapping, barren bracts, together forming the involucre. Bracts (or paleae) may also subtend the individual florets as in *Anthemis nobilis* (chamomile).

The massing together of small flowers in the umbel and capitulum has the same biological significance as in the corymb.

Cymose Inflorescences

These are usually either *uniparous* or *biparous*. In uniparous (or monochasial) forms each successive axis ends in a flower

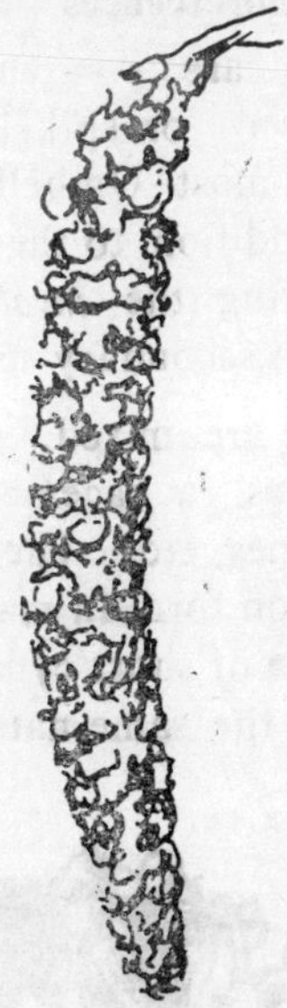

Fig. 1.48. Male Catkin of Hazel.

after producing one daughter-axis. Four different types are recognised : the bostryx. drepanium, cincinnus and rhipidium.

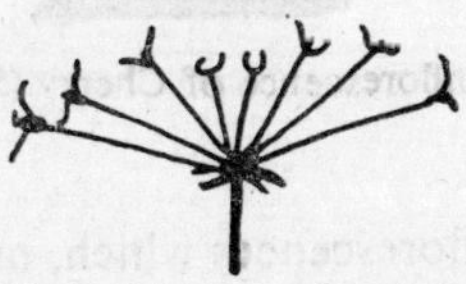

Fig. 1 49. (*a*) Diagram of a Simple Umbel.

They are *sympodial*, and sometimes resemble typical racemes. Uniparous cymes resembling racemes can be distinguished by the fact that the bracts, if they are present, are on the opposite side of the sympodial axis from the leaves. If bracts are not developed they are not easily distinguished.

In *biparous* cymes each axis ends in a flower after producing two daughter-axes. It is also called a dichasium (Fig. 1.51).

Typical examples are found in many Caryophyllaceae. Sometimes the daughter-axes are not given off at the same level, *e.g.* some *Ranunculus, Helleborus*, etc.

Compound and Mixed Inflorescences

Many inflorescences are compound, *e.g.* a raceme of racemes, (*e.g. Delphinium*), or spikelets (*e.g. Triticum*), or an umbel of umbels (*e.g.* most Umbelliferae). The *compound umbel* (Fig. 1.52), in addition to the bracts at the base of the chief branches constituting the *involucre*, may have smaller bracts at the base of each secondary umbel the *involucel*.

Many inflorescences are mixed. We may, for example, have a raceme of spikes, a raceme of capitula, a spike of capitula, a raceme of cymes, etc. The raceme of *spikelets, i.e.* the *panicle*, is a common form in many grasses (*e.g.* oat) ; in *Aesculus* there is a raceme of short cymes (cincinni). In *Syringa* the inflorescence is of the same nature, but the branching is much more copious.

Fig. 1.49. (*b*) Inflorescence of Cherry (Simple Umbel).

Special Forms

There are many inflorescences which, owing to abbreviation of axes or special crowding of the flowers, do not so readily yield to careful analysis. In *Crataegus*, for example, the in-

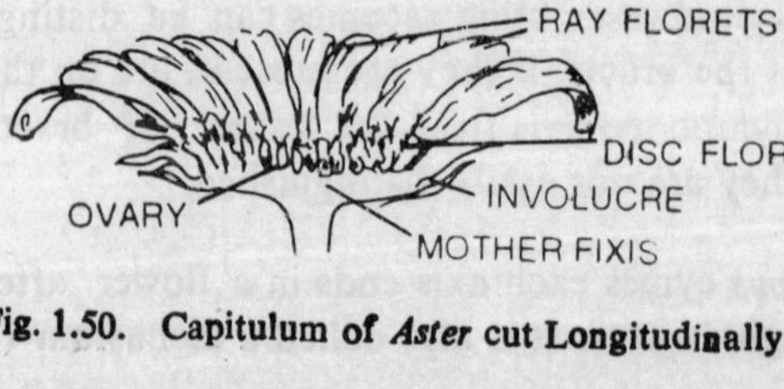

Fig. 1.50. Capitulum of *Aster* cut Longitudinally.

florescence might be mistaken for a typical corymb. Examination will show, however, that the lateral axes borne on the main axis are really cymes. It is a *corymbose cyme.*

Similarly the globular inflorescences of *Hydrangea* and of the snow-ball tree (*Viburnum opulus*) are corymbose cymes, rendered conspicuous by the petaloid calyces and enlarged corollas of flowers specialised for this purpose. In horticultural varieties all the flowers are sterile.

In the cultivated geranium (*Pelargonium*) and many species of *Narcissus* the inflorescence, at first sight, appears to be an umbel. But it will be found that the young flowers are not by any means aggregated towards the centre, and that the flowers are arranged in a number of groups. These are really cymose clusters. We may speak of the whole inflorescence as an *umbellate cymose head.* These are found in many plants. In *Narcissus* the inflorescence is protected by a membranous *spathe.*

Fig. 1.51. Dichasium of *Cerastium.*

In deadnettle (*Lamium*) and many other members of the Labiatae, the leaves are opposite and decussate, and at each node there seems to be a whorl of flowers. These *apparent* whorls are called verticillasters (Fig. 1.53). Careful analysis shows that there is in the axil of each leaf an inflorescence which is a dichasium of cincinni, *i.e.* a biparous cyme which passes on either side into a uniparous form by suppression of one of the branches at each branching. It is difficult to recognise this because the axes have been reduced and the flowers are sessile. It is easily recognised in many Labiatae where the flowers have short stalks.

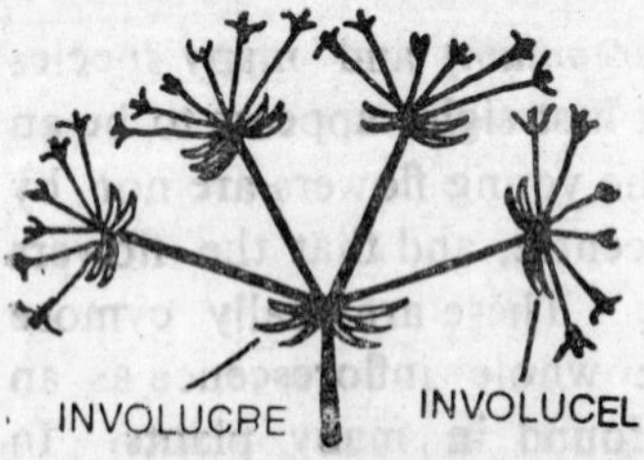

Fig. 1.52. Diagram of a Compound Umbel.

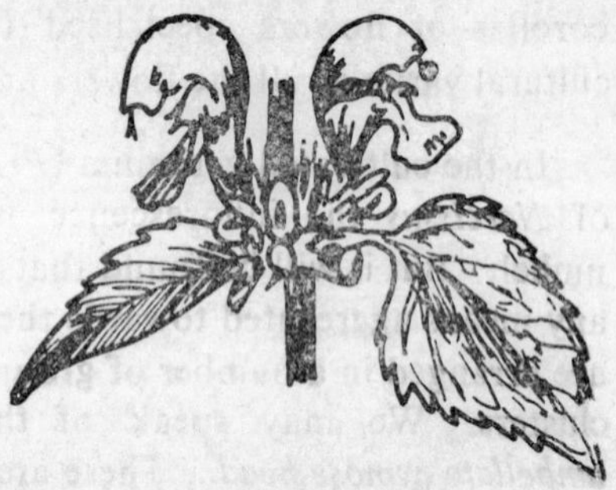
Fig. 1.53. Verticillaster of Deadnettle.

In *Dianthus barbatus* (Caryophyllaceae) and some other plants there is a copiously branched biparous cyme, in which the axes are short and all the flowers crowded together.

The cyathium (Fig. 1.54) is a peculiar inflorescence found in *Euphorbia* (spurge). There is a cup-shaped involucre, the margin of which bears a number of crescent-shaped glandular scales. Inside the cup there are several stamens; also a

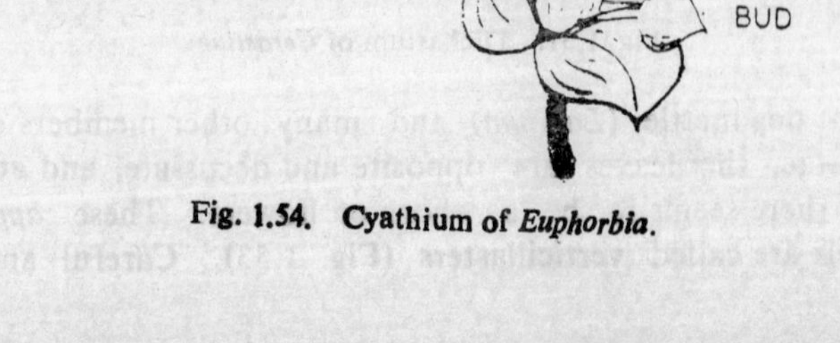

Fig. 1.54. Cyathium of *Euphorbia*.

gynaeceum borne on a stalk. Careful examination shows that each stamen is really a male or staminate flower. This is borne out by the fact that each stamen is articulated to a stalk and has a scaly bract at its base. The gynaeceum with its stalk is a female flower. The stamens are arranged in five groups, which are reduced cincinni, around the female flower. Thus an inflorescence looks like a single flower.

2

Pollination

The fertilization of the egg and the subsequent development of the seed can occur only if pollen grains have previously been deposited upon the stigma. *Pollination*, the transfer of pollen from the anther to the stigma, should be distinguished from

SPOROPHYTE ($2n$)

fertilized egg
↓
embryo
↓
mature plant

stamen { microsporangium, microspore mother cell }

pistil { integuments, megasporangium, megaspore mother cell }

GAMETOPHYTE (n)

male gametophyte
↓
microspore
↓
pollen grain and pollen tube (including tube cell, generative cell, sperms)

female gametophyte
↓
megaspore
↓
embryo sac (including the egg, synergids, polar nuclei, and antipodals)

fertilization, which is the fusion of the male and female gametes. Pollination is of two types. *Self-pollination* is the pollination of a stigma by pollen from the same flower or from another flower of the same plant. *Cross-pollination* involves the transfer of pollen from the anther of a flower of one plant to the stigma of a flower of another plant of the same or related species. The chief agents of pollination are wind and insects, but birds, snails, other small animals, and even water may carry pollen from one flower to another.

Self-pollination occurs in many plants. Among the economically important self-pollinated plants are oats, wheat, barley and rice, peas and beans, soybean, flax, cotton, and tobacco. Cross-pollination, however, is more common than self-pollination, and it probably occurs, at least occasionally, in most of the self-pollinated species. Cross-pollination brings about a more diverse combination of hereditary units of the two parents. This results in increased variability in the offspring and greater adaptability to new environments—conditions of evolutionary advantage to the species.

A more immediate effect of cross-pollination in many species is the production of more seeds or greater vigor in the offspring. A very large number of flowering plants have adaptations which prevent or reduce self-pollination. These many adaptations are usually related to pollination by insects. Prominent among them are modifications of the flower which make cross-pollination by insects possible and in some cases essential if seeds are to be formed. So numerous and varied are these adaptations to cross-pollination that ignorance of them has caused serious difficulties when new plants have been introduced. An example of this is afforded by the history of of Smyrna fig production in California. Large numbers of cuttings were introduced from Asia Minor about 1880. These grew well, but the fruits dropped from the tree before maturity. The realization that pollination is brought about by a tiny wasp resulted in the introduction of the insect, after which the industry flourished. Information on pollination is also needed by the plant breeder when he attempts to produce varieties of greater usefulness to man.

Insect pollination. A large proportion of flowering plants are insect pollinated. The insects are chiefly bees, wasps, butterflies, and moths, but beetles, flies, and other kinds also frequent flowers. Insect-pollinated flowers are usually brightly coloured or scented, sometimes both. Their pollen is heavy or sticky and is not readily carried by wind.

Many such flowers are provided with *nectaries*, specialized tissues or organs which secrete nectar, a fluid with a sugar content varying from 4 to 65 per cent. This nectar is the raw material from which honey is made. Nectaries vary greatly in form and location and may be associated with any of the floral organs. Nectar may be secreted by the surface of the receptacle, by the walls of spurlike projections of the perianth, or by hairs on the petals or ovary. In many flowers the nectary consists of a ring around the base of the ovary. In some, the stamens or petals have become reduced and modified into nectaries.

Numerous experiments, especially on the honey bee, have confirmed the general belief that insects are attracted to flowers by colour or scent. The sense of smell of the honey bee seems to be about as sensitive as that of man, but that of other insects, especially moths, is more powerful, and they may detect odors at considerable distancces. Honey bees, although partly colour blind, can distinguish colours. They are colour blind to red, which they cannot distinguish from black or dark gray. They can identify blue, but not such shades of blue as purple and violet. They also distinguish yellow, blue-green, and ultraviolet. Red flowers are usually pollinated by butterflies, which are not colour blind to red, or occasionally by hummingbirds.

Bees and other insects visit flowers to gather pollen or nectar as food for themselves or their progeny. Pollination is incidental to these activities. Pollen will usually be found adhering to the mouth parts, head, legs, and body hairs of a bee after it has visited a flower. If the bee then visits another flower, some of the pollen may adhere to the stigma, resulting in cross-pollination.

Insect pollination is of the greatest importance from the standpoint of plant evolution and reproduction. It is of economic importance also, for many kinds of plants used directly by man or as food for domestic animals depend upon insect pollination. They include more than fifty kinds of crop plants in this country, among them apple and pear, the Smyrna fig, melon and cucumber, avocado, cabbage, buckwheat, alfalfa, and many clovers. In the absence of certain insects, such plants yield neither seed nor fruit.

Floral adaptations favoring cross-pollination. In many plants with perfect flowers, the stamens and pistils mature at different times. This, of course, favors cross-pollination. Such a flower is said to be *dichogamous.* Dichogamy manifests itself in two ways. In the most common, the anthers ripen before the stigma is ready to receive pollen. In some plants the stigma matures before the pollen sacs of the stamens open. The anthers of Jacob's-ladder (*Polemonium caeruleum*), a garden ornamental, shed their pollen before the stigma of the same flower is ready to receive it ; the three lobes of the stigma are folded together at this stage (Fig. 2.1). After the pollen is shed, the lobes spread apart and may be pollinated by pollen from younger flowers.

Flowers with pistils and stamens maturing at different times are extremely common, and most families of flowering plants have some species in which this condition is present. The interval between the maturation of the stamens and the time when the stigma is receptive varies from an hour or so to several days.

The most complex example of dichogamy is found in the avocado, widely used as a food plant in the tropics and cultivated in the warmer parts of the United States. The approximately one hundred varieties of the avocado fall into two groups, termed class A and class B varieties. The flowers of class A varieties are functionally female in the morning and male in the afternoon. That is, in the morning the stigma is receptive but no pollen is shed, whereas in the afternoon the stigma has withered and the anthers open to discharge pollen.

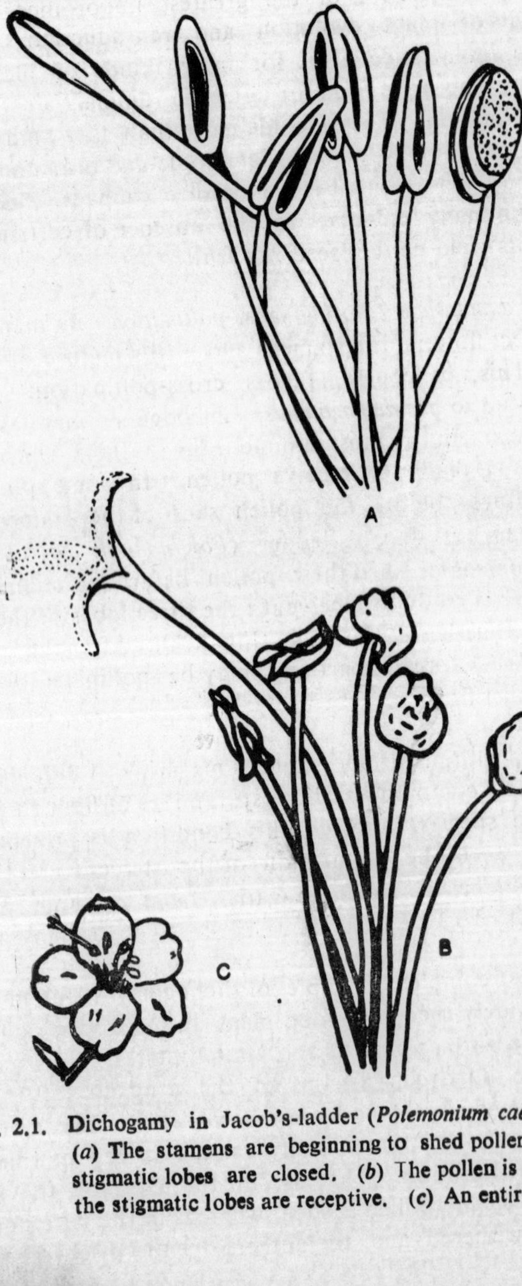

Fig. 2.1. Dichogamy in Jacob's-ladder (*Polemonium caeruleum*). (*a*) The stamens are beginning to shed pollen but the stigmatic lobes are closed. (*b*) The pollen is shed and the stigmatic lobes are receptive. (*c*) An entire flower.

The flowers of class B varieties are male in the morning and female in the afternoon. During the morning, therefore, the stigmas of class A flowers are pollinated by insects which carry pollen from class B flowers. The reverse is true in the afternoon, when the stigmas of class B varieties are pollinated by pollen from anthers of class A varieties.

This information on pollination in the avocado, obtained by several botanists in 1921-1926, is the basis of the recommendation that a grove of avocado trees should contain varieties of both class A and class B, to ensure cross-pollination. This is especially necessary in Florida. In other regions, irregularities in the time of blooming render such interplanting unnecessary, and the trees of a single variety may set fruit. Even self-pollination, however, can be brought about only with the aid of bees, for the pollen of avocado is heavy and sticky and adapted to dispersal by insects.

Insects which rove or climb over the floral parts pollinate many kinds of flowers which have no special structural adaptations to insect visitors. These visits take place by chance, and the insects bring about cross-pollination only occasionally. However, the flowers of many plants are so constructed that they are visited by one kind or at most a few kinds of insects. These insects regularly visit flowers of the same species, so that pollen is frequently carried from the anthers of one flower to the stigmas of another. Such flowers have numerous adaptations which favour cross-pollination and which make self-pollination unlikely or impossible. The stigma is commonly placed so that it is brushed by an insect visiting the flower but does not receive pollen from the anthers of the same flower. This condition commonly results when the elongated style projects well above the anthers.

In many flowers there is a close relationship between the depth of the corolla tube and the length of the sucking mouth parts of the insects which usually visit them. The nectar at the base of the corolla tube may be so deep that only a butterfly or moth with long tubular mouth parts can reach it. This is the case with a number of strongly scented white or nearly white flowers which open in the evening. Soapwort (*saponaria*), some species of tobacco, and the Jimson weed (*Datura*), which has a corolla tube 3 inches long, are pollinated only by night-flying hawkmoths. The red clover is pollinated only by bees whose mouth parts are of sufficient length to reach the nectary at the back of the flower. Flowers with a short corolla tube or exposed nectar are likely to be visited by a variety of insects, including flies.

Cross-pollination is also promoted in plants in which the length of the style differs in plants of the same species (*heterostyly*). In one kind of plant all the flowers have short styles and the anthers are above the stigma. In the other kind, the flowers all have long styles, and the anthers are located below the stigma (Fig. 2.2). The mouth parts of an insect, thrust

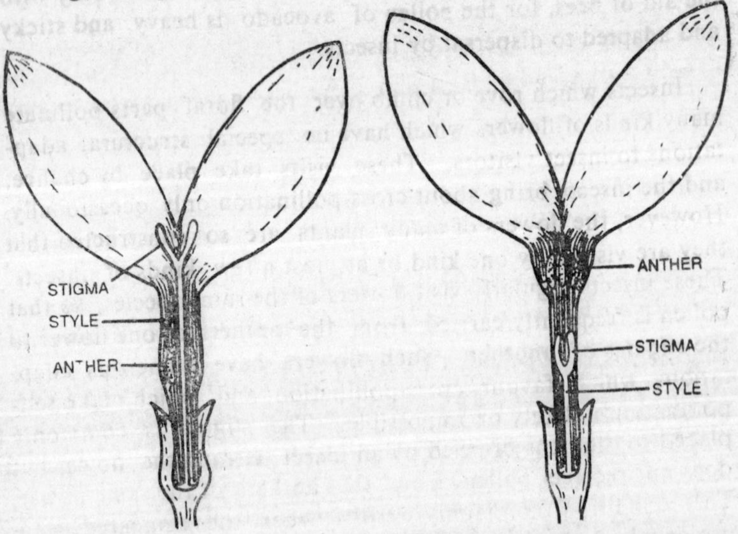

Fig. 2.2. Heterostyly in the bluet (*Houstonia caerulea*). *Left* : Long-styled flower. *Right* : Short-styled flower.

into a short-styled flower, come in contact with the anthers and become covered with pollen. If the insect should now enter a long-styled flower, the pollen is deposited upon the stigma, which is located at approximately the same level as the anthers in the short-styled flower. Similarly, the pollen adhering to an insect which has visited a long-styled flower will be deposited upon the stigma of a short-styled flower, which is at the same level in the corolla as the anthers of the long-styled flower.

Self-pollination is not precluded by this structure, for pollen may fall from the anthers of a short-styled flower onto the stigma. In the long-styled flowers, self-pollination may occur because of the elongation of the corolla tube toward the end of the flowering period. In this final growth, the ring of anthers is brought to the level of the stigma. Heterostyly is very common. It is found in a large number of plant families, including the pinks, St.-John's-worts, heaths, wood sorrels, gentians, poppies, pigweeds, and many others.

Innumerable variations in pollination mechanisms and adaptation of flowers to insect visitors are known. Two final examples illustrate how complex the relation between flowers and insects have become.

Pollination in Salvia. The salvias (sages) have a pollination mechanism whereby a sudden movement of the filaments results in the transfer of pollen to an insect visitor. This mechanism is best known in *Salvia pratensis*, a cultivated ornamental herb with blue flowers.

The flower is irregular, with a strongly developed upper and lower lip (Fig. 2.3). Nectar is secreted by a ring of glands at the base of the ovary. Only two stamens are found, concealed under the upper lip. Each filament is jointed near the base and is prolonged downward into a shell-shaped expansion which connects with a similar expansion from the other filament to form a plate which closes the mouth of the corolla tube.

When a bee pushes into the flower to obtain nectar, its mouth parts encounter this plate, which is then pushed upward and backward, causing the filaments and their anthers to tip forward. The anthers then come in contact with the back of the bee, which is dusted with pollen. When the insect withdraws, the filaments return to their former position. The stigma, located above the anthers, now matures; it grows

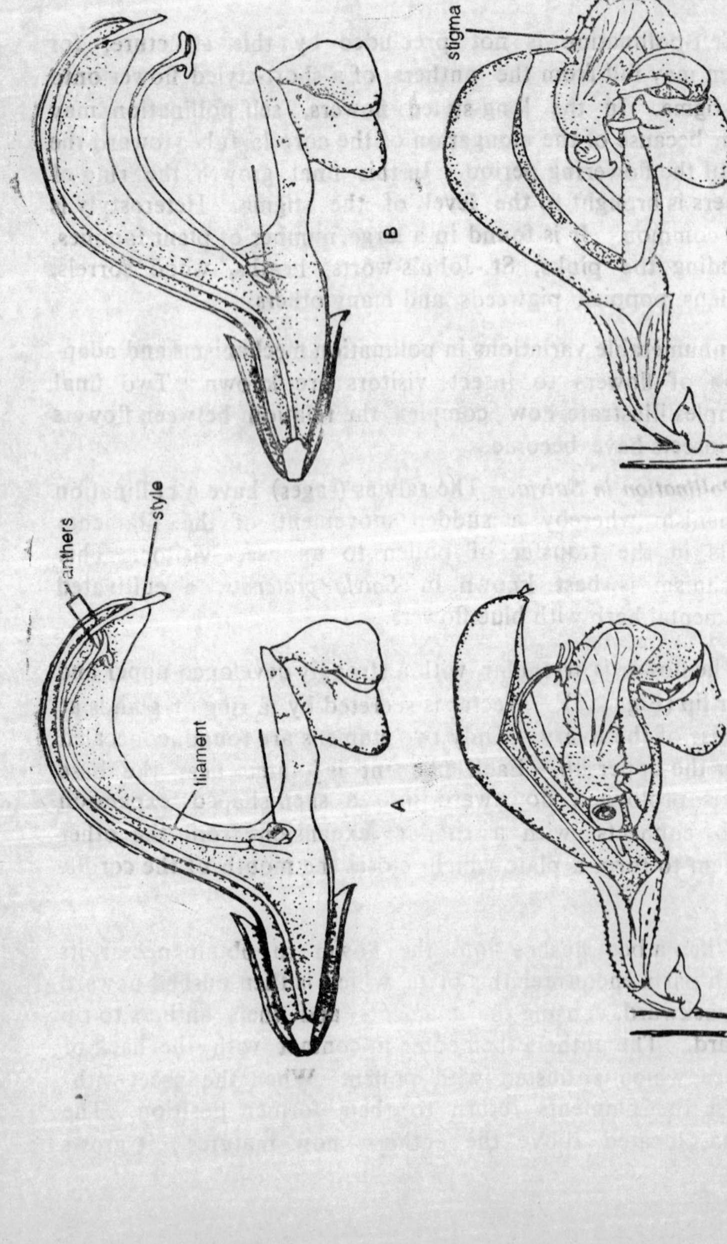

Fig. 2.3. Pollination in Salvia. (*a*) Arrow points to plate formed by fusion of bases of the two stamens. (*b*) Pressure against the plate (indicated by arrow) causes anthers to tip forward. (*c*) Bee pushes against plate, anthers tip forward and deposit pollen on back of bee. (*d*) Bee visits older flower. The stigma has now grown downward, the lobes open, and bee brushes against the stigma which receives the pollen. (Stigma, style, and ovary are shown in red.)

downward and the lobes open. If the bee then visits an older flower, it brushes against the downwardly directed lobes of the stigma, to which the pollen is transferred. Pollination of the stigma by pollen from the same flower is, of course, impossible.

Pollination in Yucca. The relationship between the Pronuba moth and the pollination of the flowers of the yucca plant is so extraordinary as to be unbelievable if it had not been verified repeatedly by qualified investigators since its discovery in 1892.

The yuccas are a group of some thirty species of lilylike plants with long, swordlike leaves. They are popularly called Spanish bayonet or Spanish dagger. Most are native to the arid parts of the Southwest and Mexico ; some are cultivated as ornamentals. In the early summer the plant sends up a tall, coarse flowes stalk bearing numerous large, white, drooping, bell-shaped flowers. The stamens, six in number, are much shorter than the pistil and also arch away from it, so that self-pollination is impossible. The style is short, and the stigma is three-lobed, with a deep chamber lying between the lobes.

In the evening after the flowers open, the silvery white Pronuba moths, about 5/8 inch long, appear in the vicinity. The moths may come from a distance, attracted by the wind-borne fragrance. Both male and female Pronuba moths take no food and live only 2 to 5 days after mating. The males and females meet within the flower and mate. The female then places herself at the tip of a stamen, where she scrapes pollen from the anther by means of highly modified mouth parts. Visits to several stamens produce a large amount of pollen, which is kneaded into a ball held just under the moth's head. The female then flies to another plant and, after boring a hole in the soft tissues of the ovary, deposits an eggs in the vicinity of the ovules.

As soon as the egg is laid, the female moves to the top of the pistil and pushes a portion of her load of pollen into the stigmatic chamber, pressing it down firmly with several vigorous thrusts (Fig. 2.4). Another egg is then deposited within the same ovary, and again a portion of the pollen mass is

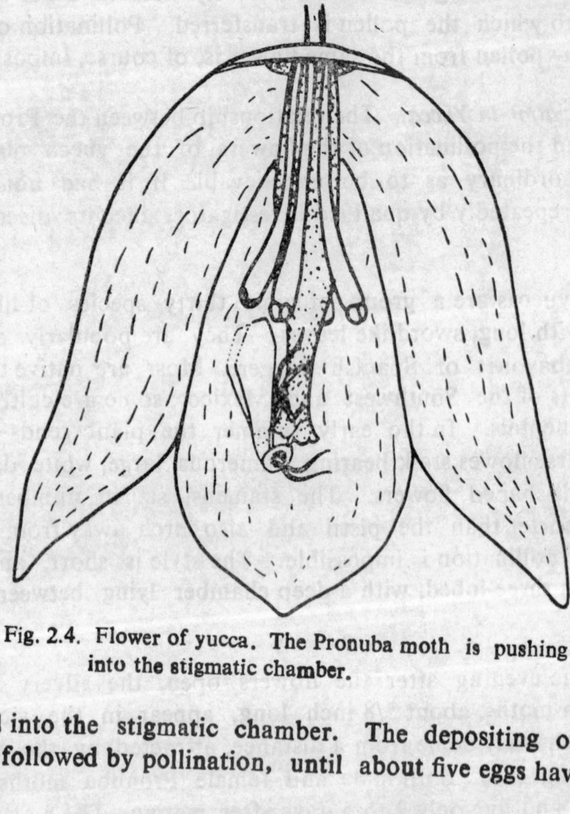

Fig. 2.4. Flower of yucca. The Pronuba moth is pushing pollen into the stigmatic chamber.

forced into the stigmatic chamber. The depositing of each egg is followed by pollination, until about five eggs have been laid.

The egg hatches into a larva, which feeds upon the seeds which have developed from the ovules. The larva matures in about a month, bores a hole in the wall of the mature ovary (capsule), and drops to the soil, where it burrows to a depth of several inches. Here it spends the winter and spring in a tough cocoon. When the yucca flowers again, the insect appears above the ground as a winged adult and the cycle is repeated.

The pollen which the female Pronuba has placed upon the stigma serves as a supply of food for the larva when it emerges from the egg. Without pollination, the ovules would not develop into seeds with their stores of readily available food. The yucca, in turn, would be unable to reproduce without the Pronuba. Each larva eats relatively few seeds, perhaps eighteen to twenty-five, and the several hundred sound seeds which remain in each fruit allow reproduction to continue.

It is difficult to avoid the conclusion that the moth behaves intelligently and purposefully in this curious relationship, perceiving in advance that her labors result in food for her young. A more rational point of view regards this relationship as the result of a long evolutionary development, in the course of which the floral structure and the mouth parts of the insect have all become modified from a simpler condition. In whatever manner the flower and the insect became mutually adapted, it is now certain that without one there could not be the other.

Self-sterility. In many plants with perfect flowers, fertilization and seed formation follow self-pollination. This condition is known as *self-fertility.* In other plants with perfect flowers, fertilization does not occur when the stigma is pollinated by pollen of the same flower. Even pollen from another flower of the same plant or from flowers of other plants of like hereditary constitution will not result in fertilization. This condition of physiological incompatibility is known as *self-sterility.*

Many species which show the phenomenon of self-sterility can be arranged in groups of individuals which fail to set seed when pollinated by other individuals of the same group but are fertile when pollinated by individuals of other groups. Several hundred such species, wild and cultivated, are known, and the list is probably far from complete. Most cases of self-sterility are due to the slow rate of pollen-tube growth when the flower is pollinated by pollen from flowers of like heredity. The rate of pollen-tube growth is determined by certain genes. If the pollen tubes carry the same sterility genes as the style, the rate of growth is so slow that the flower withers and falls away before fertilization takes place. If the genes of the style are suitably different from those of the pollen tube, the tube grows rapidly and fertilization follows.

In most ornamental plants self-sterility is of little significance because these plants are grown for their flowers or foliage. But self-sterility in such plants as red clover, apples, pears, cherries, and plums, which are cultivated for their seeds or fruits, is another matter. Some varieties of apples set fruits well if self-pollinated or pollinated by pollen from flowers of other trees of the same variety. Most apple varieties are self-sterile, and commercial orchards are therefore planted with at least two varieties whose pollen is effective in cross-pollination and whose blooming periods overlap.

Certain combinations of varieties are unsuccessful, and the grower must have a list of the varieties which are compatible when crossed. The varieties are set out in the orchard so that several rows of one variety alternate with rows of another. Bees bring about cross-pollination. Frequently hives are rented from beekeepers and placed in the orchard during the period of bloom. Even many of the self-pollinated varieties set a better crop when crossed with another variety.

The existence of self-sterility in cultivated fruits has been known since the latter part of the nineteenth century, but even today large apple orchards of one variety are sometimes planted, with the result that little or no fruit is produced. This can be overcome, however, either by grafting other varieties to established trees or by placing bouquets of apple flowers of another variety in the orchard during the pollinating period.

Pears and plums, like apples, include some self-sterile varieties. Sweet cherries are self-sterile, but sour cherries, nectarines, and peaches are not. A practical knowledge of these aspects of pollination may mean success or failure to the grower of fruit trees.

Insects that Visit Flowers

The chief flower-visiting insects are beetles (Coleoptera), flies (Diptera), bees and wasps (Hymenoptera), butterflies and moths (Lepidoptera). In connexion with the pollination of flowers, the important differences to be noticed between these insects are the size of the body, the length of the tongue (proboscis), the time of year at which each kind is most plentiful, and their habits—*e.g.*, whether they collect pollen or nector or both, whether they fly by day or in the evening. By carefully studying the structure of a flower, and noting such points as the time of flowering, the order in which the anthers and stigmas mature, the relative positions of anthers and stigmas in the open flower and any changes in position that may occur, we can often tell what kind of insect is capable of effecting cross-pollination, and whether or not self-pollination is possible.

Most flies and beetles have very short tongues, usually less than 3 mm. long. Most of the larger and longer-tongued flies, *e.g.*, gadflies, "cleggs," and horseflies do not visit flowers ; but there are some, chiefly hover-flies and bee-flies, with tongues sometimes as long as 12 mm., which are regular flower-visitors.

Flowers may be arranged in various biological group or classes according to their adaptations for insect-visitation :—

(1) *Flowers adapted for Short-tongued Insects.* These may be (*a*) flowers in which the necter is freely exposed on the surface, *e.g.*, Umbelliferae, Rutaceae, etc. ; (*b*) flowers with a very short tube, *e.g.*, moschatel, bedstraw, enchanter's nightshade ; (*c*) shallow open flowers such as stonecrop and saxifrages. Such flowers are visited by the shorter-tongued beetles and flies.

(2) *Flowers with Partially-concealed Nectar.* This group includes flowers in which the nectar can be reached only by insects with tongues at least 3 mm. in length, and which are therefore visited by the longer-tongued beetles and flies, as well as by insects of higher type. The nectar may be slightly concealed by the stamens, *e.g.*, *Ranunculus* and Myrtaceae ; by the erect stiff sepals, as in the smaller Cruciferae ; by the formation of a shallow calyx-tube, as in many Rosaceae ; by a short corolla tube, *e.g.*, the shorter-tubed Compositae, etc.

(3) *Flowers with Fully-concealed Nectar.* This type of flower differs only in degree from the last. Here the nectar can only be reached by insects having tongues about 6 mm. long, including the longest-tongued flies (chiefly hover-flies), the shorter-tongued bees, and wasps. The concealment of the nectar is effected by a further deepening of the flower, owing to the formation of a calyx-tube, to the calyx being gamosepalous or the corolla gamopetalous, or to other causes. Examples of these medium-tubed flowers are seen in some Rosaceae, Solanaceae, Scrophulariaceae, etc.

(4) *Long-tubed Flowers.* When the flower-tube becomes longer, all the shorter-tongued insects are more or less completely excluded, and the flower is adapted for, and chiefly visited by, the larger bees, butterflies and moths. Many flowers belonging to the Amaryllidaceae, Iridaceae, Orchidaceae and Scitamineae, in which the perianth nearly always forms a long tube, come under this type. Flowers like those of Papilionateae, *Antirrhinum*, and *Calceolaria*, can only be opened by large bees, and only the longest-tongued bees can reach the nectar in such flowers as *Aconitum* and *Delphinium*.

Humble and hive-bees have the most perfect mechanism (the "pollen-baskets" on the hind-legs) for collecting pollen to mix with nectar and feed their broods. Humble-bees have longer tongues than hive-bees, and are particularly skilful in finding the way to well-concealed honey.

Blue, purple, and red colours are often associated with flowers visited by bees (especially blue and purple) and butterflies (especially red), while flowers visited by other insects are usually white, yellow, or variegated ; but there are far too many exceptions to allow of a general rule.

Birds, especially small honeysuckers and humming birds, may be active as pollinators. Flowers may be visited by both insects and birds, but bird-pollinated flowers generally show brilliant colour contrasts. Examples are afforded by *Strelitzia regina*, many species of *Protea*, *Salvia* species, *Loranthus aphyllus*, and some *Acacia* species.

(5) *Butterfly-and Moth-flowers*. When the flower-tube (or at any rate the level of the nectar) is more than about 12 mm. (about half an inch) deep, the nectar is beyond the reach of bees, though they may visit the flower for pollen, or the humble-bee may bite through the tube (calyx or corolla) and thus rob the flower of its nectar. Butterflies may visit many flowers which are adapted for bees, *most* butterflies and moths having tongues of about the same length as, or a little longer than, those of bees.

Some moths, however, have far longer tongues (30 mm. or more in British species), which are (an in butterflies) carried coiled up in a spiral under the head when flying. These moths can reach nectar when it is at the bottom or a very long tube, as in *Lonicera*, which is visited chiefly by the night-flying privet hawk-moth, and the white convolvulus, which is pollinated by another species of hawk-moth (*Sphinx convolvuli*, tongue 80 mm. long). Other flowers pollinated by night-flying moths are *Oenothera*, *Nicotiana*, *Ligustrum* and *Cereus*. Moth-pollinated flowers are generally white or pale-coloured, sweetly scented, and open in the evening, usually remaining closed and almost scentless during the day.

Examples of Floral Mechanism.

(*a*) In the garden pansy (*Viola altaica*, Fig. 2.5) the authers of the five stamens are firmly joined by hairs on their edges, and the two anterior stamens bear processes, functioning as nectar-glands, which pass down into the spur of the anterior petal. A space or chamber ("pollen-box") is enclosed above the ovary, at the base of the style, by the five membranous scales borne on top of the authers. The stigma, which projects beyond the auther-scales, is dilated and hollow. It has a tuft of hairs on earch side, and below there is an opening into it, the lower edge of which is protected by a lip or flap (the "scraper").

The flowers are pendulous, and hence the pollen, which is shed on the inner faces of the anthers, and is dry and loose, not sticky as in most entomophilous flowers, falls into the

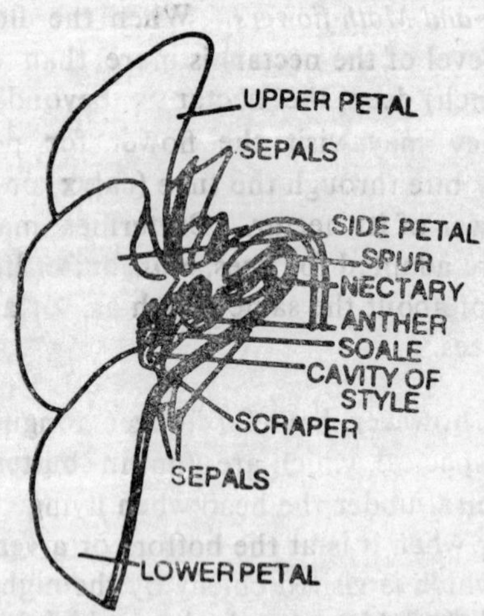

Fig. 2.5. Flower of Garden Pansy (*Viola altaica*), seen in Vertical Section, Half flower.

"pollen-box," from which it can escape only through the opening between the scales of the two anterior anthers.

The flowers are pollinated by long-tongued bees (and butterflies). When the insect enters the flower, pollen obtained from another flower may be deposited on the stigma, and cross-pollination thus effected. Pushing down into the spur of the anterior petal to reach the nectar, the insect receives a supply of pollen which has escaped from the "pollen-box". The "scraper" prevents this pollen being deposited on the stigma as the insect retires.

The conspicuously coloured centre of the flower, and the honey-guides on the lateral and spur petals serve to attract desirable insects. The entrance of small undesirable visitors is hindered by the hairs on the lateral petals and on the sides of the stigma, by the hairs lining the entrance and cavity of the spur, and by the length of the spur itself.

In *Salvia* (sage), one of the Labiatae, an interesting mechanism is found (Fig. 2.6). The corolla is bilabiate. The conspicuous lower lip attracts insects, and acts as a landing-place.

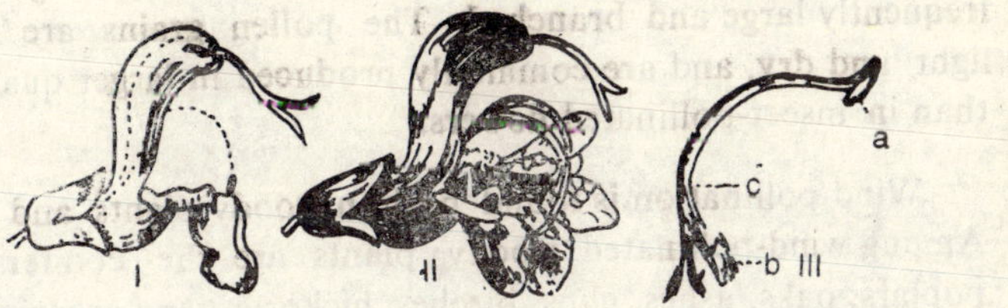

Fig. 2.6. I. Flower of *Salvia* from side ; II, With Humble bee extracting Nectar, and the anthers rubbing against its back ; III. Single Stamen.

The arched upper lip protects the stamens and style. These are only two stamens, the other two, characteristic of the Labiatae, being represented in the sage by staminodes.

The two stamens have a peculiar structure. Each has a short filament, jointed to a long curved connective (Fig. 2.6 III *c*). In some types of *Salvia* each end of the connective bears a half-anther, but in other types (*e.g.*, garden sage) the lower end of the connective is barren and flattened (Fig. 2.6 III, *b*) and the upper part of the connective is longer than the lower, the whole forming a delicate lever.

A bee on entering the flower pushes against the *united* lower ends of the two connectives in seeking the nectar, and causes the curved connectives to swing on the filaments as on hinges, so that the two fertile anther-lobes (*a*) come down and strike on the bee's back, dusting it with pollen. As the bee retires, the stamens return to their former place under the corolla-hood.

The flowers are protandrous. As the flower gets older the style curves down, and the stigma is so placed that it is touched by a bee entering the flower at this stage.

Wind Pollination. In many ways the simplest form of pollination, and probably the most primitive, is pollination by wind. The flowers of wind-pollinated plants are usually small and inconspicuous. Devoid of bright colours and nectar, they are commonly grouped in dense clusters. The perianth is often reduced or lacking. The stamens may protrude prominently from theo flower, suspended by long filaments. The stigmas are frequently large and branched. The pollen grains are small, light, and dry, and are commonly produced in larger quantities than in insect-pollinated flowers.

Wind pollination is found both in woody plants and herbs. Among wind-pollinated woody plants are the conifers, the poplars, oaks, ashes, elms, birches, hickories, and sycamore. In many woody plants the staminate and sometimes the pistillate flowers are grouped into drooping clusters called catkins. The wind-pollinated herbaceous plants include pigweeds, sorrels, docks, plantains, nettles, meadow rue, hops, grasses, rushes, and sedges. Some species may be either wind or insect pollinated. The small flowers of willows, for example, are brightly coloured and provided with nectaries. But if the pollen is not removed by insects, it may be carried away by wind. Many maples, too, are both wind and insect pollinated.

Hay fever. Large numbers of people are aware, to their sorrow, that numerous wild and cultivated plants are wind-pollinated. These people are sensitive or susceptible to the pollen of such plants, a condition known are *allergy*. The same pollen is harmless to other individuals. The allergy to plant pollens is popularly called hay fever.

In order to cause hay fever, the pollen of any given plant must have three important characteristics : it must be produced in considerable amounts ; it must be buoyant, so that it floats readily in the air ; and it must be toxic to allergic people. Thus, although nearly all hay fever is produced by wind-pollinated plants, all pollen carried by the wind does not cause hay fever. The cattails shed large amounts of pollen, as do the conifers, yet except for a few species of conifers these plants do not cause hay fever, for their pollen is not toxic.

The goldenrods, cosmos, roses, and sunflowers are popularly believed to be important sources of hayfever pollen. The pollen is toxic to some people, but these plants are insect pollinated. Their pollen is heavy and is not readily carried by the wind. Only intimate contact, such as handling roses or golden-rod, will produced in an allergic person symptoms of the disease. Usually these plants are only minor offenders in causing hay fever.

Pollination seasons. Whether or not a sensitized person will develop hay fever depends upon the kind and amount of pollen in the air. This in turn depends, except for such atmospheric conditions as rain and wind, upon the plant species which occur in a given region. The time of year when the symptoms of hay fever appear is related to the flowering period of these plants. Three hay-fever seasons are recognized in many parts of the United States. But there is great local variation in the distribution of the plants which cause hay fever and in the pollination periods of these plants.

Early Spring (March to May). The usual cause of hay fever in this period is tree pollens. Among the trees concerned are the box elder, the elms, poplars, birches, oaks, hickories, ashes and sycamore.

Late Spring and Early Summer (May to July). Grasses are the most important hay-fever plants during this period. Among the more important species are Kentucky blue grass, red top, orchard grass, timothy, sweet vernal grass, and Bermuda grass.

Late Summer and Fall (August to September). The ragweeds are outstanding pollen producers during this season.

The plants of significance in hay fever, the pollination periods of such plants, and the quantity of pollen in the air can be determined by pollen surveys for a specific region. Atmospheric pollen counts are made by exposing to the air microscope slides covered with petroleum jelly. The pollen which adheres to the slide is stained and studied under a high-powered microscope and it identified by surface features, which vary widely in pollen grains of different species. Newspapers sometimes publish daily pollen counts for the information of hay-fever victims.

Contrivances and Conditions favouring Cross-pollination

There are in flowers many arrangements and mechanism which ensure cross-pollination. Usually such arrangements and mechanism merely give chances in favour of cross-pollination without precluding the possibility of self-pollination. Sometimes, however, they make self-pollination difficult, or altogether impossible.

In plants with unisexual flowers, of course, cross-pollination is absolutely necessary if seed is to be produced. We have this condition in its extreme form in dioecious plants, *e.g.*, *Salix* (willow). There are a few plants, also, in which cross-pollination must take place if seed is to be produced, because the plants are *self-sterile, i.e.*, the flower cannot be fertilised by its own microspores ; this occurs in some specimens of *Passiflora* (passion-flower), or *Lobelia*, and of *Abutilon*. In some flowers, again, self-pollination may be rendered unlikely or difficult owing to the relative position of anthers and stigma.

A condition of much more general occurrence is that known as Dichogamy. This is a condition in which the anthers and stigma in hermaphrodite flowers come to maturity at different times, and which, when completely developed, entirely prevents self-pollination.

There are two form of dichogramy : (*a*) protandry, in which the anthers ripen first, so that when the microspores are shed the stigma of the same flower is not ready to receive them ; in this case, if the microspores are not to be wasted, they must be transferred to an older flower ; (*b*) protogyny, in which the stigma ripens first ; here the microspores must be transferred to a younger flower. Protandrous flowers are much more common than protogynous. Examples of the former are found in Compositae, Labiatae, Umbelliferae, Solanaceae, etc., of the latter in *Plantago*, *Luzula*, some Scrophulariaceae, Ranunculaceae, etc. Wind-pollinated flowers are more often protogynous than protandrous, but many are unisexual.

Anemophilous and entomophilous flowers have each special characters of their own, so that as a rule we can distinguish them at a glance. In *anemophilous flowers*, the microspores are usually dry and smooth, and produced in great abundance, as much must be wasted ; the flowers are small and inconspicuous ; there is no nectar or perfume ; and frequently the stigmas are branched and feathery, to catch the microspores. In many trees which are wind-pollinated the flowers appear in spring before the leaves, so that the microspores have free access to the flowers. In most herbaceous plants with wind-pollinated flowers, the latter are carried up on a long stem, well above the leaves, so as to expose them as freely as possible to the wind (*e.g.*, *Plantago*, Chenopodiaceae, grasses, etc.).

Much greater variety of adaptation is shown by *entomophilous flowers*. As a rule they have large, conspicuous, or highly-coloured corollas, or are arranged in conspicuous inflorescences ; they usually secrete nectar and give out perfume. The microspores are usually rough and sticky, and often not produced in any great abundance, as they are more sure of transference. The bright corollas, the perfume and nectar serve to attract insects which visit the flower in search of food.

A nectarless but otherwise insect-attracting flower is sometimes called a "pollen-flower." Examples are found in *Papaver*, *Rosa*, *Helianthemum*, *Anemone*, *Clematis*, *Hypericum*, *Ulex*, *Cytisus*, *Ulmaria*. These flowers are visited by pollen-feeding insects.

Many entomophilous flowers are further characterised by the presence of ingenious mechanical devices, which guide and control the movements of the insect and turn them to the best account. Thus, in many cases the corolla is so constructed that the insect must alight on the flower or enter it in a special way (*e.g.*, Labiatae, Papilionatae) ; the same result may be attained by the secretion of nectar into special receptacles or spurs (*e.g.* in *Viola*, Fig. 2.5). Often the insect, on entering a flower, pushes against special processes or outgrowths which move the stamens and bring the anthers in contact with its body (*e.g.*, *Salvia*, sage, Fig. 2.6) ; or the stamens may be jerked, and the microspores scattered over the insect's body. In some flowers the stamens move in response to touch by a visiting insect. This is seen in *Centaurea jacea* where the filaments contract and pull the tube formed by the syngenesious anthers backwards over the style, and the microspores are thus swept out and exposed. Curvature movements of stamens in response to a shock stimulus are seen in a number of Families such as Berberidaceae, Tiliaceae, Cactaceae and Cistaceae. Also sensitive stigmas may close over the deposited microspores as in Scrophulariaceae. Frequently spots or lines of a conspicuous colour are developed on the corolla ; these have been called "honey-guides," as they are believed to afford insects guidance in seeking out the nectar.

The general result of all these devices is that the insect receives pollen on a special part of its body, and when it enters another flower the pollen is deposited on the stigma. In many protandrous flowers this is secured by the style bending over so that the stigma is in the position formerly occupied by the stamens.

A special, but at the same time simple, arrangement for ensuring cross-pollination by insects in known as heterostyly. It is seen in primrose (Fig. 2.7). Here there are two types of flower, borne on different plants. One kind (thrum-eyed) has long stamens (with anthers in the throat of the corolla tube)

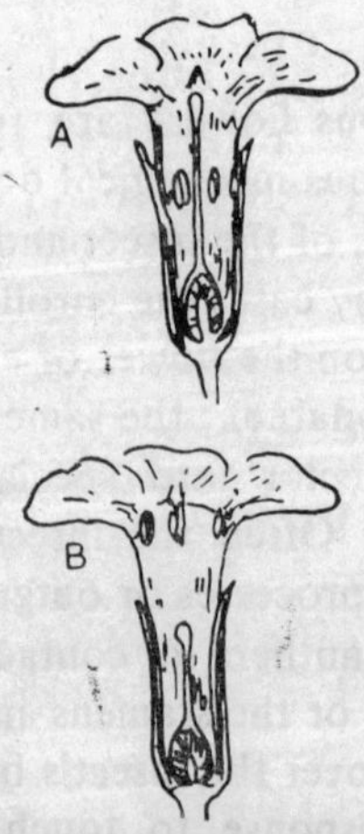

Fig. 2.7. Dimorphism in *Primula*.
A. Pin-eyed ; B. Thrum-eyed form in longitudinal section (half flowers)

and a short style ; the other (pin-eyed) has a long style and short stamens ; thus in the two types the positions of anthers and stigma are reversed. Pollination is effected by transference between these two forms and not between two flowers of the same form. This is the dimorphic form of heterostyly. In purple loosestrife (*Lythrum*) there are three types of flower combining two positions for the stamens and one for the stigma (Fig. 2.8). Such flowers are said to be trimorphip.

Trimorphism is also seen in species of *Oxalis* and dimorphism in *Turnera* and species of *Jasminum.*

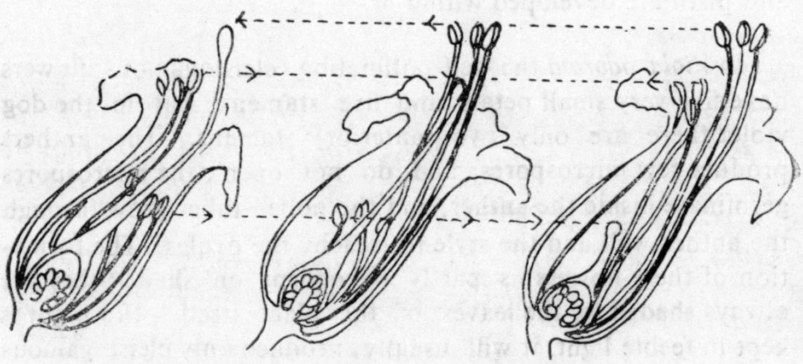

Fig. 2.8. *Lythrum salicaria,* Trimorphism. Three types of flower.

Special Arrangements for Self-pollination

In studying floral mechanisms we are too apt to forget that sels-pollination occurs regularly in most flowers were it is not precluded by dioecism, complete dichogamy, or self-sterility. Many annnual plants are commonly self-pollinated (*e.g.*, groundsel, chick-weed). They have small flowers, often without nectar or smell, and are either homogamous, that is, their anthers and stigmas mature at the same time, or so slightly dichogamous that self-pollination is secure.

Even in flowers evidently adapted for cross-pollination, if this fails, there is commonly the possibility of self-pollination. Many of them are distinctly dichogamous, but not completely so, there being usually a short period during which self-pollination becomes possible. To effect this there are sometimes special contrivances such as the curling back of the stigmas to reach the pollen on the anthers or style (*e.g.*, Compositae, Campanulaceae).

A very special adaptation for self-pollination is the production of cleistogamous flowers. These are closed flowers produced late in the year by certain plants which had previously produced entomophillous flowers, *e.g.*, *Viola odorata*, *Oxalis acetosella*, *Lamium amplexicaule* (one of the deadnettles), etc. *Commelina benghalensis* also may produce cleistogamous flowers on leafless shoots. The cleistogamous flower is small and inconspicuous. The calyx remains closed, and the stamens and pistil are developed within it.

In *Viola odorata* the self-pollinating cleistogamous flowers have five very small petals and five stamens, but in the dog violet there are only two (anterior) stamens. The anthers produce few microspores, and do not open ; the microspores germinate inside the anther, and the pollen-tubes grow through the anther-wall and the style to reach the ovules. The formation of these flowers is partly dependent on shade ; they are always shaded by the leaves of the plant itself. If a plant is kept in feeble light, it will usually produce only cleistogamous flowers.

THE ESSENTIAL ORGANS AND THEIR FUNCTIONS

The main function of the flower is the production of seeds, and for the setting of the seeds two distinct processes are necessary. They are the transference of the pollen-grains from the anthers to the stigma (pollination), and a fusion of a bit of the protoplasm of the pollen-grain with a portion of the protoplasm within the ovule (fertilisation). We know that the parts of the flower directly concerned in these processes are its essential organs, the stamens and the pistil.

For the clear understanding of these two processes, a knowledge of the essential organs is necessary. The stamens are the parts intended to produce the pollen-grains. A stamen usually consists of a filament and an anther, but it is the anther alone that is of importance. The filament is only a stalk supporting the anther, and we have many instances in which the filaments are absent, the stamens being reduced to mere anthers. To study the structural details, the stamens of Tribulus may be used, as it is typical of the anthers of a good number of plants. A transverse section of this anther reveals two distinct lobes with two cavities in each, filled with pollen. But the lobes of an old anther present only a single cavity, as the two cavities get fused. The anther lobes are held together by the connective, which is traversed by a vascular bundle. The wall of the anther consists of two layers of cells, an outer layer or the epidermis and an inner layer in which the cells have special thickenings (and hence called fibrous cells). This differentiation in the cells of the wall of the anther is intended to facilitate its dehiscence. The outer wall of the cells of the epidermis bulges slightly outwards and there is protoplasm within these cells. Until the maturity of the anther these cells remain in a turgid condition. The pollen grains take some time to develop and until they are fully formed they need protection. So the wall of the anther remains intact and does not burst, till the pollen is ripe enough for shedding. As long as the epidermal cells are turgid they exert pressure on the fibrous cells laying below the epidermis. As the anther approaches maturity the epidermal cells begin to lose water gradually. This means the gradual lessening of the pressure on the fibrous cells and, as soon as they are freed from it, they tend to expand and assume their natural size. The removal of compression on the fibrous cells leads to the curling of the anther wall. This, of course, causes the wall to break in some place where it is weak.

The pollen-grain of *Tribulus terrestris* is round and it is only a bit of protoplasm enclosed by a cell wall in which two distinct layers, an outer cutinised and an inner cellulose layer, can be distinguished. The outer layer is thickened in a peculiar manner as shown in Fig. 2.9. The inner layer is smooth and uniformly thick. There is a great deal of variation in size and form, and the sculpturing of the outer layer in the case of the pollen-grains, according to the species of the plant.

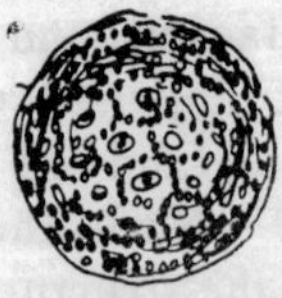

Fig. 2.9. Pollen-grain of Tribulus very highly magnified.

Those of very showy flowers are in many cases provided with spiny or other kinds of projections, or the outer layer may be sticky. In a few cases, we find some spots in the wall formed beforehand for the coming out of the pollen-tube. Some pollen-grains have special lids also, as in the case of Cucurbita.

The free end of the style or the stigma is the part intended for the reception of the pollen. It has to catch the pollen-grains and retain them, and also to assist them to germinate. The surface of the stigma is usually rough due either to papillae or hairs, and a sticky juice containing some sugar is also secreted by it. Both these conditions are useful in retaining the pollen-grains on the stigma.

Sometimes the stigma becomes very much branched and is consequently plumose, as in the case of grasses. It is said that the pollengrains are safeguarded from the attacks of bacteria, by the stigmas secreting some substance detrimental to their growth.

A stigma is not always receptive ; it becomes receptive as soon as the flower opens and generally continues to be so for sometime, this period varying with the kind of the plant.

The sugary juice secreted by the stigma seems to be necessary to stimulate the pollen-grains to germinate. It is only when the stigma is receptive that we find the sugary juice, and if it is absent it is an indication that the stigma is not receptive.

The ovule which develops into seed after fertilisation lies within the ovary. It does not develop into a seed unless a bit of protoplasm from the pollen-grain finds its way into the interior of the ovule and mixes with a definite part of the ovule. When the ovule is fully formed and ready to receive the bit of protoplasm from the end of the pollen tube, it consists of a large cell, called the *embryo-sac*, amidst a mass of cells. This mass is covered, except at the top, by two membrances called the *integuments*. The opening at the top is called the *micropyle*. The embryo-sac at this stage possesses two groups of nucleii one at the top and the other at the bottom. In each of these groups, there are three nucleii. Of the three lying at the top of the sac close to the micropyle, one nucleus is slightly larger than the others, and this is called the *egg-cell*. The other two cells seem to be helpful in directing the contents of the pollen-tube to the embryo-sac. The three cells at the other or far end of the embryo-sac do not seem to take any part in the formation of seed. Within the embryo-sac, midway between the two groups of nucleii, lies a single large nucleus, which is called the nucleus of the embryo-sac, or the *secondary nucleus*.

The pollen-grain, deposited on the stigma, emits a tube and this comes to the top of the embryo-sac finding its way through the style and the micropyle. At the end of the pollen-tube we find two nucleii, and both of them get into the interior of the embryo-sac and one fuses with the embryo-cell and the other with the secondary nucleus. This fusion of the nucleii of the pollen-grain with those of the ovule is really the process of fertilisation. It is only after this fusion that the egg-cell is capable of division and development into the young embryo plant that we find in the seed. The secondary nucleus divides and gives rise to the endosperm, after the fusion of the nucleii.

Now it is obvious, that, for the production of offsprings, the fusion of the male and female cells is essential even in the case of plants, as it is in the case of animals. Further the offsprings are likely to be better in quality when the sex cells uniting together are from different plants.

The pollen-grains of a good many of the flowering plants have no power of spontaneous movement. So they have to depend upon some external agency for the pollination of their flowers. Even in the case of plants with bisexual flowers, extraneous aid is necessary for pollination. The transfer of pollen to the stigma from the anthers of the same flower, as well as from the anthers of a different flower on the same plant, is called *self-pollination.* Pollination of the stigma of a flower with the pollen from flowers of a different plant is called *cross-pollination.*

As a matter of fact, self-pollination occurs in several plants, especially annuals. In some cases flowers are self-pollinated regularly, and such flowers are usually small and inconspicuous without smell or honey. As examples for this, we may mention the Chenopodiums. We have also certain plants wherein self-pollination is made impossible in the earlier stages, but later, just before fading, this become possible. For instance, in the flowers of *Hibiscus micranthus* and *Abutilon indicum* the style branches project a little above the anthers just when the flowers open, and so the pollen cannot reach the stigma. It is evident that pollen from a different flower has to be brought and deposited on the stigmas by some agency, either wind or insects. If the stigmas fail to receive foreign pollen, they bend down so as to come near the anthers in the same flower, to be at least self-pollinated. In both these flowers, the petals close and press on the style branches and thus assist them in coming near the anthers. In some flowers, as in *Evolvulus alsinoides,* the style lies bent away from the anthers just, at first, but later it changes its position, so as to bring the stigma nearer the anthers to make self-pollination possible.

The flowers of *Mirabilis jalapa* are also adapted for self-pollination, when cross-pollination fails to occur. These flowers open towards the evening the emit a very strong scent. Just then, both the stamens and the stigma are far exerted, the stigma alone projecting above the stamens. Usually two species of moths visit these flowers and if they are not visited by any insect, the stigma manages to get self-pollinated either by the elongation of the stamens or by the slight bending of the style, so as to reach the anthers.

The flowers of some Compositae are specially interesting in that they have special adaptations in their floral mechanism to secure self-pollination, should cross-pollination fail to take place. For example, in the plant *Tridax procumbens*, a weed found everywhere, the heads have bisexual flowers in the central portion of the disc and the ray flowers are all female. The stamens are epipetalous and the anthers are adherent, so as to form a tube. The style is bifid and the branches are pressed together, in such a manner that the receptive surfaces are in contact and so not exposed. At first, when the flowers are unopened, the style is short, but, as soon as the flowers begin to open, the anthers dehisce and the style elongates. So the pollen, lying in the anther tube, gets pushed out and remains at the top of the anther tube. The tip of the style may also carry a small amount of pollen, besides what may adhere to the hairs of the style branches on the outer surface. The style grows and after projecting a little above the anther tube, the branches separate and diverge, exposing the receptive surface. The stigmas thus exposed remain receptive for some time so that cross-pollination may occur. If the stigmas fail to receive pollen by insect visits, the style branches diverge still more and even curl round, so that the receptive surface may come in contact with the pollen adhering to the lower surface of the style branches, or with the pollen lying at the top of the anther tube. Self-pollination is thus ensured, if cross-pollination has not already taken place.

Instead of this markeshift arrangement to secure self-pollination, when cross-pollination is a matter of some difficulty, some plants produce two distinct sets of flowers, one adapted for cross-pollination and the other for self-pollination. In *Commelina benghalensis* we have a plant of this sort. The beautiful blue flowers are regularly cross-fertilised and the underground or low-lying inconspicuous flowers that remain as buds without opening are self-fertilised. These inconspicuous flowers are called *cleistogamous* flowers.

Charles Darwin, the great English naturalist, has proved by a series of experiments with different plants that cross-pollination is more advantageous to a plant than self-pollination. If we examine flowers of a number of wild plants growing in a place, we find in them numerous contrivances favouring cross-pollination to the exclusion of self-pollination. Recent workers in the field of plant-breeding ha ve also established beyond doubt, that cross-pollination is not only more advantageous to the plant than self-pollination, but it also confers on plants certain racial advantages.

A most perfect arrangement to ensure cross-pollination is to have the essential organs on separate flowers. We find a host of plants in which the flowers are unisexual. In some plants, as in Cucurbita and Ricinus, we find both male and female flowers on the same plant (monoecious), in others, as in *Cephalandra indica* and in some palms, the male and female flowers are found on different plants and not on the same individual (dioecious).

Bisexual flowers become adapted for cross-pollination to the exclusion of self-pollination, by having the essential organs active at different times. The stigmas may nature and become receptive prior to the anthers. In Cumbu spikes (*Pennisetum typhoideum*), the stigmas protrude from the spikelets and become receptive, while the anthers are still within the glumes. Such flowers and called *protogynous*. The female sexual organs in the protogynous flowers are ready for fertilisation long before the anthers are ripe in the same flower, and, therefore, the flowers have to be pollinated only by the pollen of older flowers. There are also plants wherein the stamens shed their pollen prior to the receptivity of the stigmas. The flowers are then termed *protandrous*. Protandry seems to be more common than protogyny. Flowers of several Compositae, Malvaceae and *Andropogon Sorghum* are protandrous. Pollination in this case is effected by the pollen from younger flowers.

In very many plants self-pollen is sterile, as in the case of many leguminous plants. The pollen in the flowers of certain orchids is said to act as a poison, if it falls on the stigma of the same flower. There are also instances in which the pollen from the same flower fails to fertilise, if foreign pollen falls on the stigma, soon after self-pollination. This means that foreign pollen is prepotent over self-pollen.

Another very effective arrangement promoting cross-pollination exists in the flowers of several species of Jasminum. The corolla in all cases is tubular and the stamens are epipetalous. In some flowers the style is short reaching only half the length of the tube, whereas the anthers are at or near the throat of the corolla. On the same plant we also find flowers in which the styles are longer and coming up to the throat of the corolla, while the stamens remain within the tube about midway.

In the case of land plants it is obvious that for cross-pollination some extraneous aid is necessary, because of the fixity in position of these plants and of the absence of spontaneous movement on the part of the pollen-grains. The agents usually active in this work are wind and insects and, in rare cases, water. Plants depending upon insects for fertilisation are very many and those pollinated by wind are not inconsiderable.

When plants depend upon wind for pollination, the stigmas have to wait for the pollen that may be wafted by the wind, and this is purely a matter of chance. Therefore, we should expect to find certain conditions specially favourable to this process in these plants. In the first place, large quantities of pollen should be available to ensure pollination. Winds may carry large quantities of pollen, but the pollen likely to fall on the stigmas can only be proportionately very inconsiderable. And, therefore, the flowers that are to be wind-pollinated should stand out far above the foliage leaves, so that the pollen may not be hindered from reaching the flowers. This is exactly what we find in the case of windfertilised plants. For instance in grasses which are wind-pollinated the inflorescence rises far above the level of the foliage ; in the case of trees dependent on wind for pollination, flowers appear, in most cases, at a time when the leaves have all fallen. As an example we may mention the tree, *Odina Wodier*.

The abundance of pollen is secured either by increase in the number of stamens in bisexual flowers, or by having a large number of male flowers. Sometimes the anthers become larger and produce plenty of pollen. To give a vivid idea of the abundance of pollen produced by this class of plants, we may mention the male spikes of Pandanus. A single plant is capable of producing a very large quantity of pollen. It has been computed by some American botanists that a moderate sized *Zea Mays* plant produces about 50,000,000 pollen grains. The pollen produced by these plants should also be adapted for being carried by the wind. So these grains are smooth, light and dry. The anthers are in most cases versatile, an arrangement best suited for the shaking out of the pollen very readily.

The stigmas also have certain adaptations so as to enable them to catch the stray pollen floating in the air. They are branched and plumose, thus getting a large amount of surface. All grasses and sedges, some Amaranths, Ricinus, Pandanus and Odina are wind-pollinated.

From this we see that wind-pollination involves the expenditure of a large amount of energy. As the pollen grains are composed almost wholly of protoplasm, a material most difficult to manufacture, this enormous production of pollen-grains cannot but be a drain on the resources of the plant. But this is necessary for ensuring pollination.

We have now to consider plants that are pollinated by insects. For successful cross-pollination flowers should be visited regularly and systematically by insects. Casual and erratic visits, or indiscriminateness in the choice of flowers for visits are not likely to be beneficial. Insects should have some inducement to visit the flowers. Flowers, as we know, produce pollen and honey, and both these substances are sufficient inducements to attract the insects and also ensure their visits. By the mere secretion of honey in the flowers the insects cannot be allured. To ensure their visits, it is absolutely necessary that the place where honey is secreted should be made known to them. The colours of flowers are meant to show them the place where honey may be obtained.

The infinite variety in colour and form of the flowers, as well as the different scents, is intended to attract the insect. In many plants the petals or the corollas are large and coloured so as to be very conspicuous. If the flowers are small, they are rendered conspicuous by being massed together. Sometimes instead of the petals, bracts play the same part, as in Bougainvillaea. Stamens also partake in this work, in Neptunia and Dichrostachys. There are also instances of flowers wherein the calyx becomes highly coloured, while the corolla is absent. The brightly coloured part in the flowers of Mirabilis and Boerhaavia is the calyx. Even when the petals are present the sepals become coloured like the petals, and we have such examples in Cassia and Caesalpinia.

The colour and the scent of the flowers are no doubt quite sufficient to attract the attention of the insects, but to ensure the regular visitation of insects this is not enough. Something more substantial should be offered to them to induce them to frequent the flowers very regularly. The pollen serves as food, or as material for building the hives for some insects such as the bees. Further, nectar or honey is also secreted in several flowers. Insects such as moths and butterflies live mostly on nectar. As both these substances, honey and pollen, serve as food for these insects, they come to the flowers.

Insects that visit the flowers are bees, butterflies, moths and flies. All these insects have sucking apparatus, which in the case of flies are short and in moths and butter-flies they are very long and coiled like a spring. Bees take in honey and also gather pollen ; moths, butterflies and flies drink only honey. As stated above, these insects visit the flowers, because they get food in the form of honey or pollen. Therefore, the secretion of nectar or honey is essential to ensure their visit.

In the case of insect-pollinated flowers, pollen need not be abundant. So, in these flowers, stamens are generally fewer in number. Further, the grains have usually a rough surface which often also becomes sticky. Stigmas need not have a large amount of surface as in the case of wind-pollinated flowers. The special adaptations in these flowers are so perfect that pollination is a certainty.

The nectar or honey is secreted usually by a disc at the base of the ovary. There are also instances of petals and sepals having nectaries either as glands, or as special pouches, sacs or spurs. The position of the nectary and the arrangement of the other parts of the flower are such as to make it impossible for the insects to get at the honey, without at the same time effecting pollination. It must also be remembered that, is the visit of the insect is to be of use in pollination, it is absolutely necessary that the part of the body of the insect which comes in contact with the anthers should also come in contact with the stigma ; otherwise pollination will not take place.

In flowers in which the corolla is regular and shallow any insect can get at the honey and effect pollination. For instance, in the Tribulus flower there are honey glands at the base of the stamens and between the petals ; the corolla is shallow and the stamens stand erect and so any insect, a bee, or butterfly or a fly can get at the honey. In trying to get at the honey the lower portion, or the sides of the insect must necessarily get dusted with the pollen, and as the stigma is also of the same height as the stamens it must also touch the same part of the body of the insect.

Flowers having long tubular corollas like those of Vinca and Jasminum, or papilionaceous corollas like those of Phaseolus, Indigofera and Vigna are not meant to be visited by all kinds of insects. Only butterflies and bees are able to get at the honey and in their manoenvres in search of honey they cannot help pollinating the flowers. Other insects cannot get at the honey and so they do not visit these flowers.

Colour of the petals though intended to show off well by contrast, is probably not capable of attracting insects as well as the scent of flowers and the smell of honey. The Mirabilis flower affords an example in support of this statement. This flower is visited by one or two species of the hawk moths; and there are several varieties of Mirabilis and in all of them the flowers open towards the evening and emit a strong scent. Some time after sunset the flowers that are white, light-yellow or cream-coloured alone can be seen and the magenta-coloured flowers cannot be seen. Yet the moth visits both the white and red flowers with equal ease. Insects cannot see the colour from a distance, their range of vision in this respect being limited to within a few feet. Again they are able to see some colours better than others. Bees are said to perceive blues better than yellow. Scarlet does not seem to attract them. Many of the insects visiting the flowers are usually busy with only one kind of flower at a time. For instance the common Carpenter bee was seen while sucking honey from *Bauhinia tomentosa* flowers. For over quarter of an hour it continued to hover about the same flowers, although there were several other flowers close by, that are usually visited by these bees. On another occasion this bee was thus visiting only the flowers of *Dolichos Lablab.*

Having considered the question of pollination in a general way, we shall now deal with a few flowers in a more detailed manner. The flowers of Papilionaceae are specially adapted for cross-fertilisation. At the base of the ovary there is usually a disc, which secretes plenty of honey. As the stamens are united by their filaments the honey finds its way into the staminal tube. Of the five petals, the standard is the chief attractive portion. It stands erect and is very conspicuous; in some cases besides being conspicuous it also bears some special marks which are supposed to serve as guides to the place where honey is secreted. The wing-petals invariably constitute a platform for the insects to alight and move about. The insect always sits on the wing-petals with its head directed towards the standard, whether there are special marks or not in it to direct the insect to the honey. Getting at the honey in these flowers is not at all an easy affair. It lies hidden at the base of the ovary and in the trough or cavity formed by the filaments. So it is only very intelligent insects that are likely to get at the honey.

When an insect settles on the flower and moves about, either the stamens or the stigma come out and touch the body of the insect. In some cases, as in *Crotalaria verrucosa* and *Tephrosia maxima*, when insects alight on the wing-petals and move about, pollen-dust is rejected from the pointed tips of the keel-petals first and then the stigmas come out when the insects visit the flower later.

In the flowers of *Vigna Catiang* the filaments of nine stamens are united, one stamen alone being free. At the base of the ovary and around it there is a disc secreting honey. The petals are all connected at the base by means of outgrowths, dents and foldings in such a manner that the standard stands erect, the wing-petals spread themselves out horizontally so as to serve as a platform and the keel-petal remains below. The margins of the keel-petals are united, both above and below, except at the free end where there is a small opening just enough for the emergence of the stigma. When the wing-petals are pressed down, the keel gets depressed and the stigma pops out, and when the pressure is removed, it resumes its original position. In this flower the anthers shed their pollen, before the stigma becomes receptive. Although the stigma is close to the anthers, the pollen does not reach it, because of the hairs found on the style a little below it. Even if a few grains fall on it they will not germinate, as the stigma is not receptive. When a bee or butterfly alights on the wing-petals the stigma comes out and brushes against the lower part of the body of the insect. If there is pollen already on the body of the insect the stigma gets pollinated. If on the other hand the visit is at a time when the anthers are dehiscing, the lower part of the body of the insect will be dusted with the pollen, and the same part of the body is sure to come in contact with the stigma when it goes to another flower.

In the plant *Indigofera enneaphylla* the floral mechanism is of a peculiar sort. In the flower the keel-petals have short spurs at their sides and they are intended to support the wing petals and keep them in a horizontal position. A very slight pressure on the wing-petals is enough to depress the keel-petals and separate their upper margins. As soon as the keels get depressed, the stamens and the style seem to come up with a jerk and remain outside. The essential organs cannot regain their original position, because the petals fall off as soon as the flower has received a single visit. So in this case one single visit of the insect is enough for cross pollination. Instances of the same kind of explosive arrangement are afforded by the flowers of *Abrus precatorius* and *Alysicarpus rugosus.*

The flowers of Phaseolus are very interesting in their structure and behaviour. Like other papilionaceous flowers these also depend on insects for pollination ; at any rate the arrangement of the petals and the position of the essential organs are such that self-pollination cannot take place. The keel-petals are in this flower prolonged into a beak and the beak is in the form of a spiral. The end of the spiral in the flower of *Phaseolus trilobus* and *P. Mungo* is towards the right, looked at from the front. The right keel-petal possesses a spur which supports the right wing-petal and helps it in retaining a horizontal position, so that it may serve as a platform for insects visiting this flower. There is a distinct passage between the right wing petal and the end of the keel-spiral, leading to the base of the standard where the honey is found. The left wing-petal lies higher than the right over the keel on the left side. An insect can get the honey only from the right side. On the left side the wing-petal, the keel and the standard are all close together and there is no opening. An insect coming to this flower for the sake of honey will, of course, alight on the lower wing-petal on the right hand side. As soon as it alights on the wing and begins to search for honey there will be some pressure exerted on the wing-petal. The keel will also be depressed on account of the spur which supports the wing-petal. Then the stigma will come out through the opening at the end of the keel-spiral which lies just above the wing-petal which rests on the spur of the keel on the right side. Inasmuch as the insect is on the platform, the stigma touches the back part of its body. If the stigma is receptive and if the back of the insect's body is alrealy dusted with pollen, pollination takes place. If on the other hand the insect visits a flower for the first time to start with, it will get dusted with the pollen, and the next flower visited will have its stigma pollinated.

In *Leucas aspera* and *L. linifolia* the corolla is bilabiate and the stamens are epipetalous and didynamous. The stamens and the style are enclosed within the upper lip of the corolla. Honey is secreted at the base of the ovary and when the insect alights on the lower lip of the corolla to such the honey, the stamens and the stigma are released from the upper hood of the corolla. As both the anthers and the stigma are turned downwards they will touch the back of the insect ; if the insect is already dusted with the pollen then the stigma will be pollinated, and in case the visit is for the first time the back of the insect gets dusted with pollen.

The flowers of Acanthaceae have also special devices for pollination. In many plants of this family the stamens lie within the corolla in such a way that they cannot help coming in contact with the back of the insect when it sucks honey. Even small inconspicuous flowers are regularly visited by butterflies. On one occasion in the month of October while engaged in collecting plants, the tiny flowers of *Justicia procumbens* were constantly visited by the small pretty violet butterfly, Zizera lysimon. At first the flowers of *Justicia procumbens* were not noticed ; this small butterfly was seen hovering about and settling on this plant. On close inspection it was found to suck honey from the tiny flowers of the Justicia. These plants were not at all numerous and there were stray plants scattered all over the place. The violet butterfly continued to visit these flowers persistently for over twenty minutes. The insect proved a most satisfactory guide to lead to the places where this species of Justicia was growing. Similarly another butterfly, a white one with a red border on the front wings, was seen visiting the flowers of *Spermacoce hispida* with great assiduity. The flowers of *Tephrosia purpurea* were visited by a butterfly with golden yellow wings. The flowers of Tridax were watched to see what insects visit them for honey and pollen. Seven different species of insects were found visiting these flowers and they were Zizera lysimon, Colotis amata, Catopsila pyranthe, Terias hecabe, Parnara mathias and Ceratina viridissima.

There are flowers that are exclusively pollinated by the visits of flies. The flowers of Aristolochia and Aroideae are of this kind. A very foetid odour is usually emitted by some of the Aristolochias and in one species the smell is so strong as to attract flies from very long distances (fifty to hundred yards or more). In this flower the anthers are situated on the style below the stigma and they dehisce after the withering of the stigma. Inside the tubular perianth, when it opens there are found a number of hairs all directed inwards. The flies get into the perianth tube easily. But they cannot get out of it with the same ease, on account of the hairs. This flower is protogynous and therefore it has to be pollinated by pollen from some other flower older than itself. When the flies get inside the flower they pollinate the stigma if they have already visited other flowers ; if on the other hand the visit of the insect is its first visit, it gets dusted with the pollen. When the flies get into the flower, the anthers are usually immature and the flies cannot get out of the flower, until the anthers dehisce, because the hairs dry only then. Until that time the flies will be wandering all over inside, in search of a way to escape ; and in their wanderings they get dusted with pollen. By the time the pollen is liberated the hairs disappear and the stigmas fade, and the way becomes clear for the flies to escape.

The inflorescence of Aroids are particularly interesting on account of the arrangements existing in it for pollination. The spadix bears the unisexual flowers at its lower portion only, those at the very base being female and those above male. Above or between these are a number of reduced flowers in the form of hairs directed downwards. The whole inflorescence is covered by a large spathe, the lower part of which forms a tube constricted a little above the region of anthers, or the hairs, whilst the upper portion is expanded exposing the upper free end of the spadix. Small flies attracted by the bad smell emitted by the flowers and also by the spathe, crawl down into the tubular part of the spathe. Once they are in, they cannot easily get out from the tubular part, because the neutral flowers are all directed downwards and the spathe is constricted just above these flowers, or the anthers. The flowers are protogynous as in the case of Aristolochia, and so the insect pollinates the stigma if it bears pollen from another plant. If the flower happens to be the first one that is visited by the insect, the insect is forced to stay within the spathe until the way is open for it to escape. The withering of the downward directed hairs and the dehiscence of the anthers take place together and therefore the flies escape with a coating of pollen. They cannot avoid this as the anthers are at the top of the inflorescence just at the level of the constriction or a little below it.

In the flowers of orchids we have a most perfect mechanism for securing cross-pollination although it is very complicated. As examples we shall select two species of orchids commonly met with in the plains. The orchid *Eulophia virens* grows amidst scrubs and its flowers are greenish with red streaks. There are three sepals and three petals ; the sepals are all alike, but one of the petals, the one lying on the lower side of the flower when it is open, is modified, and it is called the lip. In this flower the lip is slightly saccate at its back and the free front part is broad and in the flower it stands out as a platform for the insects to sit on. The short shallow sac secretes honey and the flowers are visited by butterflies. The anther is single and it is on the top of a column rising from the top of the ovary. The pollen in this and all orchids is collected in masses called *pollinia*, and in this flower there are two such masses in the anther. Just below the anther there is a disc-like gland to which the two pollen masses are attached by means of strap-shaped stalks. The anther in the open flower lies just above the opening leading into the honey-sac. When the butterfly sucks honey the pollen masses stick on to its head, and so the insect carries away with it the pollen masses. The next flower that is visited by this insect is pollinated by the pollen masses sticking on to its head. The stigma lies just below the anther on the front side of the column. When the insect rushes into the flower to suck honey the pollinia cannot help touching the stigma because the pollinia assume a direction which will coincide more or less, with the position of the stigma.

In Habenaria we have another very common orchid of the plains. The flowers in this orchid are all pure white and the raceme springs from the middle of two or three elliptic leaves lying flat on the ground. They are very conspicuous by reason of the racemes. Both in the morning just before sunrise and in the evenings the flowers are visited by moths. The lip is prominent below and it is prolonged at the back into a very distinct spur varying in length from quarter of an inch to one inch. The two lateral petals and the sepal lying above are united so as to form a hood. There is a narrow passage leading into the interior of the flower and the anther is situated on a column just above the passage leading into the spur. The pollinia have long stalks and the basal portions of the stalk protrude a little. When the moth gets in, the two pollinia stick on to the two eyes of the insect and when it flies to another flower the pollinia are rubbed against the stigma lying below the anthers and the masses adhere to the stigma as its surface is very sticky.

In *Calotropis gigantea* (as also in all Asclepiadeae), the pollen grains are held together in masses and there are ten pollinia. When an insect settles down on the top of the staminal column in the centre of the flower and moves about round the stigma, the pollinia adhere to the legs of the insects. When it flies to another flower the pollinium gets on to the stigma to which it will stick if it is receptive.

There are many plants whose flowers are usually-pollinated by moths. As moths move about only during the night or in the dusk, these flowers have white corollas, open in the dusk and also emit a strong scent at this time. The flowers need not be irregular as a platform is not needed, because moths do not alight and they only hover in front of the flower. We may mention as examples Ipomoea, Tobacco, Mirabilis and Jasminum. In these flowers the anthers are generally loose and are well adapted to shed pollen on to the body of a hovering moth. The hawk moths, *Herse Convolvuli* and *Cephonodes picus* visit the flowers of many of the large flowered Ipomoeas and the flowers of *Mirabilis jalapa*. The former is a very constant visitor, but the other one comes very rarely and it is extremely active and so rapid in its movements that it is impossible even to see it.

We have also plants that depend on water for pollination and they are all aquatic plants. The common water plant, Vallisneria may be selected as a type of plants fertilised by the agency of water. The flowers of this plant are unisexual and the plants are dioecious. Both the staminate and the pistillate flowers develop under water, and so, unless there is some special adaptation, fertilisation cannot take place. The male flowers are borne on short stalks almost at the base of the plant. At the time of flowering, or when the flowering season approaches, these get separated from their stalks and rise to the surface of water to float there. Simultaneously, the female flowers lying under water with spirally coiled stalks come up to the surface, in consequence of the uncoiling and growth of the stalk. The male flowers are most numerous and they float amongst the female flowers. The anthers open and the pollen is shed on the stigmas. In the case of flowers that are hermaphrodite, the flowers are raised above the water level, as in Nymphaea, so that they might be pollinated either by insects or by wind. It is only when flowers are unisexual and also submerged, pollination is effected by water currents. In all plants that are pollinated by currents of water the pollen grains are smooth, and pollination also is not a matter of certainty. So in these cases there is an abundance of pollen. During hot weather when water is low in ponds and tanks we generally see masses of small white flowers floating freely on the surface.

Sometimes the aquatics which ordinarily raise their flowers above the water, do not do so when rain is heavy. Then the flower buds do not open and are then cleistogamic.

Protection of Pollen against Rain

Microspores are much less resistant to extremes of tempe-ratue and to drying when they have been moistened and in consequence have begun to germinate. Pollen may be protected from rain in various ways. In some flowers, especially those whose pollen is exposed to rain when the flower opens, the microspores are not readily wetted, having a covering of wax or of spines, etc.

Many flowers protect the pollen by their horizontal or drooping position *e.g.,* Ericaceae, Liliaceae. In some cases the flower closes up at night or in bed weather, *e.g, Oxalis,* tulip, crocus, some Ranunculaceae ; and the same kind of closing is effected in the capitula of many Compositae by the movement of the flowers and bracts. In iris the large petaloid stigmas cover the stamens, and in many flowers the stamens are protected by a hood formed by the sepals or petals, or by both.

Germination of the Microspore

Process leading up to and ending in Fertilisation. At first the microspore is unicellular but later, even before it leaves the anther, its nucleus divides into two (Fig. 2.10, A). The larger one of these in the vegetative nucleus ; the smaller one the

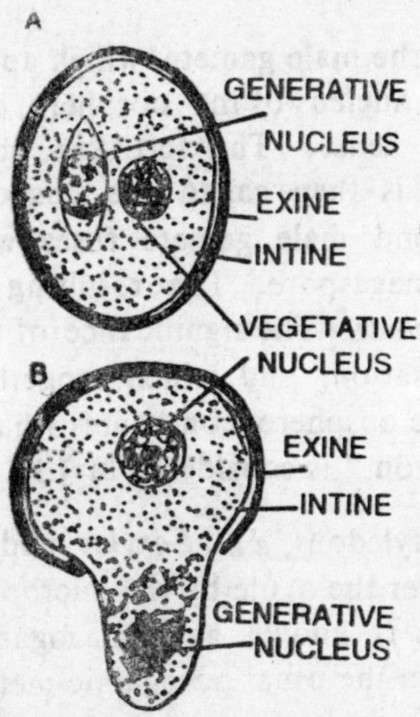

Fig. 2.10. **A.** Division of Microspore Nucleus **B,** Germination of Microspore.

generative. Both are free in the cytoplasm of the microspore. Either before or after pollination the generative nucleus divides into two male gametes. Germination and furthur development take place on the stigma, which secretes a sugary nutritive fluid.

From one of the pores in the exine a slender pollen-tube (Fig. 2.10, B) protrudes, and grows down through the tissue of stigma and style and finally enters the ovary. After entering the ovary the pollen-tube grows towards and ovule, which it enters usually by the micropyle. It penetrates the apex of the nucellus and comes in contact with the megaspore near to the oosphere and synergidae. During its growth the vegetative nucleus is found near the tip of the pollen-tube. The male gamates also pass down to the apex of the pollen-tube, but the vegetative nucleus is by this time disorganised. *One male gamete only* is concerned in the actual process of fertilisation. It passes from the pollen-tube into the megaspore and fuses with the oosphere or female gamete. The synergidae assist in the process, hence their names synergidae or "help-cells" (Gr. *συν*, with *εργον*, work). The tip of the pollen-tube swells and bursts, thus setting the male gametes free in the embryo-sac.

The fusion of the male gamete, which appears to be entirely nuclear, with the nucleus of the oosphere constitute fertilisation in the strict sense. The fertilised oosphere secretes a cellulose wall and is then called the oospore. Its nucleus is diploid. The second male gamete fuses with the secondary nucleus of the megaspore. The resulting nucleus is called the *endosperm nucleus*. The significance of this process, which resembles fertilisation, and which, together with the actual fertilisation of the oosphere, constitutes what has been called "double fertilisation," is considered in § 10.

In a few Dicotyledons, *e.g.*, *Corylus* and *Betula*, the pollen-tube does not enter the ovule by the micropyle, but by piercing the chalaza. This is known as chalazogamic fertilisation as distinguished from the usual porogamic method.

3

Fertilization

Fertilization involves the fusion of a male gamete with a female gamete. In angiosperms the female gametaphyte is seated deep in the ovarian cavity, quite away from the stigma. The pollen (male gametophyte) are, normally, held at the Stigma, and there is no device for them to reach the egg inside the female gametophyte. To effect fertilization in this group of plants the pollen grains germinate on the stigma by putting forth tubes (pollen tubes) which grow through the style and find their way into the ovules, where they discharge the sperms in the vicinity of the egg. One of the sperms fuses with the egg (forming zygote) while the other fuses with the polars or the secondary nucleus (forming primary endosperm nucleus). The distance a pollen tube has to travel in order to reach the egg depends on the length of the style which is quite variable in different species. For example, in sugar-beet it has to grow only a few millimetres whereas in corn it grows as much as 450 mm.

STRUCTURE AND DEVELOPMENT OF A STAMEN

A very young stamen in a flowers bud is small, four angled, upgrowth from the receptacle. This upgrowth will four the anther, the filament being delayed in development. In each of the four angles a pollen sac containing pollen grains will be elaborated as development proceeds. Fig 3.1 illustrates these developments as seen in transverse sections of anthers of various ages, but it must be realised that longitudinal series of cells are involved, running the whole length of the anther.

The left-hand figures represent early stages. The anther consists at first of undifferentiated parenchyma but soon cells in the four corners, lying just beneath the epidermis, become active. Divisions take place giving a group of inner cells (shaded) surrounded by flattened, actively dividing cells. The inner cells ultimately give rise to the pollen grains while the outer cells form the walls of the pollen sacs. Divisions continue in both the inner and outer cell groups and the condition seen in the top right hand figure is attained, where each corner of the anther is occupied by a young pollen sac.

Each pollen sac has a wall several layers thick, the innermost being a layer of cells known as the tapetum, which later provides food for the developing pollen grains. The cells in the centre of the sac are pollen mother cells with dense cytoplasm and large nuclei. These, at first joined closely together, soon round themselves off and float freely in a nutritive liquid derived from the tapetum. Each pollen mother cell now gives rise to four pollen grains by two successive divisions. The nucleus divides into two and then into four. Cytoplasm collects round each of the four daughter nuclei and partition walls are formed so as to give a tetrad of four pollen grains. The type of nuclear division involved is reduction division or meiosis.

As the anther reaches maturity, the tetrads of pollen grains break up and the separate pollen grains now lie freely in the cylindrical pollen sacs (Figs 3.1 and 3.2). Each young grain has at first one nucleus but this soon divides into two. One of these, surrounded by denser cytoplasm, forms the generative cell (which ultimately gives the male gametes) and the other is the tube nucleus (Fig. 3.3).

The wall of the ripe pollen grain is a double one. There is dedicate inner wall of cellulose (the intine) and a cuticularised outer one (the exine) of variable thickness. Thin places, or pores, are present in the exine. The grains vary in size from 3-300μ. Their shape may be ellipsoidal, spherical, crystal-like

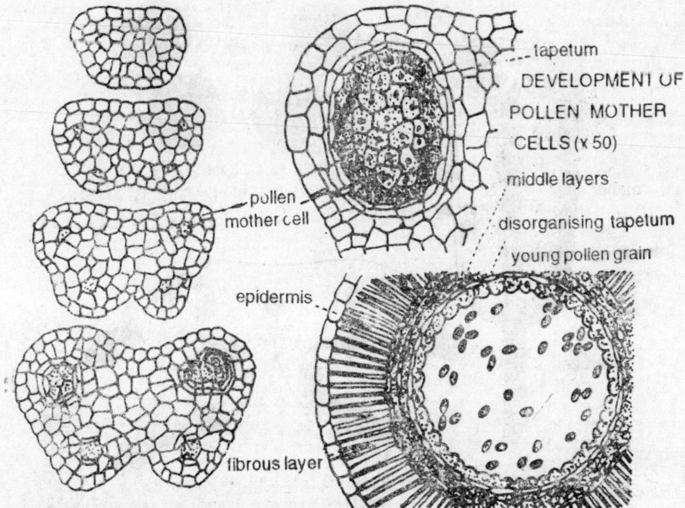

Fig. 3.1. Stages in the development of a stamen, seen in transverse sections of anther heads of different ages.

with a varying number of facets, or many other shapes. Further variation is given by characteristic sculpturing of the outer wall which may have spines, papillae, a network of ridges and so on. Many plants can, indeed, be identified merely on the basis of their characteristic pollen. It has proved possible to learn something about the changes which have occurred in our vegetation since the Ice Age by the identification of pollen grains preserved at various depths in peat deposits. It may also be mentioned that the earliest known trace of Angiosperms is in the form of pollen grains found in coal of Jurassic age in Scotland.

As a result of the various developments described above the anther now has the structure shown in Fig. 3.2. As the left-hand figure indicates, the pollen sacs are long, cylindrical structures. There are two of them in each lobe, the lobes themselves being joined by a connective containing a vascular bundle. The wall of each sac has now a highly differentiated

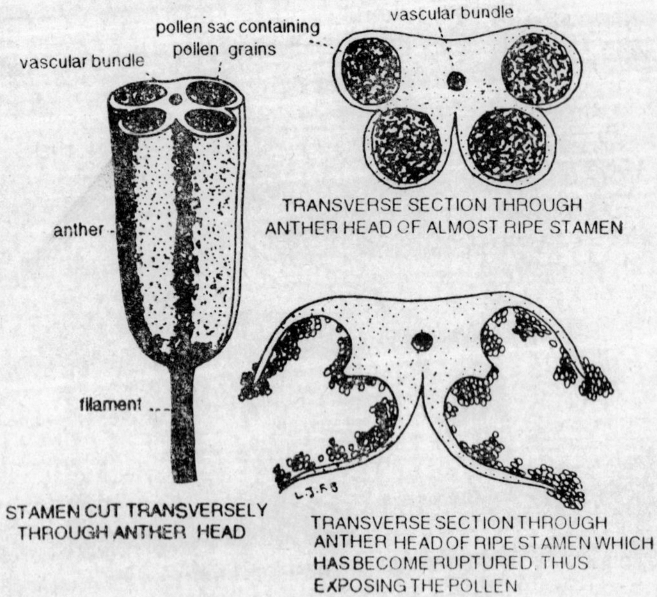

Fig. 3.2. A Typical Ripe Stamen.

structure. The cells lying just below the epidermis have enlarged and their walls have been modified by the deposition of lignin. This characteristic fibrous thickening is laid down on all the walls except the outer ones. In some anthers several layers of the wall may show this type of thickening.

The thin outer walls of the cells of the fibrous layer shrink on drying and so cause a longitudinal rupture of each anther lobe along a line of specialised cells called the stomium. This rupture is accompanied by a breaking down of the wall between each pair of sacs. The pollen grains are thus liberated (as is indicated in the bottom right-hand drawing of Fig. 3.2) and pollination can now take place.

STRUCTURE AND DEVELOPMENT OF AN OVULE

The ovule when it is ready for fertilisation is more or less egg-shaped. It consists essentially of an ovoid mass of

parenchymatous cells called the nucellus which is borne on a short stalk or funicle by means of which it is attached to the placenta in the ovary. The nucellus is invested by one, or more frequently two, integuments which are protective coats

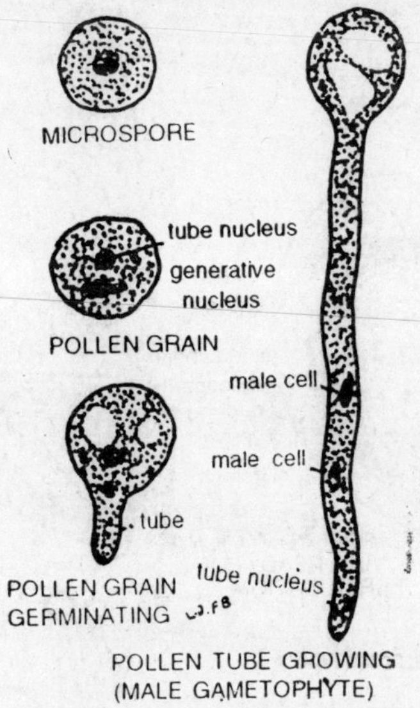

Fig. 3.3. Stages in the development of a pollen grain.

growing up from its base. These cover over the nucellus except at the extreme tip, where a narrow channel is left open as the micropyle leading down to the nucellus. Embedded in the nucellus there is a large ovoid sac bounded by a thin cell-wall. This is called the embryo-sac, since, after fertilisation, it will contain the embryonic plant.

In some plants, for example, rhubarb, the ovule is straight and the micropyle is at the end opposite to the stalk. Such ovules are said to be orthotropous. In other examples, for example, shepherd's purse, the nucellus and embryo-sac are curved so that the micropyle is placed near to the stalk. This is the campylotropous type of ovule. In the majority of ovules, however, the nucellus is straight but the entire ovule is inverted by a sharp curvature at the top of the stalk so that the micropyle lies near to the base of the stalk and also near to the carpellary wall. This is the common anatropous type of ovule (Fig. 3.4).

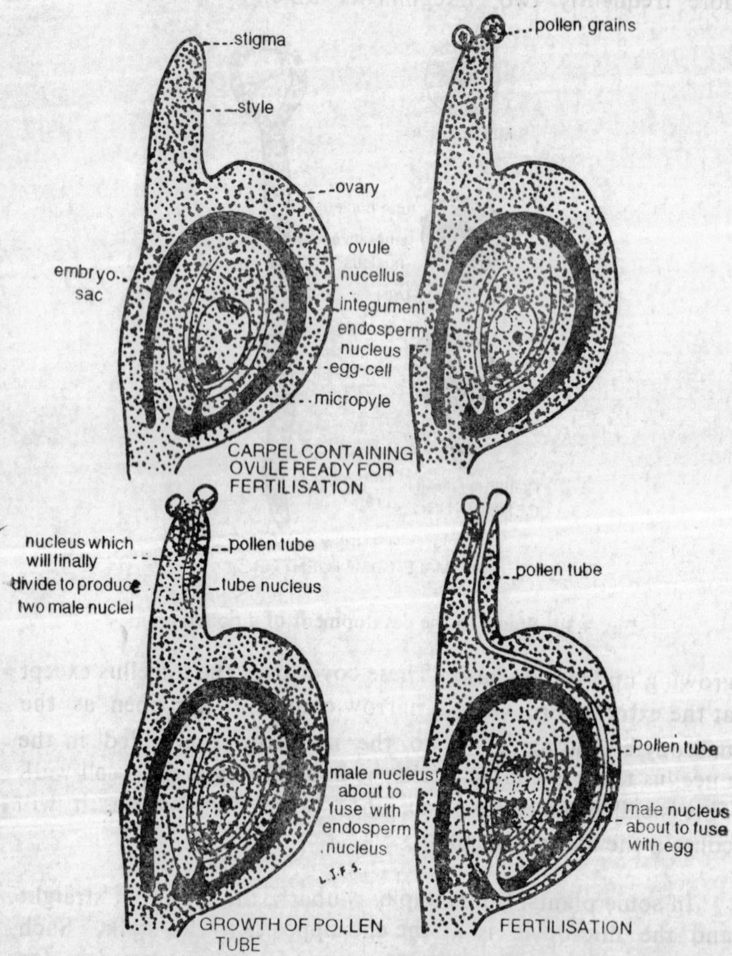

Fig. 3.4. Stages in the process of fertilisation.

The embryo-sac at the time of fertilisation contains vacuolated cytoplasm in which a number of prominent nuclei are embedded. At the end furthest from the micropyle there are three large nuclei, each surrounded by granular cytoplasm and a gel membrane ; these are called the antipodal cells (Fig. 3.5) and often appear to play a part in the nutrition of the embryo-sac. At the micropylar end there are three other large nuclei and these too are each surrounded by cytoplasm and a membrane. This micropylar group constitutes the important egg apparatus. The central cell is the female gamete called the egg cell or ovum. The two other cells, which may correspond to non-functional ova, are called the synergids. In the centre of the embryo-sac there is a conspicuous nucleus which is called the central fusion nucleus or the primary endosperm nucleus.

When the ovule is in the condition just described it is ready for fertilisation but before considering this process it is necessary to examine the way in which this structure has developed.

An ovule arises on the placenta as a small upgrowth of parenchymatous tissue which is the young nucellus. Then integuments appear as ring-like upgrowths from its base and these gradually invest the growing nucellus until only the micropyle is left at its tip. Meanwhile, if the ovule is of the anatropous type, the stalk has elongated and curved sharply at its top so as to bring the micropyle near the insertion of the stalk on the placenta.

It is during these developments that the embryo-sac is formed. A single cell of the nucellus, situated just below its tip, becomes conspicuous by reason of its larger size and dense protoplasmic contents. This is the embryo-sac mother cell which then divides first into two cells and then into a row of four cells. As in the formation of four pollen grains from a pollen mother cell, the nucleus of the embryo-sac mother cell divides by the process of meiosis to give the four potential embryo-sacs. Usually, however, it is only the innermost of these which develops further as the functional embryo-sac. The other three collapse and are crushed as the surviving embryo-sac enlarges.

When first formed, the embryo-sac has a single nucleus (Fig. 3.5). This nucleus divides by mitosis and the daughter

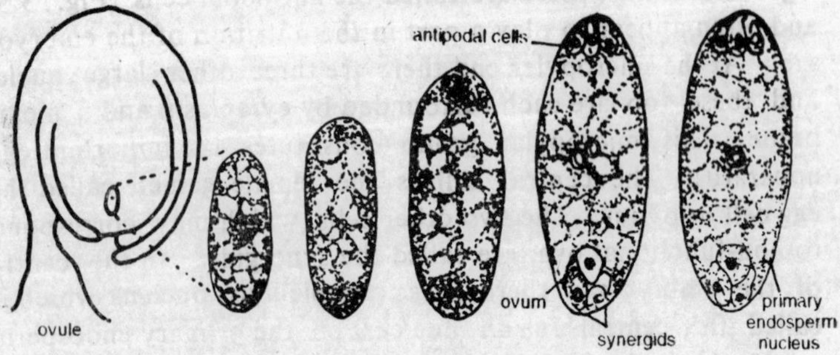

Fig. 3.5. Development of the embryo sac.

nuclei move to the two poles of the sac. Each of the two nuclei divides again twice so the four nuclei occupy the micropylar end and four the opposite (chalazal) end of the sac. One nucleus from each group of four now moves to a central position in the sac and these two fuse to give the central fusion nucleus. The three at the micropyle end form the egg-apparatus already described and the other three form the antipodal cells. The embryo-sac thus reaches the condition which is found in practically all Angiosperms at the time of fertilisation.

PROCESSES OF FERTILISATION

It is now necessary to trace the sequence of events following upon the deposition of pollen grains on the receptive stigma.

The pollen grains germinate on the stigma where there is usually a liquid secretion, often containing sugar, but sometimes other substances are also present, as, for example, malic acid in the case of members of the Ericaceae. Germination results in the growing out of the thin inner wall of the pollen grain through one of the pores in the thick exine. This results in the production of a cylindrical pollen tube (Fig. 3.3), into which the tube nucleus and generative nucleus pass. The pollen tube, tending to grow towards chemical substances produced by the stigma, towards moisture and away from oxygen, then enters the stigma and begins to grow down the style. In some examples, for example,

Viola species, there is an open canal in the centre of the style leading right down to the ovary ; in others, for example, *Rhododendron* species, there is a canal filled with mucilage. The tube grows down the moist sides of the canal or through the mucilage, as the case may be, and so reaches the ovary. It obtains water and food for this continued growth from the tissues of the style.

Having reached the ovary, the pollen tube either grows in the tissues of the ovary wall until the tip reaches the vicinity of an ovule (Fig. 3.4) when it enters the cavity of the ovary, or else it enters the top of the ovary cavity and grows down the moist inner surface in an ovule. In either case the tip of the tube enters the micropyle of the ovule, possibly directed by a fluid secreted by the synergids. It thus reaches the nucellus, and by penetrating the few overlying layers of this tissue, it comes into contact with the embryo-sac in the vicinity of the egg-apparatus.

During the downward growth of the pollen tube the generative cell has divided to form the two male gametes (Fig. 3.3). These occupy the tip of the tube, the tube nucleus having disintegrated.

Fertilisation now takes place. The tip of the pollen tube becomes ruptured and the two male gametes, now spirally coiled, thread-like bodies, enter the embryo-sac. One of the gametes enters the egg and fuses with its nucleus. The egg, so fertilised, is called the zygote and it is from this that the embryo plant develops. The second male gamete moves deeper into the embryo-sac and fuses with the central fusion nucleus. The result of this second fusion is the production of a nutritive tissue, known as the endosperm ; this is used by the developing embryo. The time elapsing between pollination and the occurrence of these fusions varies considerably. In rye it is seven hours ; in maize, twenty-four hours ; in some trees it may be a year or more.

The elucidation of the sequence of the events just outlined occupied botanists for many years. Although it was known from the time of Camerarius onwards that a fertilisation process did in fact take place, the details proved difficult to observe. In 1823, Amici, an Italian naturalist, observed the production of pollen tubes by pollen adhering to a stigma and seven years later he traced the course of the pollen tube down into the ovary and into the micropyle of an ovule. Later still (1846), the same observer showed that the ovule had an egg cell before fertilisation and that this, excited by the pollen tube, gave rise to the embryo, Hofmeister confirmed these observations in 1849 but it was not until 1884 that Strasburger observed the fusion of one of the male gametes with the nucleus of the ovum. It was still later (1898) before Nawaschin observed the fusion of the second male gamete with the central fusion nucleus in the centre of the embryo-sac.

We have now seen that fertilisation consists essentially of the fusion of a male gamete with a female gamete, the ovum, to form a zygote and ultimately an embryo. The second fusion is a subsidiary one and results only in the formation of nutritive tissue.

In the account of fertilisation given above the events have been described in relation to a single pollen grain and a single ovule. Normally, however, many pollen grains will be deposited on a stigma and they will all produce pollen tubes. If the ovary contains only one ovule, naturally only one of these tubes penetrates the micropyle and liberates its male gametes into the embryo-sac. If, however, as in the majority of ovaries, numerous ovules are present, then the production of numerous pollen tubes is a necessity if all the ovules are to be fertilised. In a flower like that of *Rhododendron*, if a transverse section is cut through a style some time after pollination has occurred, the section will show a large number of pollen tube sections embedded in the mucilage of the stylar canal.

MEIOSIS

An essential feature of fertilisation is the fusion of a male gamete with the nucleus of the female gamete. This results in the zygote or fertilised egg the nucleus of which contains the nuclear material of both the male and female gametes. If the male gamete has *n* chromosomes (a number varying with the species) and the female has *n* chromosomes the zygote will obviously possess 2*n* chromosomes. The chromosome complementss in the vegetative cells of flowering plants consist of homologous pairs, one of each pair being maternal and the other paternal in origin. It is clear that these pairs came together in the act of fertilisation which gave rise to the plant.

As the zygote develops into the embryo, and then into the new plant, all the cell divisions which take place are mitotic and thus the 2*n* number of chromosomes is maintained. If, however, this type of division occurred throughout the life-history, then the gametes would each have 2*n* chromosomes and, when fertilisation occurred, the zygote would have 4*n* chromosomes. There is, however, a compensating process in the form of another type of nuclear division known as meiosis or reduction division, by which the number of chromosomes is halved. Meiosis always occurs in the formation of the gametes or in some cell division prior to such gamete formation. The result is that the gametes themselves always have the haploid or *n* number of chromosomes and the zygote formed as a result of fertilisation will have the diploid or 2*n* number.

In flowering plants, meiosis occurs in the formation of the four pollen grains from each pollen mother cell and also in the formation of the four potential embryo-sacs from an embryo-sac mother cell. As the subsequent nuclear divisions which take place both in the pollen grain and in the functional embryo-sac are all mitotic, it follows that the male gametes and the female gamete will have the reduced or haploid number of chromosomes. All plants which reproduce sexually show the occurrence of meiosis at some point in their life-history between successive acts of fertilisation but this point varies in the different big groups of plants.

Meiosis consists of two very much modified mitotic divisions, normally giving rise to four cells, but it is essential that it should be considered as a single process with two stages.

Cells which are about to undergo this type of division are usually filled with dense cytoplasm and have a large nucleus.

The first indication that meiosis is in progress is the appearance of the chromosome threads just as in the prophase of mitosis, except that they are single and show the chromomeres more obviously along their length.

The stage where the strands are obvious as separate entities is called leptotene and this is followed by zygotene where the homologous chromosome strands come to lie together, so precisely paired that corresponding segments and even chromomeres are in juxtaposition. These paired chromosome structures are called bivalents. In each bivalent both of the chromosomes now split longitudinally (except in the region of the centromere) giving a four-stranded structure which is characteristic of the pachytene stage. Pachytene is a very short stage and is seldom seen in preparations showing meiosis ; but it is nevertheless important. In each bivalent the chromatids now contract and twist around one another but the members of each pair repulse the opposite members except at points which are called chiasmata, (sing., chiasma). These chiasmata explain cytologically the genetic problem of 'crossing over'. This stage is diplotene and it is followed by diakinesis, which is characterised simply by a greater shortening and thickening of the bivalents. The latter usually arrange themselves around the periphery of the nucleus and at this stage also the nuclear membrane and the nucleolus disappear. The above stages which constitute the prophase are shown diagrammatically in Figs. 3.6 and 3.7.

Metaphase I follows and is similar to the metaphase of a mitosis in that a spindle has now been formed and that the bivalents arrange themselves on the equatorial plate in exactly the same way as do the chromosomes in mitosis. The bivalents are placed with their two unsplit centromeres above and below the equatorial plate. Anaphase I now follows and the two centromeres of each bivalent move to the opposite poles of the cell, each pulling its pair of chromatids with it. Telophase I can be a very transient stage, and can frequently be regarded

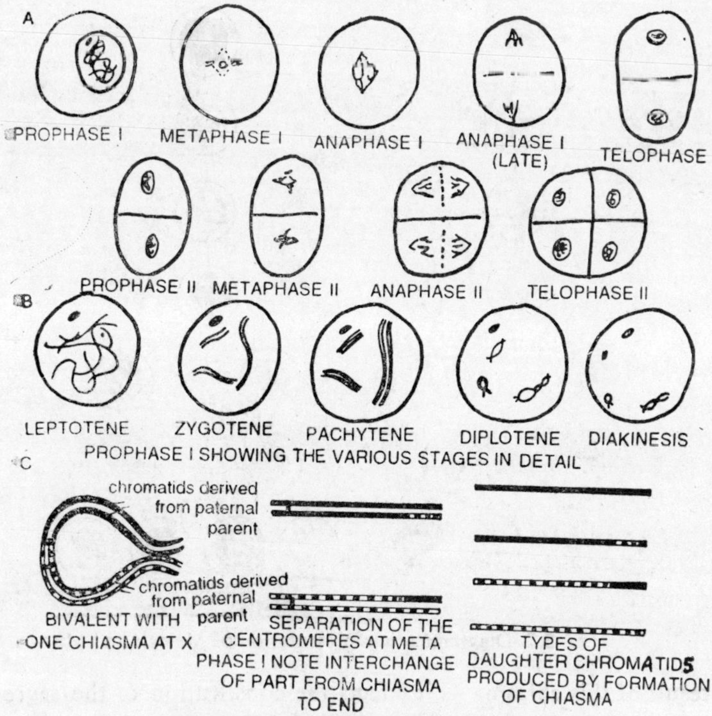

Fig. 3.6. Diagram illustrating Meiosis.
(*A. M. M. Berris*)

as non-existent since the daughter nuclei pass straight into the prophase of the second division. Metaphase II follows, each centromere (of which there are only half as many as in the original cell) moves on to the equatorial plate, taking with it the two chromatids which are attached to it. The centromeres then divide and at anaphase II the two chromatids separate. Four daughter cells are thus produced, each with half the number of chromosomes which were present in the parent cell.

One feature requires fuller treatment. At diplotene there is actual breakage and interchange of parts of chromatids, the position of the breakages usually being seen as chiasmata. Fig. 3.6 shows a chromosome which has one chiasma and the

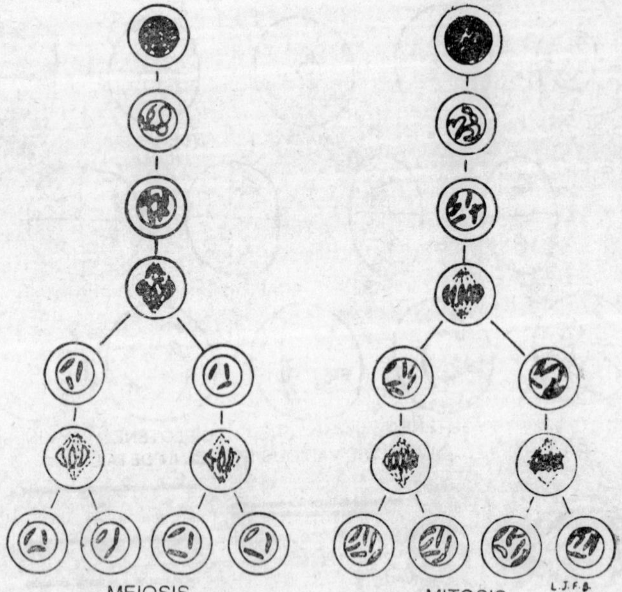

Fig. 3.7. Diagrammatic representation of Meiosis and Mitosis.

result of this chiasma as it effects the constitution of the segregating chromatids at anaphase I and anaphase II. The interchange of segments extends from the chiasma to the ends of the chromatids. This means that two of the chromatids are actually made up of both maternal and paternal material while the other two are either wholly maternal or wholly paternal in their make-up. More than one chiasma can be present in a bivalent and the interchange can take place between any two chromatids provided that one is of maternal and the other of paternal origin. Extremely complicated segregations can thus occur.

THE SEED

After fertilisation has occurred, changes take place in all parts of the ovule. The structure resulting from these further developments in the ovule is the characteristic reproductive body of the flowering plants, the seed.

Changes in the embryo-sac itself may be considered first. The fertilised egg forms a wall around itself and proceeds to divide so as to form a short chain of cells which is known as the pro-embryo (Fig. 3.8, 1 and 2). The cell nearest to the micropyle grows to a large size, while the cell at the tip of pro-embryo begins to divide in various planes, these divisions following a regular sequence. This terminal cell in fact gives rise to most of the embryo plant, the rest of the pro-embryo forming the suspensor. The large basal cell of the suspensor serves as an organ of attachment to the wall of the embryo-

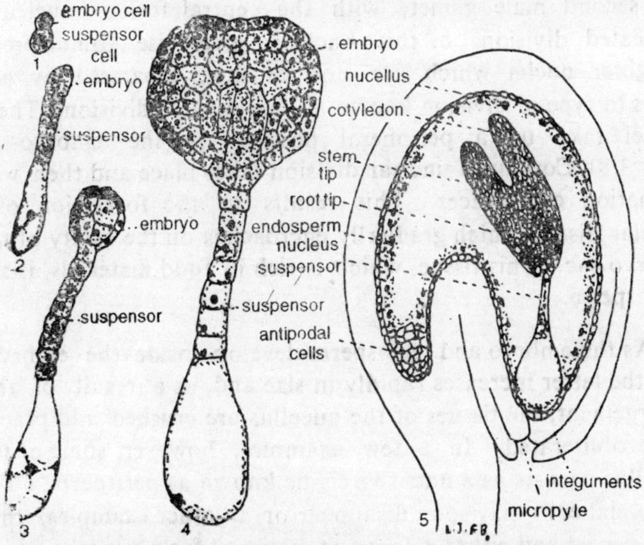

Fig. 3.8. Stages in the development of an embryonic plant from the fertilised egg.
1—4. Young developing embryo only ; 5, embryo within the ovule (now seed).

sac ; the filamentous region continues to elongate and pushes the embryo into the nutritive tissue, or endosperm, which is developing in the embryo-sac at the same time. Continued cell division in the embryo soon results in the appearance of two lobes which are the two cotyledons (or of one lobe if the plant is a Monocotyledon). The stem apex is formed between the two cotyledons (or laterally at the base of the single cotyledon in a Monocotyledon). The lower part of the embryo in both Dicotyledons and Monocotyledons forms the bulk of the first root or radicle. The tip of the latter is, however, formed by divisions in the top cell of the suspensor. The tip of the root is thus attached to the suspensor and points towards the micropyle (Fig. 3.8, 5).

Simultaneously with the development of the embryo, or even preceding this, endosperm is formed in the embryo-sac. This results from the active division of the central nucleus (which, it will be remembered, is the product of the fusion of the second male gamete with the central fusion nucleus). Repeated divisions of this nucleus give rise to numerous daughter nuclei which are not, at first, separated by cell walls (a type of division known as free nuclear division). These nuclei take up a peripheral position in the embryo-sac (Fig 3.8). Continued unclear division takes place and then wall formation commences. This results in the formation of a cellular tissue which gradually encroaches on the cavity of the embryo-sac. This tissue, which is rich in food materials, is the endosperm.

As the embryo and endosperm develop inside the embryo-sac, the latter increases rapidly in size and, as a result of this enlargement, the tissues of the nucellus are crushed and practically obliterated. In a few examples, however, some of the nucellus persists as a nutritive tissue known as perisperm. The antipodal cells may soon disappear or, in other examples, they may persist and either enlarge or divide to form a tissue which helps in the nutrition of the enlarging embryo-sac.

Changes also take place in the integuments. These usually become fibrous or woody and constitute the protective seed coat or testa.

The total result of these post-fertilisation changes in the ovule is the structure known as a seed. The testa is, as just described, derived from the integuments of the ovule. Inside the testa is the embryo. In addition there may be endosperm present, or this tissue may be absent from the ripe seed. If the embryo grows rapidly during the ripening of the seed it will use up all the endosperm to obtain the requisite food material for its growth. Such a seed is said to be non-endospermic or exalbuminous (for example, the seeds of apple (Fig. 3.9), bean, pea, etc.). On the other hand, the embryo may cease to

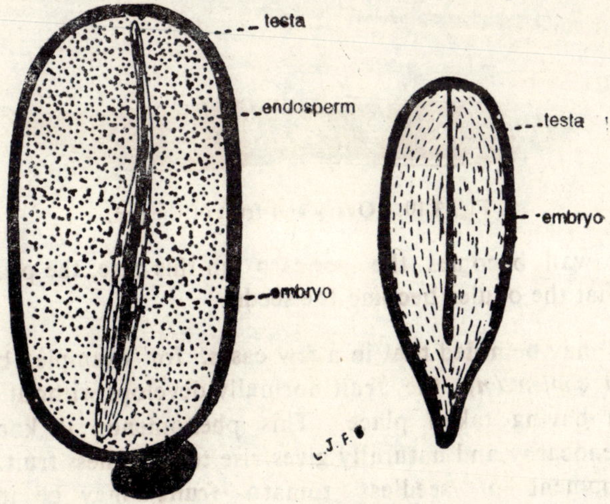

Fig. 3.9. Longitudinal section through castor oil and apple seeds.

grow before all the endosperm has been used and then the ripe seed will be endospermic or albuminous as in castor oil (*Ricinus communis*) (Fig. 3.9), wheat (*Triticum vulgare*), and coco-nut (*Cocos nucifera*).

THE FRUIT

The effects of fertilisation are not confined to the ovules but extend to the other floral parts and, in particular, to the carpels in which the ovules are borne.

The carpellary or ovary wall is stimulated to further development, possibly by growth substances secreted by the developing ovules. It may become dry and membranous, or it may thicken considerably and become either woody or fleshy. This modified ovary wall, together with the contained seed or seeds, constitutes a fruit, the modified wall being the fruit coat or pericarp. Fig. 3.10 serves to emphasise the fact that the

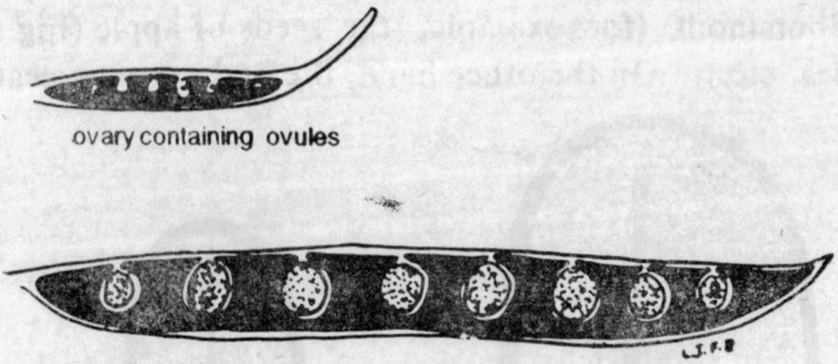

Fig. 3.10. Ovary and fruit of the pea.

ovary wall becomes the pericarp (in this case, the pea pod), and that the ovules become the seeds.

It may be noted that in a few cases, for example, banana (*Musa sapientum*), the fruit normally develops without fertilisation having taken place. This phenomenon is known as parthenocarpy and naturally gives rise to a seedless fruit. The development of seedless tomato fruits may be induced by spraying the flowers with a growth substance such as β-naphthoxyacetic acid. In this case the growth substance artificially supplied seems to act in a similar way to the growth substance produced by ovules developing after normal fertilisation.

After fertilisation the sepals, petals and stamens usually wither, although their dead remains may often still be seen, for example, at the top of an apple or gooseberry. In some examples, however, the sepals may be persistent (for example, pea, tomato). In the mulberry (*Morus* species) the perianth becomes pulpy and embeds the true fruits.

4

Production of Seeds

The embryo is the end result of the sexual cycle. These events occur in the flower, the basic structure of which is shown diagrammatically in Fig. 4.1. During flowering, pollen

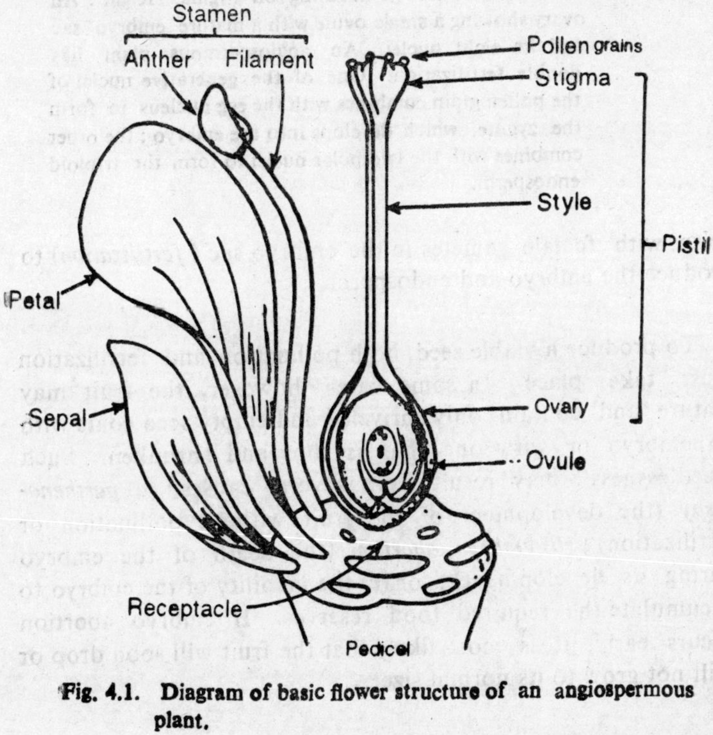

Fig. 4.1. Diagram of basic flower structure of an angiospermous plant.

is transferred from the anther to the stigma (*pollination*); it germinates and grows down the style until it reaches the embryo sac. (Fig. 4.2). Male gametes from the pollen tube

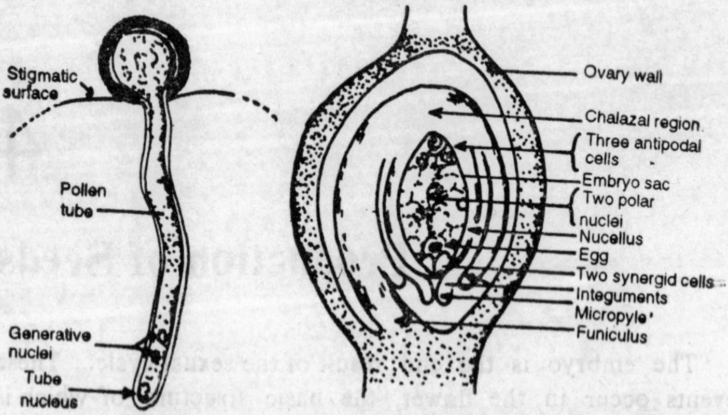

Fig. 4 2. Left : Pollen grain germinating on stigma. Right : An ovary showing a single ovule with a mature embryo sac and its eight nuclei. An angiospermous plant has double fertilization. One of the generative nuclei of the pollen grain combines with the egg nucleus to form the zygote, which develops into the embryo ; the other combines with the two polar nuclei to form the triploid endosperm.

unite with female gametes in the embryo sac (*fertilization*) to produce the embryo and endosperm.

To produce a viable seed, both pollination and fertilization must take place. In some cases, however, the fruit may mature and contain only shriveled and empty seed coats with no embryo or with one that is thin and shrunken. Such "seedlessness" may result from several causes : (*a*) *parthenocarpy* (the development of the fruit without pollination or fertilization) ; (*b*) *embryo abortion* (the death of the embryo during its development) ; or (*c*) the inability of the embryo to accumulate the required food reserves. If embryo abortion occurs early, it is most likely that the fruit will soon drop or will not grow to its normal size.

FRUIT AND SEED DEVELOPMENT

Morphological Development

The relationship between the structure of the flower and the structure of the fruit and seed in angiosperms is as follows :

Ovary............→ fruit (sometimes composed of more than one ovary plus additional tissues)

Ovule............→ seed (sometimes coalesces with fruit)
integuments............→ testa (seed coats)
nucellus................→ perisperm (usually absent or reduced ; sometimes storage tissue)

2 polar nuclei+
sperm nucleus......→ endosperm (triploid-3N)
egg nucleus+sperm
nucleus...→ zygote...→ embryo (diploid-2N)

Figures 4.3 and 4.4 show the growth relationships among the different parts of a seed and fruit during development in an angiosperm plant. The embryo develops from the zygote initially as

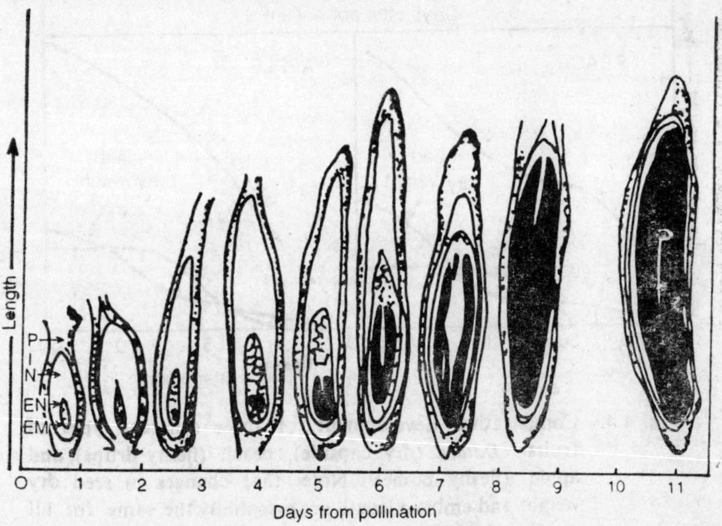

Fig. 4.3 Growth of the fruit (an indehiscent achene) and seed of lettuce, P—pericarp, I—integuments, N—nucellus, EN—endosperm, EM—embryo.

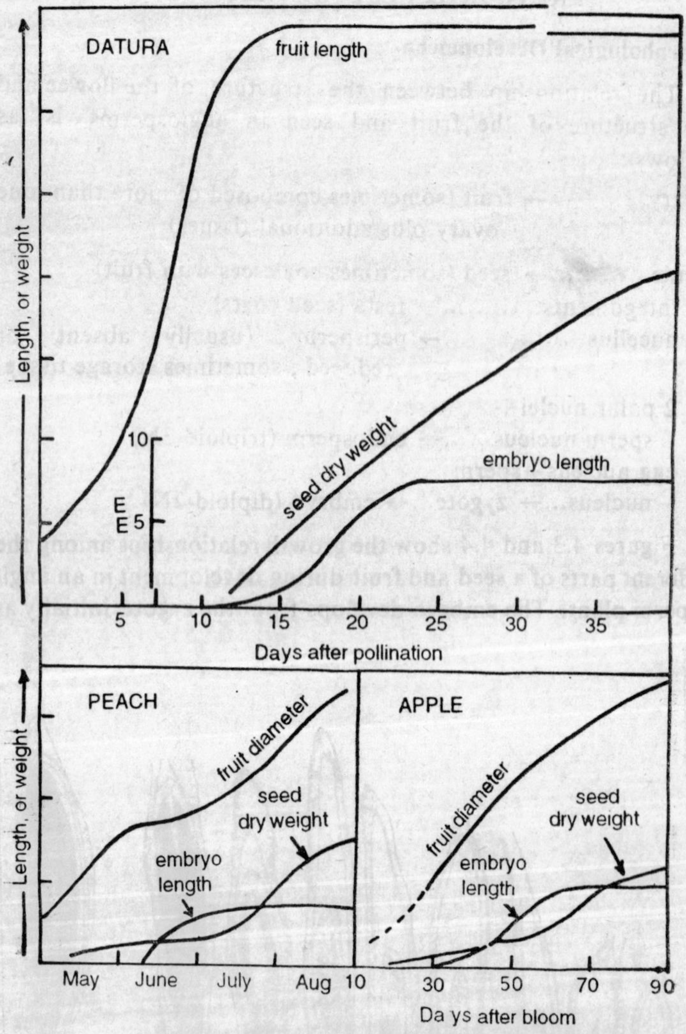

Fig. 4.4. Comparative development of three different types of fruits: *Datura* (dry capsule), peach (fleshy drupe), and apple (fleshy pome). Note that changes in seed dry weight and embryo length are essentially the same for all species shown. *Datura* data from Rietsema *et al.* apple data in part from Luckwill.

a microscopically small mass of cells embedded in the endosperm, which is in turn embedded in the nucellus. The embryo undergoes continuous morphological and physiological development. Three broad stages of development can be observed in embryos of most plants. For example, in *Capsella* and *Datura* in an eyrly *proembryo* stage, the embryo is very small (microscopic) and globular in shape ; it becomes *heart*-shaped as cotyledons begin to develop. This period is indicated by the 1- to 4-day lettuce embryo in Fig. 4.3 or the 10- to 12-day *Datura* embryo in Fig. 4.4. In the second stage the cotyledons continue to enlarge and the embryo becomes *torpedo*-shaped with its length increasing to nearly fill the seed. Five- to eight-day lettuce embryos (Fig. 4.3) and 15- to 20-days *Datura* embryos (Fig. 4.4) illustrate this stage. In the third stage. the embryos are full-sized morphologically and are increasing in weight.

The endosperm functions in a nutritional capacity for the embryo, although no vascular connections exist between the two structures. Growth of the embryo is preceded by growth of the endosperm, which digests the inner nucellar tissue as it grows. The endosperm is, in turn, digested by the developing embryo, although in some species, a large part of the endosperm remains in the mature seed.

Failure of the endosperm to develop properly results in retardation or arrest of embryo development. If embryo abortion result ; the phenomenon is called *somatoplastic sterility*. It commonly occurs when two genetically different individuals are hybridized, either from two different species or from two individuals of different polyploid constitution. It, therefore, can be a barrier to hybridization. If such embryos are excised from the developing fruit before they abort, they can sometimes be saved by being grown in aseptic culture.

Seed development in the gymnosperms is somewhat different. The ovules are exposed within the ovular cone and are not enclosed in an ovary. The ovular cone develops into the mature cone. Gymnosperms do not undergo double fertilization. The embryo is nourished by the haploid female gametophyte and is not dependent upon a triploid endosperm. Consequently somatoplastic sterility is not a problem.

Various external agents can prevent embryo development, even though the fruit itself continues to develop. Numerous kinds of insects attack the developing seed and fruit. This is particularly serious in forest trees. The developing seed of carrot and other Umbelliferous plants is attacked by *Lygus* bugs which can penetrate the fruit and feed on the embryo. Adverse weather, such as frosts during early fruit development, sometimes kills the embryo, but the fruit itself continues to develop. Growth tensions, causing the hardening endocarp to split or crack, can result in abortion of stone fruit seeds. This condition is associated with excessive vigor, reduced crop density, root restriction, or tree injuries resulting in girdling.

Accumulation of Food Reserves in the Seed

The accumulation of storage materials in the seed can be measured by changes in dry weight of the seed, although in the earlier part of the fruit development period, increase in weight may occur because of increase in size. Later, when the seed has attained its full size, the increase in dry weight is a measure of this accumulative process. These reserve materials originate as carbohydrates produced by photosynthesis in the leaves and translocated to the fruits and seeds, where they are converted to complex storage products—carbohydrates, fats, and proteins. This process largely takes place in the final developmental periods of fruit growth.

This accumulative process must take place properly if high-quality seeds are to be produced. Such seeds should be plump and heavy for their size. Since the initial growth of the seedling depends upon these reserve materials, heavier seeds should result in better germination and produce more vigorous seedlings. If conditions interfere with this storing process so that few reserve materials accumulate, the seeds will be thin, shrunken, and light in weight. The more severe this condition is, the less the seeds can survive storage periods, the poorer is the germination, and the weaker are the seedlings that are produced.

Ripening and Dissemination of Seeds

Specific physical and chemical changes take place during ripening which lead to the senescence of the fruit and the dissemination of the seed. One of the most obvious changes is the drying of the fruit tissues. In certain fruits, this leads to dehiscence and the discharge of the seeds from the fruit. Changes may take place in the colour of the fruit and the seed coats, and softening of the fruit may occur. An immature fruit is invariably green because of the presence of chlorophyll, but when the fruit ripens, the chlorophyll breaks down and may disappear, causing other colours to become evident as particular pigments are exposed.

The Mature Seed

Botanically, in the angiosperms the seed is a matured ovule enclosed within the ovary, or fruit. Seeds and fruits of different species vary greatly in appearance ; size ; shape ; location and structure of the embryo ; and presence of storage tissues (Fig. 4.5). These points are useful in identification. From the standpoint of seed handling, it is not always possible to separate the fruit and seed, since they are sometimes joined in a single unit. In such cases, the fruit itself is treated as the "seed," as in wheat or corn.

Parts of the Seed

The seed has three basic parts : (*a*) embryo, (*b*) food storage tissues, and (*c*) seed coverings.

Embryo

The embryo is a new plant resulting from the union of a male and female gamete during fertilization. Its basic structure consists of an *axis* with growing points at each end, one for the shoot and one for the root, and one or more seed leaves (*cotyledons*) attached to the embryo axis. Plants are classified by the number of cotyledons. Monocotyledonous plants (such as the grasses or onion) have a single cotyledon, dicotyledonous plants (such as the bean or peach) have two, and gymnosperms (such as the pine or *Ginkgo*) may have as many as fifteen.

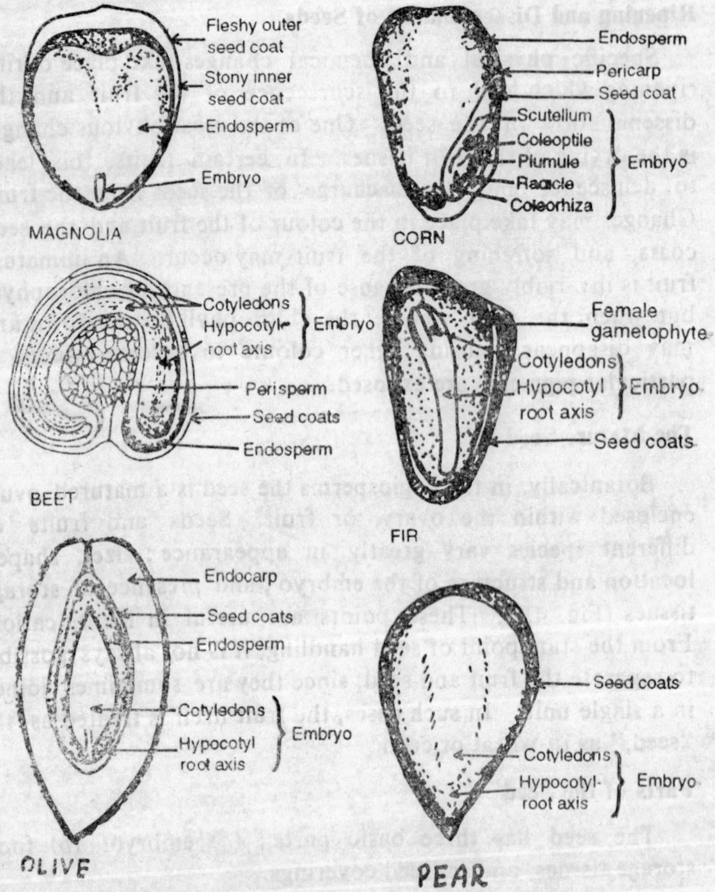

Fig. 4.5. Seed structure of representative species.

Storage Tissues

The storage tissues of the seed may be the cotyledons, the endosperm, the perisperm, or, in gymnosperms, the haploid female gametophyte. Seeds in which the endosperm is large and contains most of the stored food are referred to as *albuminous seeds* ; seeds in which the endosperm is lacking or reduced to a thin layer surrounding the embryo are referred to as *exalbuminous* seeds. In the latter case, the stored food is usually within the cotyledons, the endosperm being digested by the embryo during development. The perisperm, originating from the nucellus, occurs in only a few families of plants, such as the Chenopodiaceae and the Caryophyllaceae. Usually it is digested by the developing endosperm during seed formation.

Seed Coverings

The seed coverings may consist of the seed coats, the remains of the nucellus and endosperm, and sometimes parts of the fruit. The seed coats, or *testa*, usually one or two (rarely three) in number, are derived from the integuments of the ovule. During development the seed coats become modified, and at maturity they present a characteristic appearance. Usually the outer seed coat becomes dry, somewhat hardened and thickened, and brownish or otherwise coloured. On the other hand, the inner seed coat will usually be thin, transparent, and membranous. Remnants of the endosperm and nucellus are found within the inner seed coat, sometimes making a distinct, continuous layer around the embryo.

In some plants, parts of the fruit remain attached to the seed so that the fruit and seed are commonly handled together as the "seed." In certain kinds of fruits, *e.g.*, achenes, caryopsis, samaras, and schizocarps, the fruit and seed layers are contiguous. In others, such as the acorn, the fruit and seed coverings separate but the fruit covering is indehiscent. In still others, such as the "pit" or stone fruits or the shell of walnuts, the covering is a hardened portion of the pericarp, but it is dehiscent and can usually be removed without much difficulty.

The seed coverings provide mechanical protection for the embryo, making it possible to handle seeds without injury, and thus permitting transportation for long distances and storage for long periods of time.

Seed Production of Herbaceous Plants

To keep a variety genetically unchanged, control of the seed source is necessary. With no control, the genetic characteristics of the variety may become somewhat different from that initially developed by the plant breeder. Contamination may result from chance-pollination with plants of a different genotype, or it may result from a mixture with other crop seed of lesser value or with weed seed. Foreign seed similar to the pure crop seed cannot be easily removed with seed-cleaning equipment.

METHODS OF MAINTAINING PURITY

Isolation

Isolation is used (*a*) to prevent contamination by cross-pollination with a different but related variety, and (*b*) to prevent mechanical mixing of the seed during harvest. It is primarily achieved through distance, but it also can be attained by enclosing plants or groups of plants in cages, enclosing individual flowers, or removing male flower parts and then employing artificial pollination.

Not as much isolation is needed in the production of self-pollinated plants as in the production of cross-pollinated plants. As a general rule, two or more closely related varieties of self-pollinated plants can be planted close to each other without one contaminating the other. The principal reason for separation in this case is to prevent mechanical mixing of seed during harvest. The minimum distance usually specified between plots of self-pollinated seed-producing plants is 10 ft (see Table 4.1).

Seed of different varieties must be kept separate during harvest. Careful cleaning of the harvesting equipment when a change is made from one variety to another is required. Likewise, sacks and other containers used to hold the seed must be carefully cleaned to remove any seed which may have remained from previous lots.

Isolation is a primary factor in the production of plants cross-pollinated by wind or insects. The minimum distance between varieties depends upon a number of factors, such as the degree of natural cross-pollination, the pollination agency, direction of prevailing winds, and the number of insects present. The required distances specified by seed-regulating agencies varies with the class of seed produced. That is, greater distances should be used with seeds in which a high degree of genetic purity is desired (Table 4.1). The minimum distance usually specified for insect-pollinated plants is ½ mi., although for others, such as the onion, 1 mi. is specified. Some seed producers have indicated that a minimnm of ½ mi. should be used, preferably 1 mi. or more, depending upon the location of colonies of bees.

The distance that should separate varieties of wind-pollinated plants varies with different kinds of plants. The distance usually specified for corn is 40 rd (⅛ mi). In hybrid seed corn, this distance may be reduced by planting border rows of the pollinating variety. With beets, ½ to 1 mi. is recommended, and to produce stock seed 2 mi. is preferable Seed-producing fields of two different varieties should not be in the line of prevailing winds.

Cross-pollination takes place between certain varieties and not between others. In general, any variety can contaminate any other variety of the same species; it may or may not contaminate varieties of a different species but in the same genus; and rarely will it contaminate varieties belonging to another genus. Since the horticultural classification may not indicate taxonomic relationships, the seed producer should have some familiarity with botanical relationships among the varieties he grows.

Roguing

The existence of off-type plants, plants of other varieties, and weeds in the seed production field are other sources of contamination which should be eliminated. Although a low percentage of such plants may not seriously affect the performance of any one lot of seed, their continued presence will lead to deterioration of the variety over a period of time. The removal of such plants is referred to as "roguing".

Off-type plants may arise because some recessive genes are present in a heterozygous condition even in highly homozygous

Table 4.1. Minimum Isolation Requirements for Seed Production of Certain Species of Field and Vegetable Crops.*

Type of Pollination	Species	Seed Class Foundation	Registered	Certified
Self-pollinated	barley, oats, wheat, rice, peanut, soybean, lespedeza, field pea, garden bean, cowpea, flax	Fields should be separated by a definite boundary at least 10 ft wide.		
	grasses (self-pollinated and apomictic species)	10 ft	10 ft	5 ft
Self-pollinated but to a lesser degree than those in the above list	cotton (upland type)	100 ft from varieties which differ markedly.		
	cotton (Egyptian type)	1320 ft	1320 ft	660 ft
	papper	200 ft	100 ft	30 ft
	tomato	200 ft	100 ft	30 ft
	tobacco	150 ft or by four border rows of each variety. Isolation between varieties of different types should be 1320 ft.		

Cross-pollinated by insects	alfalfa, birdsfoot trefoil, red clover, white clover	1320 ft	660 or 1320 ft†	165 ft
	sweet clover	5280 ft	660 or 990 ft †	165 or 330 ft †
	miliet	1320 ft	1320 ft	1320 ft
	onion	2280 ft	5640 ft	1320 ft
	watermelon	2640 ft	2640 ft	1320 ft
Cross-pollinated by wind	hybrid field corn	660 ft (may be reduced if field is surrounded by specified numbers of border rows and the varieties nearby are of same colour and texture).		
	grasses	1320 ft	660 ft	330 ft

*As recommended by the International Crop Improvement Association.

†First number if plot is less than 5 acres ; second number if plot is more than 5 acres.

varieties. Recessive genes may arise by mutation, a process continually occurring at a low rate. The effect of a mutant recessive gene controlling a given plant characteristic may not be immediately observed in the plant in which it arises. The plant becomes heterozygous for that gene, and in a later generation the gene segregates and the character appears in the offspring. Off-type individual plants should be rogued out of the seed production fields before pollination occurs. Regular supervision of the fields, by trained personnel is required.

Volunteer plants arising from accidentally planted seed or from seed produced by earlier crops is another source of contamination. Fields for producing seed of a particular variety should not have grown a potentially contaminating variety for a number of preceding years.

Testing

Varieties being grown for seed production should be periodically tested for genetic purity to make sure that they are being maintained in their true form. Many seed companies maintain test gardens for this purpose under the supervision of trained personnel.

Genetic Shifts

Sometimes a seed-propagated variety that is produced initially in one given environment may undergo a *genetic shift* if it is grown for several consecutive generations in a different environment. Such a change takes place because those seedling plants better adapted to the new environment survive in greater amounts than others. Physiological changes affecting yielding ability, disease resistance, and environmental adaptation are difficult to detect without testing. To minimize the opportunity for such shifts to occur in seeds of forage crops, only one generation of seed production is allowed in regions of mild winters if the crop seed is to be used for planting in regions with cold winters.

Seed Certification

Genetic purity in commercial seed production is often maintained through a system of *seed certification.* Such programs exist in most states of the U.S. through the cooperative efforts of public research, extension, and regulatory agencies in agriculture and a state Crop Improvement Association. The Crop Improvement Association is usually a legally designated seed-certifying agency. Its members include growers as well as other interested individuals, involved in the production of "certified seed." These individual state organizations are coordinated through the International Crop Improvement Association, with members in the United States and Canada.

The principal objectives of seed certification are to maintain and to make available crop seeds, tubers, or bulbs and sometimes turfgrasses, which are of good seeding value and true to variety. To accomplish these purposes, the seed-certifying agency determines the eligibility of particular varieties. It sets up production standards for isolation, presence of off-type plants, and quality of harvested seed, and makes regular inspections of the production fields to see that the standards are maintained. Supervision of seed processing is also involved. The International Crop Improvement Association has recommended minimum standards for many crops (Table 4.1), although these regulations are set within the individual state organization. Since specific requirements for certification may vary from state to state, the local state regulations should be examined. These usually specify the varieties which can be certified, the amount of isolation required of any particular crop, required inspections, and standards for the seed quality after harvest.

The mechanics of maintaining genetic purity are facilitated by utilizing certain classes of seed which are maintained at different levels of purity. These classes are given below, as defined by the International Crop Improvement Association.

Breeder's seed. Breeder's seed is seed or vegetative propagating material which is directly controlled by the originating or, in certain cases, the sponsoring plant breeder or institution, and which provides for the initial and recurring increase of foundation seed.

Foundation seed. Foundation seed, including *elite* in Canada, is seed stock so handled as to most nearly maintain specific genetic identity and purity, and that may be designated by an agricultural experiment station. Production must be carefully supervised or approved by representatives of the station. Foundation seed is the source of all other certified seed classes, either directly or through registered seed.

Registered seed. Registered seed is the progeny of foundation or registered seed that is so handled as to maintain satisfactory genetic identity and purity, and that has been approved and certified by the certifying agency. This class of seed should be of a quality suitable for the production of certified seed.

Certified seed. Certified seed is the progeny of foundation, registered, or certified seed that is so handled as to maintain satisfactory genetic identity and purity, and that has been approved and certified by the certifying agency.

Certified seed is the class grown in largest volume to be placed on the market and sold to growers. Bags of certified seed have attached, with a metal seal, a blue tag distributed by the seed-certifying agency as evidence of the genetic identity and purity of the seed contained therein. Registered seed is used to produce the plants from which certified seed is to be harvested. Bags of registered seed are labeled with a purple tag or a blue tag marked with the word "registered." Likewise, foundation seed is labeled with a white tag or a regular certified seed tag with the word "foundation."

Vegetable and Flower Seed Production

In the U.S. vegetable and flower seed is largely produced by commercial seed companies who either grow the seed themselves or contract with private growers to produce seed for them. The company preserves genetic purity by specifying the particular variety an individual contractor will grow, by supervising the fields during production, and by maintaining test gardens.

Although such seed is not necessarily produced as certified seed, similar procedures are used to maintain genetic purity. Essentially the same classes of seed are used, although they may be designated by different names. The primary source of seed of the variety, the foundation seed, is usually maintained by the seed company. Trained personnel have the responsibility for maintaining the foundation seed at the highest level of purity. Seed provided by the company to the contractor-grower for growing his crop is known as *stock seed.* The seed actually harvested from the fields of the contractor-grower will be placed on the market and sold commercially by the seed company.

The Production of Seed for Hybrid Varieties

Hybrid varieties have become an increasingly important category of cultivated plants within recent years. These are the progeny produced by the repetitive crossing of two or more parental lines that are maintained either (*a*) by seed, such as inbred lines, or (*b*) asexually, such as clones. To produce commercial hybrid seed, the parental lines must be grown side by side so that cross-pollination takes place between them. The seed produced (the F_1 progeny of the cross) is the seed used to grow commercial crops. This cross must be repeated every time the seeds are produced.

Hybridization between two inbred lines is known as a *single-cross.* Combining two single crosses produces a *double-cross,* the usual case with hybrid corn varieties.

Isolation of the seed-producing fields should be even more stringently observed than in other cross-pollinated crops. With enforced cross-pollination, the possibility of pollen contamination is greater.

Hybrid plants were first utilized in the production of field corn, and almost all the field corn grown in the United States is of this type. Likewise, hybrid varieties of sweetcorn and popcorn have been developed. Hybrid varieties of other plants have not been developed as rapidly as those of corn, the limiting factor in most cases being the development of methods whereby the hybrid variety can be produced on a large scale at a moderate cost. Nevertheless, a number of vegetable hybrids have become available as have hybrids of some flower plants (for instance, petunia).

Methods of Cross-Pollination

(*a*) Plants of monoecious species (where male and female organs are in different flowers) readily adapt to hybrid seed production (Fig. 4.6). To produce hybrid corn seed, the parental lines are planted in rows, one male row to three female rows or a similar combination. Production of pollen is not allowed to take place in seed-producing rows, and thus pollen to fertilize the female flowers (the silks) will come only from the male parental lines. This is done either by removing the male flowers (tassels) by hand or by using pollen-sterile plants as described later.

(*b*) Controlled hand pollination is sometimes practiced in producing seeds from particularly selected parents. Although used mostly in plant breeding, it is also used to produce some commercial hybrid seeds. The specific technique used depends upon flower structure and the pollination characteristics of the individual species, but it follows certain principles.

Accidental contamination by unwanted pollen can be prevented by washing the hands and instruments with alcohol before handling pollen or flowers. Pollen should be collected from flower buds immediately preceding bloom and before the anthers open (dehisce). This eliminates the chance of contamination by unwanted pollen. Anthers are extracted by pulling them from the flowers with tweezets, by squeezing the flower between the fingers, or by rubbing the buds across a wire screen. Catkins (walnut, birch, aspen) or staminate cones (pine) will open on drying and will shed large quantities of pure, dry pollen if placed on a sheet of paper in a warm room. Anthers are dried on a sheet of paper until their opening can be observed under a magnifying glass. Pollen and anthers can be screened through a fine sieve to remove extraneous material; they are then stored in a glass vial or bottle.

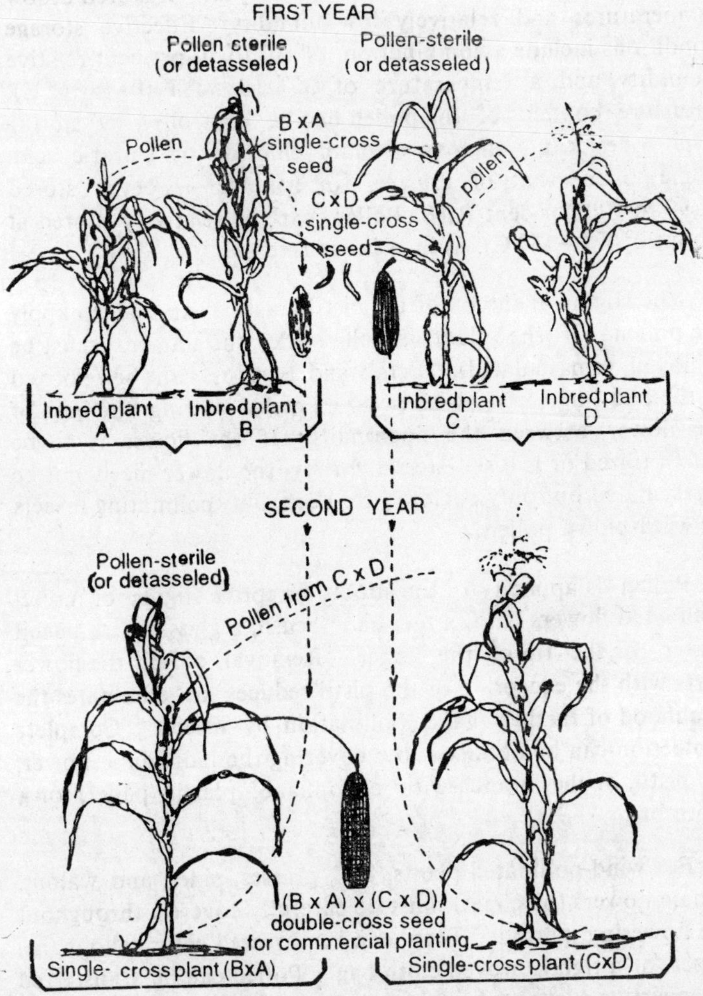

Fig. 4.6. The production of hybrid corn. The seed used to grow the crop of hybrid corn is usually a combination of four inbred lines. The single cross is made between the original inbred lines. Seeds from two single crosses are planted to produce the double cross.

Most kinds of pollen will remain viable for only a few days or weeks at warm temperatures, but many kinds can be preserved for several months to several years if stored at low temperatures and relatively low humidity. Effective storage conditions include a combination of 10 to 50 per cent relative humidity and a temperature of 32° to 50° F (0° to 10° C). Moisture content of the pollen can be controlled by storing over a desiccant, such as calcium chloride or sulfuric acid. Some pollen—that of grasses, for instance—is better stored at 90 to 100 per cent R.H. Pollen can be relatively stored at about 0° F (−18° C), as in a home freezer.

The stigma at the tip of the pistil must be exposed to apply the pollen. If the plant is self-fertile, the stamens must be removed (emasculated). Petals and stamens can be removed at the bud stage either with tweezers or by cutting the base of the flower between the fingernails. If the flower is to be self-fertilized or is a self-sterile variety, the flower need not be emasculated but only enclosed to keep out pollinating insects or wind-blown pollen.

Pollen is applied to the sticky, receptive stigma of insect-pollinated flowers with a fine-hair brush, a glass rod, a pencil eraser, or the tip of the finger. Removal of all the flower parts with the exception of the pistil reduces or eliminates the likelihood of further chance pollination by insects. Complete protection can be obtained by covering the individual flower, the pistil, or the branch with cellophane, plastic, paper, or a cloth bag.

For wind-pollinated plants, *e.g.*, grasses, pines, and walnut, female flowers (or cones) must be carefully covered throughout the flowering period. They can be covered with cellophane, paper, or a tightly woven cloth bag. Pollen can be transferred by inverting a paper bag containing pollen over the exposed flower; or dry pollen can be blown into the bag with an atomizer or inserted with a hypodermic needle.

The flower or branch on which it is borne should be carefully labeled to maintain a record of the pollen used. The usual method is to identify the cross as follows : "seed parent × pollen parent."

(*c*) Mass production of hybrid seed of most species is possible only by utilizing or developing some system of self-sterility that prevents self-fertilization. Enforced hybridization results when a self-sterile, seed-producing line is grown with an appropriate pollen-producing parental line and some vector for cross-pollination is present. The major difficulties in utilizing this procedure in commercial seed production are (1) discovering suitable self-sterility genes, (2) incorporating them into useful varieties (a job of plant breeders), and (3) maintaining the necessary self-sterile genotype in the parental lines. If parental lines can be economically propagated vegetatively (or by apomixis) then self-sterility can be maintained intact with no difficulty. Seed-propagated lines, however, are more difficult to manage and a thorough knowledge of inheritance in the self-sterility system is required. Methods must be used to insure that only the desired self-sterile genotype is produced when needed or procedures must be used to identify the unwanted pollen-producing individual plants early enough in their development so that they can be rogued out of the seed-producing fields.

Actually, complete cross-pollination might not be necessary. Burton describes a system utilized in the hybrid-seed production of 'Gahi-1', a partially cross-pollinated variety of pearl millet (*Pennisetum glaucum*) used for forage. Two or four selected parental lines are grown together, and the seed is harvested from the entire block and mixed thoroughly. About 75 per cent of the offspring are hybrids. When planted at high density, the more vigorous hybrid seedlings crowd out the weaker inbred seedlings. At lower planting densities, mixtures of hybrid and inbred plants result. Later in this chapter, a similar procedure is described for some tree hybrids.

Several naturally occurring systems of enforced cross-pollination exist and are utilized in the seed production of some hybrids. Self-incompatibility has been used in the production of some cabbage hybrids. Some grass hybrids of Pensacola Bahiagrass (*Paspalum notatum*), for instance, have been produced by interplanting self-incompatible (but cross-compatible) parental clones. The plants spread vegetatively to occupy the field and all seeds are harvested together as hybrids. Dioecious plants can also be utilized effectively in this way. 'Mesa' buffalograss (*Buchloe dactyloides*) is produced by interplanting vegetatively propagated male and female parents in isolated fields.

Genes for male sterility that inhibit pollen production have been discovered in many crop plants and have been incorporated in the breeding lines of a number of important crop species. Male sterility is inherited in several ways in different species and not all systems are effectively used in producing hybrids. The most significant one is the cytoplasmic-genetic male-sterility system which was first utilized for the production of onion hybrids (Fig. 4.7). It is also used in the commercial production of hybrid seed of field corn, sugar beets, sorghum, and pearl millet. Since it can be used for other crop plants as well, its use will undoubtedly be extended. The system involves the combination of a cytoplasmic factor S (as opposed to a normal cytoplasm N) and a recessive gene ms (as opposed to a dominant fertility-restoring gene Ms). A male-sterile plant has a genotype of S ms ms, whereas a male-fertile plant can have any of the following : N ms ms, N Ms ms, S Ms Ms, or S Ms ms. The male-sterile line is maintained by continuous backcrossing to a pollen parent of N ms ms. Two examples of the inheritance of these factors are the following :

(*a*) S ms ms × N ms ms
(seed) (pollen)
↓
S ms ms
(all male-sterile)

(*b*) S ms ms × N Ms ms
(seed) (pollen)
↓
½ S Ms ms
(male-fertile)
½ S ms ms
(male-sterile)

Use of F_1 Seed for Propagation

From the standpoint of continued propagation, one important point should be emphasized in regard to hybrid varieties. The F_1 hybrid plants are heterozygous, and seeds from them will not continue to produce plants with the characteristics of the F_1 plants. The effect of hybrid vigor is greatest in the F_1

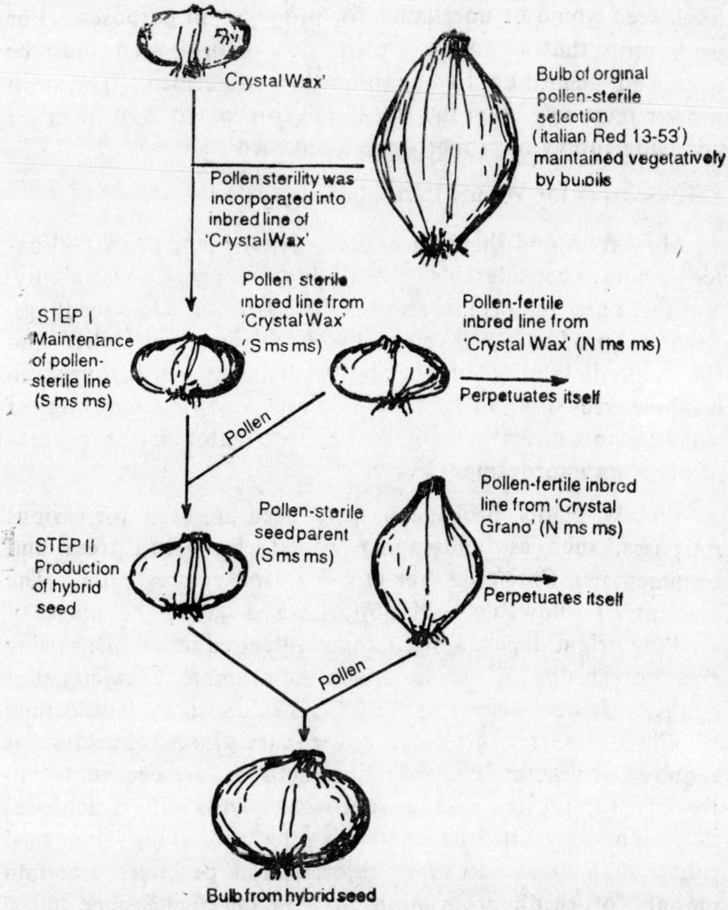

Fig. 4.7. Hybrid onion seed production utilizing pollen sterility to facilitate cross-pollination is illustrated by the onion hybrid 'Crystal Wax' × 'Crystal Grano'. The factors of pollen sterility, S ms ms, discovered in 'Italian Red 13-53', were incorporated by plant breeders into an inbred line of 'Crystal Wax' by backcrossing. The pollensterile line is maintained by crossing it with a pollen-fertile inbred line of 'Crystal Wax' which has the factors N ms ms.

Commercial hybrid seed is produced by planting together this pollen-sterile line of 'Crystal Wax' with a pollen-fertile inbred line of 'Crystal Grano', the resulting hybrid seed being used to grow the commercial crop of onion bulbs.

generation and declines sharply in succeeding generations. Such seed would be unreliable for propagation purposes. For each crop that is to be grown, new hybrid seed must be produced according to the procedure described. The crop grower must rely upon the seedsman to provide him with an adequate supply of properly produced seed.

Seed Sources for Woody Perennials

Most trees and shrubs are heterozygous and cross-pollinated and have considerable potentiality for genetic variability. Yet there are important reasons for growing tree seedlings. Seedling production involves two problems : variability and the unpredictability of the ability of many seed sources to produce trees of known characteristics. Growing seedlings of woody plants directly involves the propagator in the process of effecting improvement in tree crops.

Woody plants propagated from seed are used for various purposes, such as landscaping, rootstocks (fruit trees and ornamentals), Christmas tree growing, and reforestation. The amount of allowable variation within a group of plants of seedling origin depends upon their intended use. In growing trees and shrubs for uses in which the ultimate characteristics of the individual plant should be known, such as landscaping or Christmas tree growing, either careful seed selection is required or vegetative propagation methods are needed. Uniformity is important among root-stock plants and is achieved in producing fruit tree rootstocks by preserving the seed source as a clone. In some reforestation practices a certain amount of seedling variation may be beneficial since initial planting density is usually high. Close spacing promotes good stem form, and competition among plants gradually eliminates the weakest trees. On the other hand, modern forestry practices are more and more involved with the growing of carefully selected trees of good timber quality.

PRINCIPLES OF SELECTION

Seed Origin

Seed origin refers to the particular climatic and geographical locality where the seed is obtained. The original geographic source of seed is referred to as *provenance*. It is particularly important in reforestation and is second in importance only to choice of species. The significance of seed origin was first recognized during the nineteenth century when extensive plantings of forest trees from unselected seed were made in Europe. Observers noticed that trees grown from imported seeds were generally inferior to trees grown from seeds of local origin. In one instance, thousands of acres of Scotch pine (*Pinus sylvestris*) were eradicated because of their inferior characteristics. After that many European countries enacted laws covering the importation and use of forest tree seeds, in part requiring the labeling of seed as to origin.

Origin is important when plants of single species grow over a wide range of ecologically different areas in nature. Variation in morphology, physiology, adaptation to climate and soil, and resistance to diseases and insects may exist the represent specific races, or *ecotypes*, that are characteristic of a particular locality. Variation may also exist in a continuous gradation, or *cline*. A locality may be defined by its latitude, longitude, and elevation. Seeds of a given species collected from one locality may produce plants completely unadapted to a different locality. For example, seed collected from trees in warm climates or at low altitudes is likely to produce seedlings that will be injured when grown in colder regions even if the species is the same and the plants are similar in appearance. Although the reverse situation—collecting seed from colder areas for growth in warmer regions—would be more satisfactory, it might result in a net reduction in growth due to the inability of the trees to fully utilize the growing season.

Distinct racial variation has been demonstrated to occur in the United States in Douglas fir, ponderosa pine, lodgepole pine, red pine, eastern white pine, slash pine, loblolly pine, shortleaf pine, and white spruce. Evidence shows that similar variations occur in other native tree crops, although details are lacking. On the other hand, some seed sources have been shown to be superior to others and preferable to most local sources—for example, the East Baltic race of Scotch pine ; the Hartz Mountain source of Norway spruce, the Sudeten strain of European larch, the Burmese race of teak, Douglas fir from the Palmer area in Oregon, ponderosa pine from the Lolo Mountains in Montana, and white spruce from the Pembroke, Ontario area.

An important principle in seed origin is to use local seed where possible unless adequate research has demonstrated that another source is better. In 1939 the United States Department of Agriculture adopted the following policy regarding the use of seed stocks and individual clones obtained by them for forest, shelter-belt, and erosion-control plantings. It serves as a guide for the selection of woody plant seed from natural sources.

1. Only seed of known locality of origin or nursery stock grown from such seed is to be used.
2. Evidence for the place and year of origin is to be required of the vendor when seeds are bought.
3. An accurate record of the following data is to be required of all shipments : (*a*) lot number ; (*b*) year of seed crop ; (*c*) species ; (*d*) seed origin as to state, county, locality, and range of elevation ; and (*e*) proof of origin.
4. Whenever available, local seed from natural stands is to be used unless it is demonstrated that another source can give desirable plants. "Local seed" means seed from an area subject to similar climatic influences, and this may usually be considered to mean within 100 mi. of the planting and within 1000 ft of its elevation.
5. When local seed is not available, seed should be used from a region having as nearly as possible the same climatic characteristics, such as length of growing season, mean temperature of growing season, frequency of summer droughts, and latitude.

Seed-collection zones designating particular areas that have a specific climatic and geographical basis have been established in a number of areas in the U.S. For example, California has 14 such zones and 5 additional sub-zones. Similar zones are established in Washington and Oregon and in the central states area of the United States.

Heit has shown that choice of seed sources is highly important in growing conifers and various hardwood species of trees in the nursery—for instance, for landscape purposes or for Christmas trees. These differences were demonstrated in nursery test plants. For example, in his tests Douglas fir had at least three recognized races, *viridis*, *caesia*, and *glauca*, with various geographical strains within them. The *viridis* strain from the West Coast was not winter-hardy; strains collected further inland were winter-hardy and vigorous, whereas those from Montana and Wyoming were very slow growing. Trees of the *glauca* (blue) strain from the Rocky Mountain region were winter-hardy but varied in growth rate and appearance. Similarly, differences among sources occurred in Scotch pine, mugho pine, Norway spruce, and others.

Seed-Tree Selection

Although local sources of seed of forest species can produce the best adapted plants for a given site, selection of *individual seed trees* may be necessary to improve quality in other characteristics. In forest trees there is a great deal of evidence that the quality of the individual seed trees is a good index of the quality of its offspring. The more knowledge the collector has about the transmission of specific characteristics from parent to offspring, the more accurate selection will be. Parent tree selection for many characteristics, such as stem form, branching habit, growth rate, resistance to diseases and insects, presence of surface defects, and certain other lumber qualities, can be effective.

Selected trees with a superior phenotype are referred to by foresters as "plus" trees and are often not cut for lumber. Such individual trees have value for their seeds either for natural reseeding or as seed sources for other areas. They can be multiplied and maintained as a clone in an established seed orchard.

Seeds of many woody plants used in nurseries are collected wherever the species is found growing locally, as in orchards, forests, woodlots, streets, and roadsides. A particular seed source should be selected not only for convenience of collection but, more important, because seed obtained from it possesses desirable genetic potentialities. Consequently an initial step in selecting a seed source is to evaluate the characteristics of the seed trees themselves. Although plants possessing the desired characteristics are most likely to pass them on to their offspring, this assumption cannot be safely made without progeny testing.

A second step in selecting a seed source is to evaluate the characteristics of surrounding trees that could cross-pollinate the seed trees. Most woody tree species are cross-pollinated; in general, better quality seeds will likely result thereby. Consequently it is undesirable to collect seed from an isolated plant or from those growing near others of related species, such as they might be in an arboretum. Cross-pollination from unwanted sources can nullify the potential value of a given seed source by producing excessive and unpredictable variability. For example, prior to World War I, "Old French" pear seedlings were grown from seed obtained in Europe from *Pyrus communis* trees. Much variability resulted from hybridization with *Pyrus nivalis*. Similarly, the "Oriental pear rootstock" species obtained during that same period were from seeds obtained from Asia with little knowledge about the characteristics of either the seed tree or the pollen source. Plants grown from these sources have turned out to be a heterogeneous group with many species and hybrids. The differences among flame eucalyptus (*Eucalyptus ficifolia*) plants arising from separate collection of a particular source in Australia have been explained by the fact that seeds collected from the center of the block were likely to breed true and produce upright, uniform groups of trees; those from the edge of the block were likely to be hybrids with surrounding *E. calophylla* trees, and consequently produced many dwarfed and off-type plants.

The most desirable procedure is to collect seeds from groups of the same kind of plants with desired characteristics growing in a *pure stand*. Seedling plants grown from such sources, although not necessarily homozygous, would be most likely to reproduce the characteristics observed among the parent seed trees.

Control of both seed and pollination source can be achieved by collecting from trees of clonal varieties growing in orchards with a known pollination arrangement. Such a procedure has been used extensively to produce rootstock seeds in fruit growing. For example, to obtain true *Pyrus communis* seedlings, *e.g.*, the so-called "French pear", seeds can be obtained from trees of 'Winter Nelis' growing as scattered trees for cross-pollination in 'Bartlett' ('Williams Bon Cretien') pear orchards. All seedlings would be 'Winter Nelis'×'Bartlett' hybrids. Likewise, 'Lovell' peach seeds or 'Royal' apricot seeds collected from fruit-drying yards come from known clones that most likely are grown in a solidly planted block of trees where self-pollination occurred.

Progeny Testing

A true genetic value of a seed source can only be established through a *progeny test*. A representative sample of seeds is planted, and the resulting progeny are grown under test conditions that will identify essential characteristics or demonstrate presumed superiority to other sources, or both. Foresters refer to seed-source trees with a superior genotype, as demonstrated by a progeny test as "elite" trees.

Progeny testing has been used to identify fruit-tree clones that will transmit high yielding characteristics—as in tung (*Aleurites*)—or important rootstock qualities, such as nematode resistance. Heit has used nursery tests to identify and characterize specific seed sources for landscape and Christmas tree uses. Progeny testing is an essential phase of identifying seed sources of superior "elite" forest trees. Here the procedure is particularly difficult, since the tester is working with a wild species with no standard cultivars for comparison, requiring a long-term evaluation of perhaps 20 years or more.

Nursery Selection

Desired individuals can be obtained and variability reduced to some extent through *nursery selection.* This procedure requires that variability involve identifiable characters of vigor or appearance, or both. Paradox hybrid walnut seedlings can be identified in nursery plantings of *Juglans hindsii* seedlings. "Blue" seedlings of the Colorado spruce (*Picea pungens*) appear among others having the usual green form. Differences in fall colouring of *Liquidamber* and *Pistacia* trees among seedlings necessitates fall selection of individual trees for landscaping. Variation among seedlings grown for rootstocks may be reduced by grading out the weak and small seedlings, a usual practice in nurseries in the United States.

PROCEDURES IN SELECTION OF WOODY PLANT SEED

Seed-Collection Areas

A *seed-collection* area is one where the trees have promise as a potential seed source have not been progeny-tested. This should be a pure stand of uniform trees large enough to insure adequate cross-pollination and separated from other plants having undesirable characteristics. Such areas are not managed or treated specifically for seed production.

Seed-Production Area

This refers to an area enclosing a group of trees that has been set aside permanently as a seed source. Use of such an area for a seed source applies primarily to forest trees, but the procedure could be utilized in any woody plant seed enterprise. Seed trees within the area are evaluated for the characteristics desired. Off-type trees or those that do not meet minimum standards are rogued out and eliminated. Other trees or shrub species that would interfere with the operations may be removed. It may be desirable to remove additional trees to provide adequate space for tree development and seed production. An isolation zone at least 400 ft wide from which off-type trees are removed should be established around the area. Trees may or may not have been progeny tested prior to establishment of the area.

Seed Orchards

These are orchards or plantations established specifically for seed production, usually from seed trees of acceptable origin and quality, and preferably from progeny-tested seed trees. Seed orchards are used directly or indirectly in the production of fruit-tree rootstock seeds and to a limited extent in the production of forest trees. This procedure will undoubtedly increase as progeny-tested seed sources become available. Seed orchards are useful for maintaining rootstock seed trees involved in virus-control programs.

There are three general types of seed orchards. They may be made up of (*a*) seedlings produced from selected parents through natural or controlled pollination ; (*b*) clonal seed orchards in which selected clones are propagated by grafting, budding, or rooting cuttings and grown in the block ; and (*c*) seedling-clonal seed orchards in which certain clones are grafted into branches of some of the trees. The choice of which type should be used will depend upon the species that is being grown.

Details of setting up a seed orchard vary with the species. A site should be selected for maximum, efficient production. For most native species enough different tree selections should be included in a suitable arrangement to insure cross-pollination and to decrease effects of possible inbreeding, If the species is dioecious, pollen-producing trees must be included. Fruit-tree rootstock clones that are self-pollinated can be planted as solid blocks. An isolation zone at least 400 ft wide should be established around the orchard. The size of this zone may be reduced if a buffer area of the same kind of tree is established around the orchard.

Hybrid Seed Production

Production of F_1 hybrid seeds is used to grow some tree crops. The principles are the same as those described earlier in the chapter. These seeds result from crosses between species or within species. Hybridization between the parental plants to produce the desired F_1 offspring may be achieved by various means. (*a*) Hybrids may be produced by hand pollination. This procedure is used mostly to produce F_1 hybrids for testing purposes. Mass production of seeds by this method would normally be too expensive, although hybrids of *Pinus rigida* × *P. taeda* have been produced this way in Korea. (*b*) Hybrids may be produced in seed orchards by interplanting the parental trees (1). Hybrids of *Larix decidua* and *L. leptolepsis* have been produced in seed orchards in Europe. In Idaho, a western white pine (*Pinus monticola*) seed orchard was established in the late 1950's containing 13 parental clones selected by progeny testing. Their F_1 offspring showed an average of 30 per cent resistance to blister rust. The F_2 generation showed an even higher average level of resistance in tests and will eventually be utilized as a seed source. (*c*) Another method is illustrated by the production of Paradox hybrid walnut seed, a procedure probably limited to certain kinds of plants. Such seed is obtained by collecting from individual Northern California black walnut (*Juglans hindsii*) seed trees where natural cross-pollination is known to take place with pollen from nearby English walnut (*Juglans regia*) trees. Seed-source trees are identified by progeny testing. Individual Paradox seedlings are identified in the nursery by vigor and leaf characteristics. (*d*) Seed from F_1 seed trees gives F_2, or second-generation hybrids. Collection of seed from many such sources, particularly if interspecific hybrids are involved, produces a group of seedlings that is highly variable in regard to vigor and other characteristics. As a group these may fail to reproduce the derirable attributes of the F_1 hybrid generation, although some individuals may be highly desirable. Such variability is undesirable if the hybrid plants are to be used for rootstocks or for landscaping where uniformity and predictibility of the final form is important. On the other hand, in forest trees, some variability in vigor commonly associated with second-generation hybrids may have advantages. In tree plants with relatively high initial seedling density, competition among plants will favour the most vigorous and tend to eliminate the weaker, less desirable plants. Consequently, planting seeds from F_1 hybrids (the F_2) may be an economical way to obtain a natural stand of vigorous hybrids, F_2 hybrids of selected lines within species may be quite uniform and desirable as in the blister-rust-resistant, white pine hybrids described in an earlier paragraph. The value of any F_1 hybrid tree as a seed source should be established by progeny testing, the same as should that of any other seed source.

5

Germination of Seeds

A seed consists of an embryo and its stored food supply, surrounded by protective seed coverings. When the seed separates from the plant on which it was produced, the seed is *quiescent*; that is, there is no external evidence of activity within the seed. The resumption of active growth by the embryo, resulting in the rupture of seed coverings and the emergence of a new seedling plant capable of independent existence, is known as *germination*.

In order for germination to occur, three conditions must be fulfilled. *First*, the seed must be *viable*; that is, the embryo must be alive and capable of germination. *Second*, *internal conditions* within the seed must be favourable for germination; that is, any physical or chemical barriers to germination must have disappeared. *Third*, the seed must be subjected to favourable *environmental* conditions, the essential factors being available water, proper temperature, a supply of oxygen and sometimes light. Although in any one seed each of these conditions may have an effect distinct from the others, the beginning of germination may be more often determined by the interactions among them.

The Germination Process

The germination process involves a complex sequence of biochemical, physiological, and morphological changes in which certain stages can be recognized. The *first stage* begins with the imhibition of water by the dry seed, the softening of the seed coverings, and the hydration of the protoplasm. This process is largely physical and occurs even in a non-viable seed. As a result of water absorption the seed swells, and the seed coats may break. The *second stage* begins with the initiation of cellular activity and includes the appearance of specific enzymes and a rise in respiratory rate. A plant growth hormone—gibberellin—(Fig. 5.1) is believed to play a key role in the initiation of germination. This effect has been demonst-

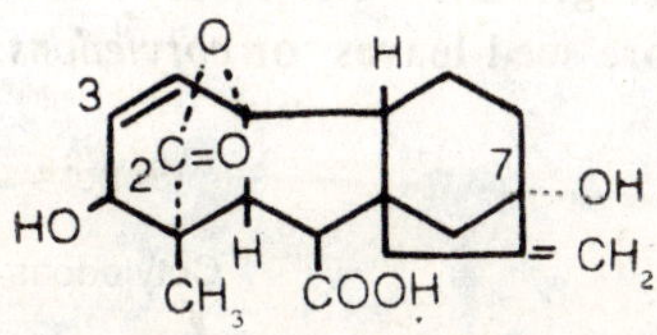

Fig. 5.1. Chemical structure of gibberellic acid (GA_3), a plant growth hormone associated with seed germination.

rated most clearly in seeds of cereal grains, such as wheat and barley. In these seeds the important structures are the *embryo*, *endosperm* (nonliving storage tissue containing mostly starch), and the *aleurone* (the outer layer of endosperm one or two cells thick). When the dry seed imbibes moisture, gibberellin appears in the embryo and is translocated to the aleurone layer where it activates enzymes. One of these enzymes, alpha-amylase, moves to the endosperm, causing the starch to be converted to sugar. Other enzymes appear in the aleurone, weakening the seed coats and allowing the root tip to burst through. Cell elongation and emergence of the radicle are events associated with the onset of germination. Cell division may also occur at an early stage, but this seems to be independent of cell elongation.

A *third stage* is the enzymatic digestion of complex insoluble reserve materials, mostly carbohydrates and fats but sometimes proteins, to soluable forms which are translocated to the active growing areas. The *fourth stage* is the assimilation of these substances at the meristematic areas to provide energy for cellular activity and growth and for conversion into new cell components. In the *fifth stage*, the seedling grows by the usual processes of division, enlargement, and differentiation of cells at the growing points. The seedlings depends upon reserves in the seed for continuing development until such time as the leaves can function adequately for photosynthesis. In summary, germination takes place in the following steps: imbibition, enzymatic and respiratory activity, digestion, translocation, assimilation, growth.

As germination proceeds, the structure of the seedlings soon becomes evident. (Fig. 5.2). The embryo consists of an axis bearing one or more seed leaves, or *cotyledons*. The growing

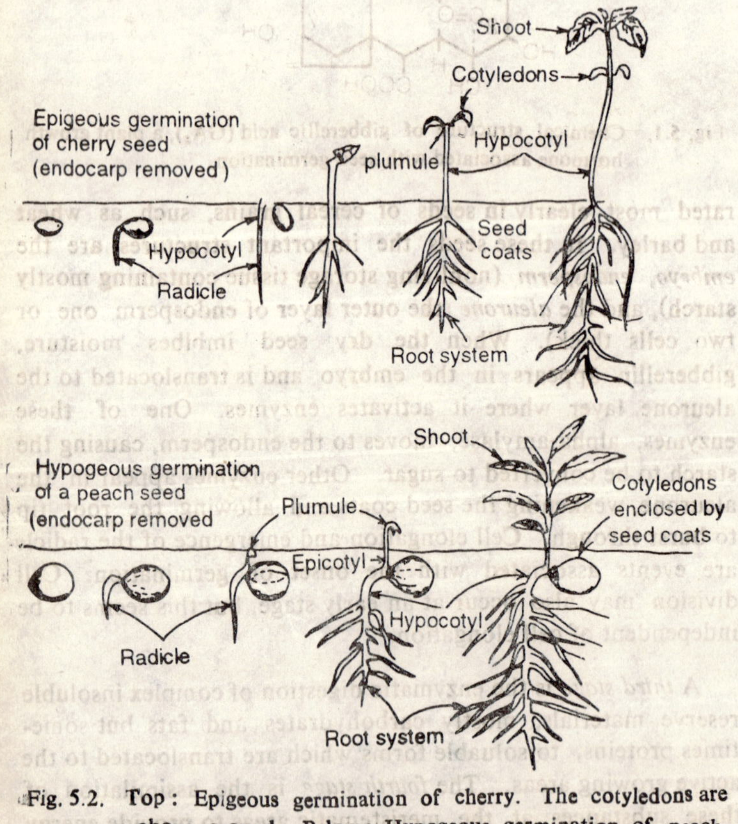

Fig. 5.2. Top: Epigeous germination of cherry. The cotyledons are above ground. Below: Hypogeous germination of peach. The cotyledons remain below ground.

point of the root, the *radicle*, emerges from the base of the embryo axis. The growing point of the shoot, the *plumule*, is at the other end of the embryo axis, above the cotyledons. The seedling stem is divided into the section below the cotyledons—the *hypocotyl*—and the section above the cotyledons—the *epicotyl*.

The initial growth of the seedling follows two patterns. In one type—*epigeous* germination—the hypocotyl elongates and raises the cotyledons above the ground. In the other type—*hypogeous* germination—the lengthening of the hypocotyl does not raise the cotyledons above the ground, and only the epicotyl emerges.

Viability of Seeds

A supply of viable seed is essential in successful propagation from seed. However, the distinction between a live and a dead seed is not necessarily sharp, and is often characterized by necrosis or injuries in localized areas of the seed. Viability is represented by the *germination percentage*, which expresses the number of seedlings which can be produced by a given number of seed. Germination should be prompt, and growth of the seedling vigorous. This is the seed vitality, or germinating power, and may be represented by *germination rate*. In seeds of low viability, a low germination percentage and a low rate of germination are often associated. A reduction in seed viability and seed vitality may result from incomplete seed development on the plant, injuries during harvest, improper processing and storage, or aging. With long storage, reduced viability is usually preceded by a period of declining vitality.

Seeds with a low germination rate or embryos that produce weak and abnormal seedlings are less able to withstand unfavourable environmental conditions in the seed bed than more vigorous seedlings. Weak and injured seedlings are likely to succumb to attacks by disease organisms, or they may lack the strength to emerge if planted too deeply or if adverse seed bed conditions exist. Consequently field survival is apt to be less than a predetermined germination percentage would indicate.

Measurement of Viability

If one measures the time sequence of germination of a given lot of seeds or the emergence of seedlings from a seed bed, one usually finds a pattern that is illustrated by the germination curve shown in Fig. 5.3. There is an initial delay

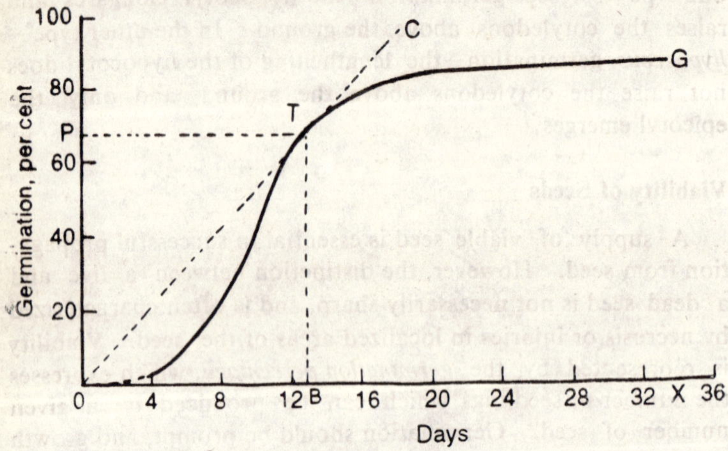

Fig. 5.3. Typical germination curve of a seed sample. After an initial delay, the number of seeds germinating increases, then decrease. Such a curve can be used to measure germination. See text for explanation. Redrawn from Czabator.

in the start of germination, then a rapid increase in the number of seeds that germinate, followed by a decrease in the rate of appearance. When viability is less than 100 per cent, the exact end point may be difficult to ascertain.

Measurement of germination involves two factors—the *germination percentage* and the *germination rate*. Germination percentage should involve a time element, indicating the number of seedlings produced within a specified length of time. Germination rate can be measured by several methods. One determines the number of days required to produce a given germination percentage. A better method, that has been used for many years, calculates the average number of days required for radicle or plumule emergence as follows :

$$\text{Mean days} = \frac{N_1T_1 + N_2T_2 + \ldots N_xT_x}{\text{total number of seeds germinating}}$$

N values are the numbers of seeds germinating within consecutive intervals of time ; T values indicate the time between the beginning of the test and the end of the particular interval of measurement. Kotowski has used the reciprocal of this formula multiplied by 100 to determine a *coefficient of velocity*.

Czabator has suggested another measurement that includes both rate and percentage—*germination value* (GV). To calculate GV, a germination curve, as shown in Fig. 5.3, must be obtained by periodic counts of radicle or plumule emergence. The important values on the curve are T—the point at which the germination rate begins to slow down—and G—the final germination percentage. These points divide the curve into two parts—a rapid phase and a slow phase. Peak value (PV) is the germination percentage at T divided by the days to reach that point. Mean daily germination (MDG) is the final germination percentage divided by number of days in the test. Then GV=PV×MDG.

Regulation of Germination

A seed enclosed within the fruit normally will not germinate while still attached to the plant on which it is produced. Certain highly specialized plants—for example, the mangrove, a swamp-growing tree—are exceptions in that they produce *viviparous* embryos that germinate on the tree. These grow into massive seedlings, with a foot-long, javelin-shaped root ; such seedlings eventually fall and become embedded in the mud below. Also, wet weather during harvest in certain seasons may result in premature sprouting in some grain crops, such as wheat. The tendency toward *vivipary* is inherited as a defective characteristic and is automatically selected against.

Ripening of seeds involves the development of internal mechanisms that function to regulate the time of germination so it will coincide with periods having environmental conditions most likely to favour survival of the seedlings. During maturation most seeds lose moisture to a level below that required for germination. The seed can remain in this dry state for long periods of time and can be handled with relative economy, ease, and safety. If the embryo is merely quiescent and no other germination block exists, the seed can germinate immediately whenever sufficient water is absorbed and suitable temperature and aeration conditions prevail. On the other hand, most freshly harvested seeds have additional germination regulating mechanisms that exist within the seed, and germination may fail to take place even under seemingly favourable environmental conditions.

Propagators of cultivated plants have long recognized these germination delaying phenomena and have learned to cope with them through the adoption of appropriate pre-germination and handling procedures arrived at through trial and error. Furthermore the domestication of many seed-propagated cultivars has undoubtedly included selection for easier propagation. Terminology based on these empirical methods has depended to a large extent upon the particular group of propagators. On the other hand, investigations of the fundamental nature of these phenomena have been carried out only in comparatively recent times, and an understanding is still far from complete. Such investigations have been extended from the cultivated plants to their relatives in nature, where such phenomena are not only much more prevalent and complex but play an important role in ecological adaptation.

Seeds may be grouped into the following broad categories on the basis of their response to particular environmental conditions and to methods of handling.

Group I. Seeds in which the covering is hard and moisture-impermeable and in which the embryo thus cannot absorb water. The group also includes seeds with coats so hard as to resist embryo expansion.

Group II. Seeds which have a dormant embryo that responds to pre-germination chilling.

A. Seeds requiring a single chilling period.

B. Seeds requiring a warm period (for root or embryo development) prior to the chilling period.

C. Seeds requiring two consecutive chilling periods, separated by a warm period.

Group III. Seeds which combine an impermeable seed coat with a dormant embryo.

Group IV. Seeds that contain chemical inhibitors that can be removed by leaching.

Group V. Seeds that are dormant when freshly harvested but become germinable upon dry storage (post-harvest dormancy).

A. Seeds in which germination is promoted by light.

B. Seeds in which germination is inhibited by light.

C. Seeds in which germination is inhibited by high temperatures.

The term *dormancy* has broad applications in plant physiology to refer to lack of growth in any plant part due either to internally or externally induced factors. Seed technologists, on the other hand, have used the term in a somewhat narrower sense to refer to nongermination resulting from conditions within the seed (other than nonviability). The Association of Official Seed Analysts, in the "Rules for Testing Seeds," further distinguishes between "hard seeds" and "dormant seeds". *Hard seeds* include those (groups I and III, above) which cannot absorb moisture because of the impermeable seed coat. *Dormant seeds* include those (Groups II, IV, and V, above) which are able to absorb moisture but fail to germinate because of restrictive influences within the seed that block some physiological reaction in the embryo and prevent the initiation of germination.

Conditions that exist within the seed at the time it matures on the plant to prevent germination have been called *primary dormancy.* The physiological changes that occur within the seed so that germination can take place are referred to in horticultural literature as *after-ripening.* Dormancy may be transitory, lasting only for a few days, disappearing with dry storage (in the case of seeds in group V, above) ; on the other hand, dormancy removal may require complex and time-consuming treatments. Once a seed has passed through an after-ripening period it may again become dormant if the imbibed seed is subjected to particular unfavourable environmental conditions. This is referred to as *secondary dormancy.*

Germination controlling mechanisms are important in nature because they contribute to natural survival and to the dissemination of plants to different areas. Such mechanisms are particularly important in plants growing in desert or cold regions where environmental conditions may not be favourable for germination immediately following dissemination of the seed. For instance, in the temperate zone, spring-ripening seed on such trees as elm, maple, or willow germinates immediately, and the seedling establishes itself the same season. If fall-ripening seeds of trees growing in the same region germinated immediately, the seedlings would probably be killed during the winter. Seeds of many plants of the latter type require winter chilling for germination, and consequently do not germinate until the following spring.

Variation among individual seeds of a given lot within a particular species may distribute germination over a period of years. If those which germinate at a certain time are lost, other seeds remain to germinate at a later and perhaps more favourable time.

With cultivated plants, prevention of premature germination on the plant or during seed harvesting is beneficial. On the other hand, expensive and time-consuming pre-germination seed treatments required to obtain good, uniform germination are undesirable. If reliable pre-germination treatments are unknown, germination may be unsatisfactory and reduced stands may result. Consequently the development of cultivated crop plants has involved the selection of cultivars which do not have complex germination problems. Most difficulty in germination regulation in propagation is found in seeds of trees and shrubs of native plants or those recently introduced into cultivation. Particular difficulty is experienced in getting quick and reliable results in testing seed for viability, especially when this is attempted shortly after the seed is harvested.

INTERNAL CONDITIONS AFFECTING GERMINATION

Seed Coverings

Impermeability to Water. The maturation of seeds of many species results not only in reduced water content but also in deposition of particular substances in the seed coverings to produce impermeability to water. This is a genetic characteristic of certain plant families, but it can also be modified by particular environmental conditions and methods of handling. A hard, impervious seed coat has value in prolonging storage life, since the internal parts of such seeds, once dry, are held in sealed storage until the seed coverings become modified. The effective seed covering may be only the hardened outer seed coat, or it may include various parts of the pericarp which either are attached to the seed or surround it. A number of plant families contain species whose seeds have hard coats ; these include the Leguminosae, Malvaceae, Cannaceae, Geraniaceae, Chenopodiaceae, Convolvulaceae, and Solanaceae. The plant group most commonly associated with hardseededness is the legumes, in which are found clover, alfalfa, and similar agricultural plants, as well as locust (*Robinia*), *Acacia*, and other woody legumes.

Seed coats can sometimes be modified during seed development or by the treatment the seed receives at harvest or during storage. For instance, white clover seeds are hard when ripened in hot, dry weather but are soft when ripened during rainy weather. If seeds are harvested before they are completely ripe and before the covering has become hard, and then kept from drying out, hardseededness may never develop. Further changes in the degree of hardness can take place during storage. For instance, in studies with *Symphoricarpos*, as the length of time after harvest increased, the length of the sulfuric acid treatment required to soften the coats also increased. In some cases, seeds which normally do not have hard seed coats, such as beans, develop hardseededness with dry storage and when planted, fail to germinate unless they are pretreated.

Seeds with hard coverings may be germinated by any method which artificially breaks or modifies the seed coat, provided that another type of dormancy is not also present. In nature, softening of the seed coat comes about through agencies of the environment : mechanical abrasion, alternate freezing and thawing, attack by microorganisms in the soil, or passage through the digestive tract of birds or other animals. Fungus activity takes place at moderate temperatures—50° F (10° C) or higher—but does not occur to any appreciable extent at lower temperatures. For effective seed coat decomposition, the seeds must be held moist at warm temperatures in the soil. Addition of nitrates to the medium has increased seed softening, presumably because it stimulates fungus activity.

Mechanical Resistance to Embryo Expansion. In most seeds, once water is absorbed, the expanding pressure of the embryo during germination is sufficient to rupture the seed coats. However, the seed coverings can be so hard as to offer mechanical resistance to embryo expansion. In the olive, for instance, the seed is surrounded by a thick, bony, indehiscent endocarp which is both mechanically hard and water-impermeable. The hard seed covering of *Symphoricarpos* is indehiscent, but water can be absorbed through the micropyle. Other seed structures, such as the "pits" of stone fruits or the shells of walnuts and other nuts, are extremely hard and may produce temporary resistance to germination. These structures are dehiscent, however, and water can be absorbed through the layer separating the two halves. Seeds of certain weeds, *e.g. Amaranthus retroflexus* and *Alisma plantago*, also have hard coats of this type.

The force required to overcome resistance of freshly harvested seeds of black walnut (*Juglans nigra*) and hickory (*Carya sp.*) was shown to be considerably greater than that of the calculated breaking force of the embryo during germination. However, shell resistance decreased markedly during moist, warm storage, provided that the medium was unsterilized.

Some structures within the seed, such as the endosperm, are believed to mechanically restrict embryo expansion in some kinds of dormant seeds Germination occurs when the expanding force of the embryo increases or when the restrictive force of the covering decreases, or both.

Restriction of Gaseous Exchange. Physical restriction of the movement of oxygen to the imbibed embryo by the inner membranous seed coat or an enclosing nucellus or endosperm layer has frequently been cited as a factor in seed dormancy. Evidence for this is that freshly harvested dormant seed, which are light-sensitive or temperature-sensitive, can be made to germinate promptly by excising the embryo or by physically or chemically altering the seed coat. Increasing the oxygen content of the surrounding atmosphere will stimulate germination of the intact seed in many cases. The benefits of after-ripening in dry storage have been attributed to increase in permeability.

The regulatory function of these seed coverings in dormancy may be real, but the mechanism does not seem to be simple. Other factors, in addition to physical restriction, may be involved. For instance, differences in oxygen requirements of different parts of the seed or at different stages of after-ripening may be involved. Cocklebur (*Xanthium*) seed illustrates this phenomenon. Two seeds are present in the bur, one germinating readily, the other delayed for one or two years. This difference was first attributed to a difference in inner seed coat permeability to oxygen uptake. It has since been found that the embryos contain two water-soluble chemical inhibitors that will not pass through the seed covering. However, the inhibitors can be removed by leaching the excised embryo with water or by subjecting the seed to an increased concentration of oxygen.

Dormant Embryos

Seed with a Single, Low-Temperature Requirement for Germination. Plant propagators have known since early times that moist seeds of certain fall-ripening tree and shrub species of the temperate zone must be chilled as they overwinter in the ground before they will germinate in the spring. This requirement led to the horticultural practice termed *stratification*, in which seeds are placed between layers of moist sand or soil in boxes (or in the ground) and exposed to chilling temperatures either out-of-doors or in refrigerators. A more accurate term for this procedure, used by some propagators, is *moist-chilling*.

Conditions required for after-ripening changes to take place within these dormant seeds include: (*a*) temperatures just above or slightly below freezing; (*b*) imbibition of moisture by the seed; (*c*) aeration; and (*d*) a certain length of time. After-ripening is considered to be most effective at approximately 35° to 45° F (2° to 7° C), although changes also occur at higher and lower temperatures. De Haas and Schander have shown that the lowest temperature at which such changes occur in apple seeds is about 23° F (−5° C). Moist, dormant seeds of many species that are exposed to a high temperature (greater than the effective chilling range) before after-ripening is complete not only fail to germinate but also develop secondary dormancy. Abbott has shown that in apple seed at 62° F (17°C) the processes leading to after-ripening and those resulting in secondary dormancy are in equilibrium. He suggested that this *compensation temperature* should be used to test germination of dormant seeds. Compensation temperatures occur in dormant seeds of other plants, but the exact temperature probably varies with different species and possibly with the stage of after-ripening.

The response of partially after-ripened lots of apple seeds to germination temperatures is illustrated in Figure 5.4. If the seeds are placed at relatively low temperatures, the germination rate is slow but the germination percentage high. At higher temperatures germination rates are faster, but the germination percentage decreases in proportion to the increase in temperature; nongerminating seeds revert to secondary

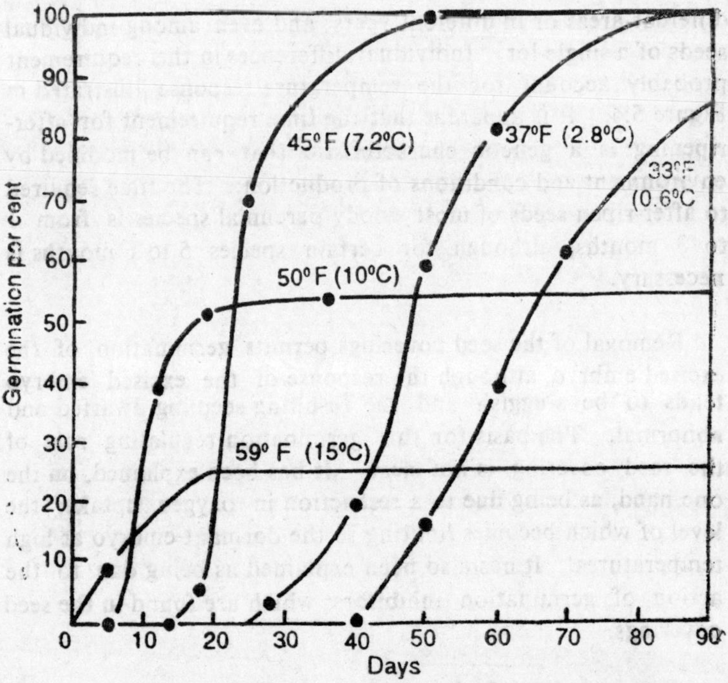

Fig. 5.4. The effect of temperature on germination of apple seeds previously stratified for 65 days at 37° F (3° C). The abscissa represents the days following stratification.

dormancy. The temperature conditions most effective for germination may be similar to those in natural environment in which soil temperatures gradually increase with the advance of spring. Late planting or an early hot period can inhibit germination and produce secondary dormancy.

Moisture in the seed is necessary during pregermination chilling. Drying will stop after-ripening causing secondary dormancy, or it can injure the seeds. Reduced aeration will slow or stop after-ripening and has been used experimentally to induce secondary dormancy.

The time required for moist-chilling (stratification) varies among species, among clones within a species, among different seed lots of the same kind which were grown in different areas or in different years, and even among individual seeds of a single lot. Individual differences in this requirement probably account for the temperature response illustrated in Figure 5.4. It is apparent that the time requirement for after-ripening is a genetic characteristic that can be modified by environment and conditions of production. The time required to after-ripen seeds of most woody perennial species is from 1 to 3 months, although for certain species 5 to 6 months is necessary.

Removal of the seed coverings permits germination of the excised embryo, although the response of the excised embryo tends to be sluggish and the resulting seedling dwarfed and abnormal. The basis for this germination-regulating role of the seed covering is not clear. It has been explained, on the one hand, as being due to a restriction in oxygen uptake, the level of which becomes limiting to the dormant embryo at high temperatures. It has also been explained as being due to the action of germination inhibitors which are found in the seed coverings.

Flemion showed that incompletely after-ripened peach seedlings remain dwarfed for many years if held at high temperatures but will resume normal growth after being subjected to 6 weeks of chilling. Later, Pollock demonstrated that peach seedlings were dwarfed when the apical meristem of the excised embryo was exposed to germination temperatures between 73° and 81° F (23° and 27° C) but normal when exposed to higher or lower temperatures. Dwarfing can be offset by exposure of the seedlings to long photoperiods or continuous light (provided that this action is taken before the apical meristem becomes fully dormant) or by repeated application of gibberellic acid. Treatment of nonchilled seeds of some peach varieties with thiourea will produce prompt germination, but the resulting seedlings will be dwarfed.

The biological mechanism whereby embryo dormancy is overcome within the seed is not well understood. Many changes that occur within the embryo during after-ripening appear to indicate merely the increasing ability of the seeds to germinate. Such changes include increasing water-absorbing power, increasing enzyme activity, increased acidity, and gradual changes of the insoluble complex storage materials into soluble simpler substances. Pollock has measured the increase in fresh weight, number of cells, and respiratory activity in the embryo axis of cherry seeds held for 16 weeks at 50° F. At this temperature, phosphorus was increasingly found in the energy-rich compounds (nucleotides) involved in the metabolic activities of the cell, whereas in seeds held at higher temperatures, the phosphorus accumulates as inorganic phosphate.

Growth-inhibiting and growth-promoting substances occur within various parts of the seed. Luckwill found inhibitors in dormant apple seeds in the endosperm, integuments and embryo in the ratio of 30 : 13 : 1, respectively. The inhibitors disappeared during chilling, and growth-promoting substances appeared. Inhibiting substances have likewise been shown to be produced in the cotyledons of *Fraxinus* seeds during the period in which the dry seeds imbibed water. Lipe and Crane identified an inhibitor extracted from the integuments of peach seed as dormin (or a compound similar to it). Dormin is a dormancy-inducing substance which has been extracted from Sycamore buds and leaves ; this is also known as abscisin II or as *abscisic acid*. Similarly, an inhibitor occurs in the integuments of nonchilled birch (*Betula*) seeds. Dormin and abscisin II have been shown to induce dormancy in seeds. In each of these cases the effect of the inhibitor is decreased by gibberellic acid and, to some extent, by leaching with water. Although these inhibitors tend to diminish during chilling, there is also an increase in the content of growth-promoting substances prior to and during germination. Consequently it is difficult to ascertain specific cause-and-effect relationships. Nevertheless, there is considerable belief among plant physiologists that dormancy of the kind represented in these seeds is controlled through a balance among naturally occurring inhibitors and promoters, interacting with environmental factors such as temperature and possibly oxygen supply. Gibberellin has been cited as a major part of the promoter complex, and dormin as a part of the inhibitor complex.

Seeds with a High-Temperature, Followed by a Low-Temperature, Requirement. Several cases exist in which a single low-temperature treatment is inadequate to produce germination unless the moist seed is previously subjected to warm temperatures. In nature, ripe seeds fall to the ground, are exposed to warmth and moisture in late summer and fall, and are then subjected to chilling in winter. With this sequence, germination occurs the following spring ; without it germination may be delayed until the second spring. In one group of plants—wild ginger (*Asarum canadense*), several lily species (*Lilium auratum, canadense, japonicum, rubellum, superbum,* and *szovitsianum*), herbaceous and tree peony (*Paeonia*), and a number of *Viburnum* species—only the radicle grows at high temperatures. Low temperatures must follow to after-ripen the epicotyl and cause the shoot to grow. Two to 6 months (in some *Viburnums,* 17 months) at high temperatures are required for root growth before a chilling period—of 1 to 4 months—is provided.

Another group of plants with similar requirements include those in which the embryo at the time of fruit ripening is incompletely developed ; it increases in size during the warm period, then after-ripens during the chilling period. Ash (*Fraxinus excelsior*) seeds have this requirement. Holly (*Ilex sp.*) seeds consist mostly of endosperm—the embryo being small and hardly differentiated—and require a warm period for the embryo to develop. In both ash and holly, however, the problem is complicated by seed coat dormancy conditions as well.

Seeds with a Low-Temperature, Followed by a High-Temperature, Followed by a Low-Temperature, Requirement. Still more complex requirements occur in a number of other plants, mostly wild species of the temperate zone—blue cohosh (*Caulophyllum*), lily-of-the-valley (*Convallaria*), Solomon's seal (*Polygonatum commutatum*), bloodroot (*Sanguinaria canadensis*), false Solomon's seal (*Smilacina racemosa*), and *Trillium erectum* and *T. grandiflorum.* Their requirements are as follows : a cold period of 1 to 6 months to after-ripen the radicle, a warm period of 1 to 6 months for the radicle to grow, and then a second cold period of 1 to 6 months to after-ripen the epicotyl.

Combinations of Two or More Types of Seed Dormancy

Seed germination in a number of species is complicated by the presence of more than one type of dormancy, the usual combination being that of seed coat dormancy combined with a dormant embryo. Such seeds have been recognized by nurserymen as being among the most difficult to handle because of the long period of time required to produce germination. This group has been referred to as "two-year" seeds, because seeds planted in the spring remain dormant and will not germinate until the next spring. If planted in the fall, they will not emerge until the second spring following planting.

Treatment of seeds with double dormancy involves the combination of procedures which overcome each type of dormancy present. In nature, decomposition of the seed coverings takes place by various agencies of the environment, as has been previously discussed. Warm stratification for several months, during which time micro-organisms act upon the seed covering, followed by cold stratification, is an effective method of handling. This explains the fact that seeds planted in the spring or summer fail to germinate until the following spring. The period of handling can be shortened by treating the seed coats artificially and then stratifying. In some cases giving the seeds a short, warm stratification period betweed the time of seed coat treatment and cold stratification will assist considerably in the elimination of dormancy. If seeds of some woody plant species are dried following harvest, a hard seed coat develops, and germination does not take place until the second spring following harvest. If, on the other hand, the seeds are fall planted without drying, germination takes place the following spring.

Chemical Inhibitors

Substances which inhibit germination can be extracted from a variety of plant parts, including seeds, fruits, leaf sap, bulbs, and roots. Some of these naturally occurring germination inhibitors have been identified as specific chemical substances : ammonia (from beet seed), hydrogen cyanide (from amygdalin), ethylene (from ripe fruits), essential oils, unsaturated organic acids, alkaloids, (nicotine, cocaine, caffein), unsaturated lactones—coumarin and others. The existence of such substances in plants does not necessarily mean they have an essential role in controlling seed germination. Nevertheless, there is increasing emphasis on the biological role that naturally occurring inhibitors play.

During fruit and seed development many different chemical materials accumulate in the fruit, seed coverings, and embryo. It is not surprising then that most fleshy fruits, or juices from them, strongly inhibit germination. This occurs, for instance, in citrus, cucurbits, stone fruits, apples, pears, grapes and tomatoes. Likewise many dry fruits and fruit coverings, such as the hulls of guayule, *Pennisetum ciliare*, and wheat (*Triticum*), and the capsules of mustard (*Brassica*), and others produce inhibition. Undoubtedly these substances play an important biological role in preventing premature germination while the seeds are present on the parent plant. Where the fruit covering remains with the seed, the inhibiting effect can persist for a considerable period even after the seed is separated from the plant.

Inhibitors in seeds appear to play a direct germination controlling role in some plants. For instance, seed inhibitors are important in the ecology of certain desert plants. Such substances have been referred to as chemical rain gauges in that a heavy rain (which would insure survival of the seedling) is necessary to leach out enough of the inhibitor to allow germination to take place. A light soaking is insufficient.

Inhibitors appear to the involved in germination control of many freshly harvested seeds, although their exact role is difficult to assess. Part of the evidence is that removal of the seed coat makes germination possible. Inhibitors have been repeatedly demonstrated to occur in various seeds when freshly harvested. Such materials either disappear in dry storage or leach out at germination. Dormancy has been induced experimentally in lettuce seed by treating it with an inhibitor—coumarin—and subsequently overcome by the same treatments required for naturally dormant seed. A probable role of inhibitors was described in the discussions or gas-restricting seed coats and dormant embryos.

ENVIRONMENTAL CONDITIONS AFFECTING GERMINATION

Water

Imbibition of water by the seed is the first step in the germination process. The two most important factors which affect water uptake by seeds are : (*a*) the nature of the seed and its coverings and (*b*) the amount of available water in the surrounding medium. Seeds have great absorbing power, owing to their colloidal nature. In storage, seeds can absorb water from the surrounding air. Different kinds of seeds vary greatly in the amount and rate of water absorbed, either in storage or during germination. The rate of water uptake is also influenced by temperature, a higher temperature favouring an increased rate. The seed covering also plays an important role in water uptake. In some seeds, it is so impermeable to water that germination will not occur until the seed covering has been altered in some manner.

The moisture supplied to the germinating seed may affect both the germination percentage and the germination rate. Germination percentage tends to be equal over most of the range of available soil moisture from field capacity (FC) to permanent wilting percentage (PWP). Differences between species become evident as the moisture content of the soil approaches dryness (PWP). Some seeds germinate only above PWP; others can germinate below PWP. For instance, vegetable plants can be grouped according to their moisture requirements for germination as follows :

Group 1. Seeds that will germinate in soils with moisture from permanent wilting percentage (or a little above) to moisture content above field capacity.

cabbage	sweet corn	muskmelon	pepper
turnip	squash	cucumber	onion
radish	watermelon	tomato	carrot

Group 2. Seeds that will germinate in soils from intermediate moisture content to above field capacity.

snap bean	pea	endive	beet
lima bean	lettuce		

Group 3. Seeds that will germinate only in soils near field capacity.

celery

Group 4. Seeds that germinate well at lower moisture contents but show reduced germination near field capacity.

spinach New Zealand spinach

Rate of emergence from a seed bed is influenced particularly by the available moisture supply. From a point approximately halfway through the range from FC to PWP there is a decline in rate, as illustrated in Figure 5.5.

Excess soluble salts in the germination medium may inhibit germination and reduce seedling stands. Such excess soluble salts may originate from the soil and other materials used in the germination medium, the irrigation water, or excessive fertilization. Since the effects of salinity become more acute when the moisture supply is low and the concentration of salts thereby increased, it is particularly important to maintain a high moisture supply in the seed-bed where the possibility of salinity exists. Surface evaporation from subirrigated beds can result in the accumulation of salts at the soil surface even under conditions in which salinity would not be expected. Planting seeds several inches below the crown of a sloping seed-bed can minimize this hazard.

Maintaining an adequate continuous moisture supply can be difficult, because germination takes place in the upper surface of the germination medium, which is subject to

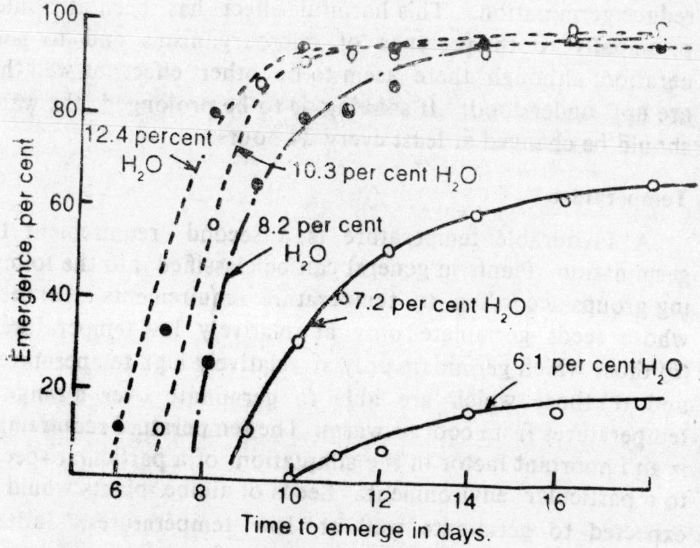

Fig. 5.5. The effect of different amounts of available soil moisture on the germination (emergence) of 'Sweet Spanish' onion seed in Pachappa fine sandy loam.

fluctuations in temperature and in moisture supply. The problem is greater with the necessarily shallow planting of small seeds or where the rate of germination is slow. Careful watering, deeper planting, and the application of a mulch are ways to maintain a more uniform water supply. On the other hand, excess watering accompanied by poor drainage can be deleterious, because it reduces aeration in the germination medium and favours "damping-off".

Soaking seeds before planting is sometimes utilized to initiate the germination process and to shorten the time required for seedlings to emerge from the soil. Such a treatment may be advantageous with seeds normally slow to germinate, with seeds which are hard and dry, or when certain dormancy conditions exist. However, if the seeds ordinarily germinate without difficulty, there is little need for soaking. Seeds which have imbibed water are easily injured and more difficult to plant.

Prolonged soaking can result in injury to the seed, and reduce germination. This harmful effect has been attributed principally to the presence of microorganisms and to poor aeration, although there seem to be other effects as well that are not understood. If soaking is to be prolonged, the water should be changed at least every 24 hours.

Temperature

A favourable temperature is a second requirement for germination. Plants in general can be classified into the following groups according to temperature requirements : (*a*) those whose seeds germinate only at relatively low temperatures ; (*b*) those which germinate only at relatively high temperatures ; and (*c*) those which are able to germinate over a range of temperatures from cool to warm. The temperature requirement is an important factor in the adaptation of a particular species to a particular environment. Seeds of alpine plants would be expected to germinate well at low temperatures. Indeed, studies have shown that some alpines, for instance *Camassia leichtlini* and *Lewisia rediviva*, germinate only below 41° F and 50° F, respectively. On the other hand, seeds of most tropical plants require high temperatures.

Temperature largely determines the time of year at which seeds will germinate out-of-doors. As a result, seedlings of many self-sown plants, particularly annual weeds, appear only at certain times of the year, depending upon whether they are warm-season or cool-season plants.

The temperature requirements for germination of seeds are generally considered in relation to three points : minimum, maximum, and optimum. The designation of these temperatures for any one species is somewhat difficult, because temperature affects both the germination percentage as well as the germination rate. Furthermore, temperature requirements of dormant seeds may change with the age of the seed and the progress of after-ripening. Germination percentage can be rather constant within the range of temperatures at which germination takes place. The rate of germination, on the other hand, is affected more ; an increase in temperature invariably increases the germination rate. Above the optimum, rates decrease as injury occurs.

Minimum temperatures are those below which germination will not occur. For "cool-season" plants, this temperature is approximately 40° F (4.5° C) or less, although for some, such as endive, lettuce, onion, parsnip, and spinach, it approaches freezing. Germination rates are invariably very low at these low temperatures. Minimum temperatures for "warm-season" plants are in the range of 50° to 60° F (10° to 15° C). Below these temperatures, either the seeds fail to germinate, or "chilling injury" can result in seedling abnormalities. In lima bean, Pollock has shown that there is a temperature-sensitive period during the initial imbibition in which injury to the embryo axis occurs if seeds are exposed to temperatures below 60° F (15° C).

The maximum temperatures are the highest at which germination will occur. The upper limits of soil temperature for the survival of most vegetable seeds is between 86° F (30° C) and 104° F (40° C). Seeds placed at 113° F (45° C) were killed within 24 hours. During hot weather, direct sunlight striking the soil surface may raise the temperature to a level injurious to plant tissue. Also, moisture is rapidly lost. This so-called "heat injury" affecting young seedlings resembles the "damping-off" caused by pathogenic organisms. For may plants shading is necessary.

Short exposures can be made to temperatures much higher than those which would be used for germination. In one study, exposure of seeds of a number of different species to 120° F (49° C) for 1 hour did not decrease the germination percentage even though the seeds had imbibed water. At higher temperatures, the amount of injury increased but depended upon the kind of seed, the duration of exposure, and the moisture content of the seed. Dry seeds could withstand brief exposure to temperatures of 212° F (100° C), but above 250° (121° C) all were killed.

Such heat treatments are important in freeing planting material of harmful organisms. The temperature should be high enough to kill the organisms but not so high as to impair germination. Heat tolerance is greater with seeds of low moisture content and with those that are fresh, uninjured, and of high viability.

Seeds of some species—for instance, lettuce, celery, endive, delphinium, and numerous flower species—remain dormant at temperatures over 75° F (24° C). This temperature sensitivity is associated with freshly harvested seeds and tends to disappear with after ripening in dry storage. It has been called *thermodormancy*. Germination can be induced by subjecting the seeds to lower temperatures or to dry storage. In laboratory testing, chilling moist seeds at 41° to 50° F (5° to 10° C) for 5 days will induce prompt germination when the seeds are subsequently shifted to higher temperatures. Seed germination of some woody species is also inhibited by high temperatures.

Optimum temperatures are those most favourable for germination. This should be the range wherein the highest percentage of seedlings will be produced at the highest rate of germination. On this basis, the optimum temperature for seeds of most plants is between 80° and 95° F (26.5° and 35° C).

Germination is often much better if the seeds are subjected to daily alternating temperatures rather than a constant temperature. Commonly used alternations are 59° F (15° C) or 68° F (20° C) for 18 hours and 86° F (30° C) for 6 hours. The effective temperature alternation has been shown to change as the seed after-ripens in dry storage, and the effect may disappear entirely.

The above discussion refers to the effect of temperature on the initiation of germination itself. In addition, temperature affects other influencing factors in the seed-bed, such as the activity of pathogenic organisms. Temperature likewise affects seedling growth following germination. Usually a temperature somewhat lower than that which is optimum for germination is best for growing the seedlings.

Oxygen

Respiration may be summarized as follows :

sugar +	oxygen →	carbon dioxide +	water +	energy
$C_6H_{12}O_6$	$6O_2$	$6CO_2$	$6H_2O$	673 kcal

This process takes place in seeds as long as the seed is alive. In a nongerminating dry seed the respiration rate is low, and little oxygen is utilized. During germination the respiration rate increases, oxygen uptake increases, and carbon dioxide is given off at an increased rate. The biochemical mechanisms that control seed germination are not well known, but oxygen is utilized in separate metabolic processes that may change with the progress of germination.

Oxygen is involved in some way with the initial triggering reactions of germination, and aerobic conditions are necessary for after-ripening reactions. Consequently limitation in the oxygen supply at the initial stages can inhibit germination. In seeds, fats are transformed into sugars, in which form they are translocated. Since a fatty acid molecule contains less oxygen than a sugar molecule, additional oxygen must be supplied. Energy to drive the cellular mechanisms and to transform specific materials into growth processes is provided through the oxidation of the reserve substances in the seed. Metabolic pathways in seeds appear to be essentially the same as in other plant tissues. Part of the energy is evolved as heat and can be measured by appropriate instruments.

Reduced aeration can decrease germination. One of the effects of a hard crust in a seed-bed is to limit oxygen diffusion and thus inhibit seedling emergence. Limitation in oxygen supply in poorly drained seed-beds or after heavy rains or excessive irrigation can result in the pore spaces of the soil being filled with water.

Seeds of land plants vary considerably in their ability to germinate under water, but those of the following species germinate well : timothy, *Agrostis nebulosa*, *Axonopus compressus*, Canadian bluegrass, Bermuda grass, lettuce, wormwood, carrot, celery, portulaca, iceplant, snapdragon, *Melissa officinalis*, bittersweet, and petunia. Seeds of water plants and rice normally germinate under water. The ability of rice to germinate at low oxygen pressures has been attributed to the presence of an anaerobic, energy-liberating system within the seeds. Cattails (*Typha latifolia*) give poor or no germination in air but germinate promptly in water or when the oxygen supply is reduced.

Light

Light can play an important role in seed propagation both because of its effect upon the initiation of germination and because of its controlling influence upon seedling growth.

Light and Germination

The fact that visible light can stimulate or inhibit germination of seeds of some plants has been recognized since the middle of the nineteenth century. Plants whose seeds have an absolute requirement for light, with viability being lost in a few weeks without it, include mistletoe (*Viscum album*), strangling fig (*Ficus aurea*), and *Areuthobium oxycedri*, all epiphytic plants. Seed germination of another group, including celery, lettuce, tobacco, most grasses, and many conifers is favored by light. Seeds of another group, *e.g.*, some species of *Phacelia*, *Nigella*, *Allium*, *Amaranthus*, and *Phlox*, are inhibited by light.

Germination control is exerted through a photochemically reversible reaction involving the response of a pigment, known as *phytochrome*, to light of particular wavelengths :

$$\text{phytochrome}_{red} \xleftrightarrow[\text{in far-red light}]{\text{in red light}} \text{phytochrome}_{far\text{-}red}$$

(inhibits germination)	(instantaneous reaction) in darkness (slow reaction)	(promotes germination)

Exposure of the imbibed seed to red light (6400 Å to 6700 Å) causes the phytochrome, in it to change to phytochrome$_{fr}$, which is linked in some way to reactions inducing germination. Exposure to far-red light (7200 Å to 7500Å) produces an instantaneous change to the alternate form, phytochromer, which inhibits germination. In the dark, in the presence of oxygen, and at favourably low temperatures, the latter change takes place slowly. In plants under natural light the red wavelengths dominate the far-red so that the phytochrome is converted to the active $P_{far-red}$ form.

Light-inhibited seeds are less common, but evidently their response is to the same phytochrome system. Blue light may also be involved, but the reactions are not known. Light control appears to involve the seed coats or the endosperm layer, and the light requirements disappear if the embryos are excised.

Knowledge of light effects has been obtained by experiments with particularly sensitive seeds under controlled conditions. The role of light as a germination-controlling mechanism in propagation and in nature is not always clear, since response varies with the kind of seed, the conditions of seed production, and accompanying environmental conditions. In general, a light requirement is found in freshly harvested seeds and disappears with dry storage ; light stimulation will not occur if the temperature is too high. The light requirement can be partially replaced by alternating temperatures, by potassium nitrate, or by kinetin, as well as by gibberellic acid and thiourea.

In the soil, light penetration depends upon the length of the rays. Red light penetrates to a depth of about an inch in sandy soil. Most light-favoured seeds are small. Seed germination occurs with those siturated close enough to the surface for seedlings to emerge but deep enough so that the water supply will not be limiting. At greater depths there is complete darkness, or only far-red rays penetrate ; in this case the seeds remain dormant until the soil is disturbed.

Photoperiod can affect germination of seeds of some species. Eastern hemlock (*Tsuga canadensis*) and birch (*Betula*) for instance, are "long-day" seeds, but germination is also promoted by low temperatures. Photoperiodic control can partially replace a seed-chilling requirement and vice versa. These light-temperature requirements for seed germination evidently are involved in the ecological adaptation of the species. Seeds of some other species, such as *Nemophila, Nigella, Veronica persica,* and *Eschscholtzia californica* are inhibited by long days.

Light and Seedling Growth

Light affects the growth processes in the seedling and has an important role in its emergence through the soil. In the darkness of the seed-bed, the hypocotyl or epicotyl is elongated and etiolated, and the leaves are unexpanded. There is a hook at the end of the plumule which pushes through the ground. When the seedling shoot emerges into the light, shoot elongation decreases, the plumular hook straightens out, leaves expand, and normal growth takes place.

In the early germination stages, the seedling utilizes the reserve supply of the seed. Later growth depends upon the production of carbohydrates and other materials resulting from photosynthesis in the leaves. Light of a relatively high intensity is necessary to produce sturdry, vigorous plants.

An extremely high light intensity, on the other hand, can result in high temperatures which may produce heat injury to the seedling. Consequently excessively high as well as excessively low light intensities need to be avoided. Partial shading is necessary for many plants during their early seedling growth.

Supplementary artificial light to increase the length of the photoperiod is sometimes provided in growing seedlings, particularly during winter when the day length is short and cloudy weather often prevails. Artificial light has been used as the sole illumination in indoor propagating structures for germinating seeds. The principal problems in artificial lighting in relation to plant growth are to get a sufficiently high intensity without heating and to obtain a proper balance of light quality.

Light intensities of 700 to 1200 foot candles are satisfactory, although lower intensities can be used for some shade plants. The length of the photoperiod should be 12 to 18 hours ; below 12 hours the amount of photosynthesis becomes limiting, while photoperiods longer than 18 hours may produce poor results with some plants. Where low light intensities are used, lower temperatures should also be used. With higher light intensities, higher temperatures are permissible.

A certain amount of blue light is necessary to produce stocky seedlings ; red light alone produces tall, "leggy" plants. Satisfactory artificial light sources are fluorescent lamps of the daylight type, 3500° K, or 4500° K white types, or a mixture of the two, or the "cool-white" alone.

Seedling Production Indoors

The production of seedlings within some type of indoor shelter is an important method of propagating plants. Seeds are sown in special germinating media and the seedlings transferred to pots, flats, or special beds and later transplanted into their permanent location. Seeds may also be sown directly into pots or other containers. (See Fig. 5.6).

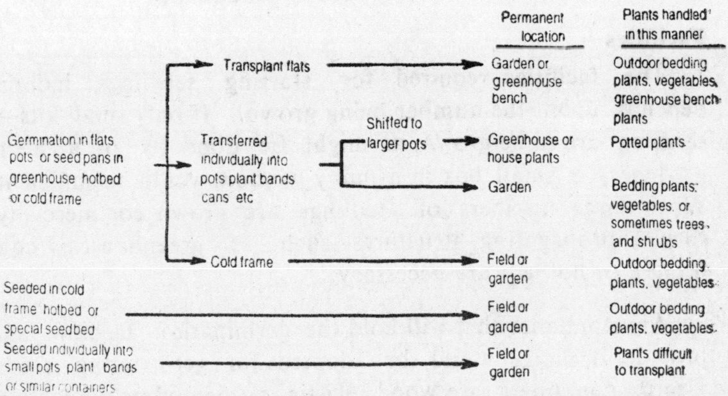

Fig. 5.6. Methods of handling seedlings in relation to indoor culture.

This technique is used, when outdoor growing conditions are unfavorable, to produce plants for pot plants or benching in greenhouses or to start plants for transplanting later to out-of-door locations. It is extensively used to raise vegetables for early production, to grow plants which might not otherwise be adapted to an area having a late spring and a short growing season, and to produce bedding plants for landscaping purposes. It may also be used in the nursery production of trees and shrubs.

Starting seedlings indoors for outdoor growing may enable the propagator to obtain higher germination percentages because he can control environmental conditions and reduce loss from disease, insects, and other adverse conditions.

To obtain these benefits certain facilities are required, and somewhat more labour may be necessary in the handling of seedlings through to the transplant stage than would be required with direct seeding. It also requires rather close attention to details and a certain degree of skill and experience. On the whole, growing seedling indoors is somewhat more expensive than direct seeding and would be justified only with valuable plant material or expensive seeds ; or when seedlings are difficult to start under more rigorous outdoor conditions, or to obtain the benefits of early season production.

Facilities

The facilities required for starting seedlings indoors depends upon the number being grown. If only small lots of seedlings are to be grown, as might be done by an amateur gardener, a small box in a sunny window would be sufficient. When large numbers of seedlings are grown commercially, special propagating structures such as greenhouses, cold frames, or hotbeds are necessary.

Any container that will hold the germination medium and provide drainage could be adapted for germinating seeds. Useful containers are wood, plastic, or metal flats ; individual flower pots ; bulb pans ; or plastic freezer boxes. Seeds of plants which are difficult to transplant bare-root—for instance, cucumber or melon—could be planted individually in plant bands. Polyethylene plastic bags can also be used for germinating seeds.

SEEDLING PRODUCTION

Germinating Media

There are a number of materials and mixtures of materials which are satisfactory for seed germination.

Clean sterilized containers should be used. The container is completely filled with the germination medium, which is worked carefully into the corners to eliminate air pockets. Excess soil is removed with a straight board drawn across the top of the flat. The soil is then tamped firmly with a flat block of wood to a level approximately ½ in below the rim of the container. If fine seeds are to be sown, if may be desirable to water before planting.

Sowing

The method of sowing seeds depends on the size of seed. Containers with the germination medium may be presoaked, drained, and then enclosed in polyethylene to hold moisture until planted. Depth of covering varies with the kind of seed and the germinating conditions. Very fine seed may merely be dusted onto the surface. Larger seeds may be covered to one to two times their minimum diameter, assuming that good moisture conditions prevail in the medium during the first week after planting. During the initial germination period a uniform, but not excessive moisture supply is essential in the appear layers of the medium. Containers should be sprinkled lightly after sowing and then covered. A plastic or glass cover will prevent drying out, but it should be shaded from the direct rays of the sun. Some seeds, however, must be exposed to light. Placing the planted containers under intermittent mist controlled with a time clock is an excellent method of handling.

The temperature of which the flats are placed depends upon the germination requirements of the seeds. The best temperature for seeds of most plants is around 68° F (20°C). For some, 55° F (13°C) is better, and for seeds of warm-season and tropical plants temperatures up to 86° F (30°C) are preferable. Higher temperatures (over 86° F) result is rapid moisture loss and inhibition of germination in some seeds.

Handling of the Seedlings

The primary objectives from seedling emergence until the plant is placed in its permanent location are to control damping-off and to develop stocky, vigorous plants capable of being transplanted with little check in growth. After seedling emergence, the cover over the flat is removed and the seedlings subjected to increased but not intense light. Moderate to low temperatures produce stocky growth. High temperatures lead to undesirable succulent growth, excessive moisture loss, and development of damping-off organisms. General recommendations are a 60° to 65° F. (15° to 19°C) temperature during the day and 5° to 10° F lower at night for cool season plants, and 70° to 75° F (22° to 24° C) during the day, and 10° to 15° F lower at night for warm-season plants. Seedlings subjected to adequate light produce short, stocky plants rather than spindly, elongated plants, which result if they are grown in weak light. However, full sunlight during the early stages of growth should be avoided because of injury from high temperature.

The moisture supply should be constant but not excessive. Over-watering can result in poor aeration, and the resulting high humidity contributes to damping-off. When root growth is limited in the initial germination stages, frequent light waterings are necessary to keep the upper layer of the germination medium moist. As the root system develops the frequency of watering can be reduced, but the medium should be wetted completely. Watering early in the day so that plants dry before evening is desirable to reduce fungus development. Covers of closed frames should be opened at intervals to lower the humidity as the seedlings become larger.

If damping-off becomes a problem, it can be controlled to a certain degree by watering at intervals with a commercial fungicidal solution, although the success with damping-off can be controlled by this means is variable.

Transplanting Seedlings

When planted closely in the medium, the seedlings soon become overcrowded and a check in growth will occur if they are not promptly transplanted. This operation comes as soon as the first two to four true leaves have developed and the seedlings are large enough to handle. A seedling should be transplanted—with wider spacing—either into another flat or individually into pots, plant-bands, or similar containers.

A transplant flat is prepared as was done for the germination medium. Seedlings are lifted carefully and inserted 1 to 2 in. apart into holes made with a wooden dibble in the transplant medium. The soil is then pressed firmly about the seedlings to eliminate any air pockets about it, and the flat is then thoroughly watered. For the first few days while the seedlings are becoming re-established, the flats should be shaded, held at low temperature, and kept from drying. When the seedlings have grown so that they are well developed and have begun to crowd, they are again transplanted either into their permanent location in the field or into pots or other containers.

Success in transplanting depends to a large extent upon the previous handling. When the permanent location is in the greenhouse, environmental conditions are controlled and are not different from those to which the seedling has been exposed during development. When movement is from a greenhouse or hotbed to the open field, the operation is somewhat more critical and requires that the plant be "hardened" prior to its shift to the open field "Hardening" involves a checking of growth resulting in the accumulation of carbohydrates which makes the plant better able to withstand adverse conditions. The process can be brought on by withholding the supply of moisture, reducing the temperature, and gradually shifting from the greenhouse or hotbed to the environment of the permanent location. A cold frame or lathhouse is useful for this purpose. Flats of seedlings can be mcved into them and allowed to remain for 7 to 10 days prior to placing in the field.

Before being moved into the field, the plants should be watered thoroughly, Planting is done in the field by hand or, in some cases, by transplanting machines. During transplanting, it is desirable to retain as much soil about the roots as practical to avoid disturbing the root system. Afterward, the plants should be thoroughly watered, and if practical a temporary shade provided. For the first few days, until the plants have become established, they should be watched for wilting, and watered as necessary.

The use of 'booster" or "starter" solutions containing nitrogen, phosphorus, and potassium shortly before or after transplanting is sometimes beneficial in established plants in the field. Such solution should be used with caution, since if the soil is low in moisture at transplanting time, high concentrations of fertilizer can be injurious. This effect may vary with the soil condition.

Direct Seeding

Direct sowing of seeds in permanent plant locations is a technique of seed propagation often preferred to the use of special seed-beds. In general, direct seeding is more economical, since it utilizes no special plant-growing facilities and involves no individual handling of seedlings. With many plants it is the only practical method to use. It is a standard method of growing field crop; and many commercially grown vegetables. Much forest regeneration is done through direct seeding practices. Aerial broadcasting is used for large acreages. Amateur vegetable and flower gardeners may find direct seeding preferable to other methods, since it is simple and does not make demands on time and facilities. The lower germination percentages often obtained with direct sowing can be offset by increasing the rate of sowing and thinning any excess plants after germination. This procedure is feasible where the cost of the individual seed is not a factor. Of course, if much thinning is necessary, the cost of production again increases.

Direct seeding results in continuous and rapid development of the seedling, with no check in growth associated with transplanting. The total time between seeding and harvest is less than by the other methods. Some plants, such as corn, cucumbers, or beans, are difficult to transplant, and are almost always planted directly in the field.

Seed-Bed Preparation

Proper seed-bed preparation is the keynote for successful field sowing. The requirements of a good seed-bed are (*a*) to have an initial supply of moisture to carry the seeds through the germinating and early seedling period, (*b*) to have a physical condition that allows moisture to be supplied continuously to the seed, and (*c*) to have good aeration. A problem in seed-bed preparation is to obtain the proper balance between the extremes of low moisture-high aeration and high moisture-low aeration. This can be best obtained by a medium-textured loam which can supply moisture continuously without excessive drying. A light sandy soil dries out rapidly, whereas a heavy clay soil gives poor drainage and inadequate aeration, and tends to form a hard surface crust as it dries. Soil conditioning can be provided in the field by plowing under a cover crop or animal manure, allowing sufficient time before sowing for decomposition to take place. Incorporating peat moss, leaf mold, or other organic material may be feasible on a small scale or for high value crops. The ideal seed-bed would be one in which the soil is moist but not wet, finely pulverized to a depth of 6 to 10 in, particularly in the upper 3 or 4 in., and firmed to eliminate large air spaces which would increase evaporation and water loss.

The method of preparation depends upon the size of the operation and upon the crop being grown. For large-scale operations, implements such as plows, disks, harrows, and land-levelers are necessary. For smaller operations, the rototiller type of implement is convenient, since it loosens and pulverizes the soil in one operation. For small plots, spading or forking is the time-honored system. The soil should be spaded to a depth of 6 to 10 in. and then vigorously raked.

If it is dry, irrigation several days before preparation is necessary. Working the soil when it is too wet or too dry destroys its structure and leads to difficulty in aeration and water penetration later. Determining the optimum soil condition is often difficult, but as a rule of thumb, if the soil forms a tight ball when squeezed with the hand, it is too wet. It should crumble when squeezed.

Seed-bed preparation may involve soil treatments to destroy harmful insects, disease organisms, and weed seeds. The modern development of specific chemicals and effective methods of application has made this phase of operation important, if the crop has sufficient economic value to justify the cost. This cost is offset to some extent by reduction in later costs of cultivation and weed control.

Seed Planting

The time of planting is determined by the temperature requirement for germination. Cool-season crops are planted early in the spring when temperatures are low ; conversely, seeds requiring higher temperatures should not be planted until after the soil warms up. The time of planting is also determined by the date at which production from the plant is desired. For instance, planting small numbers of vegetable seeds at consecutive time intervals throughout the early spring and summer to produce a continuous supply from a home garden would be more desirable than a single planting to produce the entire crop at one peak harvest.

One of the problems in field planting is to determine the rate of sowing to obtain a desired stand of plants. The final stand of the plant is determined in order to produce the maximum yield of a high-quality, plant product. Too low a stand will reduce yields ; too thick a stand will decrease size and quality. Once the desired final stand is determined, the required rate of sowing can be calculated if the germination percentage, the purity percentage, and the number of seeds per pound (or ounce) are known :

$$\text{Pounds (ounces) of seed required per unit area} = \frac{\text{Number of plants per unit area desired}}{\text{Number of seeds per pound (ounces)} \times \text{percentage germination} \times \text{percentage purity}}$$

This figure is the minimum number of seeds to use. In practice, it is often desirable to increase the rate of sowing to offset expected losses in the seed-bed. The seedlings are thinned to a desired spacing after emergence. The depth of planting depends upon the size of the seed, the conditions of the seed-bed, and the environment at the time of planting. With too-deep planting, germination may fail because the seedling is not able to emerge before the food reserves in the seed are exhausted. In addition, poor aeration may occur at lower depths. Proper depth is also related to moisture supply. Too-shallow planting will place the seed in the top soil layers, which are subject to drying. In light sandy soils, or during periods of warm weather, deeper planting is best, whereas in heavier soils and during periods of cool, cloudy weather, shallower planting is permissible. A general rule is to plant seeds three to four times their thickness, being sure this depth is well within the moist layer of the seed-bed.

Various mulching materials have been used on seed-beds. Paper of special manufacture, and clear or black polyethylene or polyvinyl plastic film have been used. These materials are laid over the surface and either removed at the time of seedling emergence or prepared with perforations for the seedling to grow through. A later innovation is a black asphalt emulsion which can be sprayed on the surface of the ground ; it is nontoxic to plants and remains in place until disturbed through cultivation. Other materials include perlite or vermiculite treated with polyvinyl acetate. These materials benefit seedling stands by preventing crust formation, conserving moisture, controlling weeds (in some cases), and controlling seed-bed temperatures. Temperature control depends upon the type of material ; clear plastic and asphalt emulsions increase day temperatures more than black plastic but lose heat more rapidly at night.

Growing Seedlings in the Nursery

Growing trees and shrubs by seeds is an important part of nursery production. Some nurseries propagate seedlings almost entirely, dealing very little in vegetative methods. This is true in the production of trees and shrubs for reforestation, roadside planting, erosion control, and similar purposes. Nurseries producing fruit trees or ornamental shrubs and trees use seedlings extensively as rootstocks upon which to bud or graft selected clones or varieties.

The discussion which follows gives a general picture of the operations involved in the handling of seeds under nursery conditions. Details of operation vary from nursery to nursery, depending upon such factors as the kind of plant, the size of the operation, the purposes for growing, the inclinations of the operator, and the location of the nursery. Nursery management is a specialized field involving many more problems than the actual propagation of plant material. (See Figure 5.7).

Time of Planting

Seeds may be planted in the fall or in the spring depending upon their germination requirements and upon the management practices of the nursery. Most tree species can be placed into three general categories in regard to their seed-germination requirements.

Group I includes species (apple, pear, *Prunus, Rosa multiflora,* and yew) whose seeds require moist-chilling (stratification) and that, upon the completion of after-ripening, are sensitive to high germination temperatures. Germination temperatures of 50° to 62° F (10° to 17°C) would be optimum. High soil temperature results in damping off, seedling injury, and a tendency to revert to secondary dormancy. Such seeds could be effectively planted in the fall and sprouting would not take place until early in the spring. If they are kept in stratification storage overwinter, planting should take place as early in the spring as practical, but after frost danger is over.

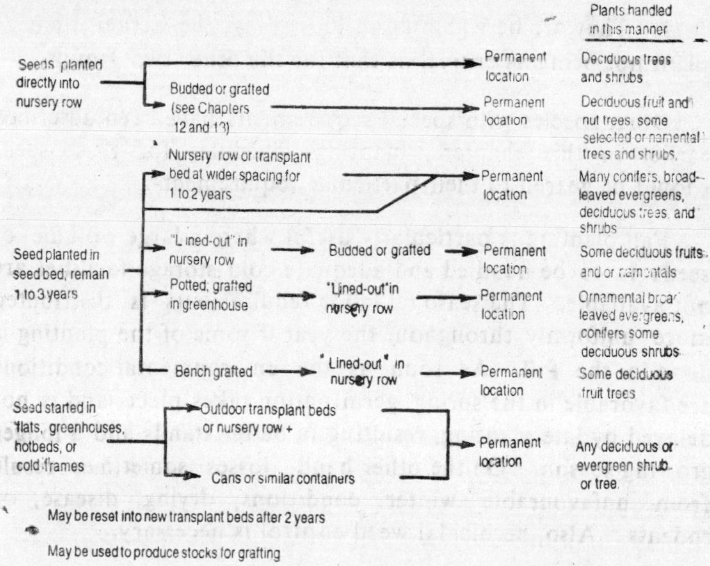

Fig. 5.7. General schedules for the propagation of seedling and grafted nursery stock.

Group II includes most fir, pine, and spruce species and many deciduous hardwood species whose seeds require varying amounts of moist-chilling (unless exposed to light) but that, after this is over, do not germinate in the spring until after the ground becomes warm. Optimum germination temperatures are 68° F to 86° F (20° to 30° C). Seeds of these species do not germinate in the cool fall season and not until late in the spring. Such seeds can be handled without difficulty either by fall planting or by moist winter storage and spring planting.

Group III includes those species (black locust, Colorado and Norway spruce, Chinese and Siberian elm, European larch, bristlecone pine, and Douglas fir) whose seeds are not temperature sensitive and are able to germinate over a range of 60° to 90° F (15° to 32°C). Some of these species require moist-chilling or have hard coats requiring special treatment. If fall-planted, seeds of this group will usually germinate prematurely and the seedling plants can be injured by winter cold. They are best planted in the spring, the actual time to plant not being as critical as that for the other two groups.

Other species with special requirements have been described earlier in this chapter. Planting schedules for these seeds should be geared to their particular requirements.

Fall planting is particularly useful where a large volume of seeds is to be handled and adequate cold storage facilities are not available. The seasonal labor requirements is distributed more uniformly throughout the year if some of the planting is done in the fall. As soon as the environmental conditions are favorable in the spring, germination takes place, and is not delayed by late planting, resulting in better stands and a longer growing season. On the other hand, losses sometimes result from unfavourable winter conditions, drying, disease, or rodents. Also, herbicidal weed control is necessary.

Seed-Bed Culture

Seedlings of many species are grown in special out-of-doors seed-beds during the first 1 or 2 years in the nursery. A common size is 3½ to 4 ft. in width with a walkway between seed-beds, the length varying with the size of the operation. In some cases, sideboards are placed alongside the seed-bed after sowing to maintain the shape of the bed and to provide support for glass frames or lath shade.

Seed may either be broadcast over the surface of the bed or drilled into closely spaced rows. Regulating the density of planting is one of the major problems in planting a seed-bed. For economical use of space, seeds should be planted as closely together as feasible. On the other hand, over-crowding leads to greater difficulty with damping-off and reduces vigor and size of the seedling resulting in thin, spindly trees and small root systems. Seedlings with these characteristics do not transplant well.

The optimum density depends on the species and on the purposes of propagation. In other words, if a high percentage of the seedlings are to reach a desired size for field planting or for use as rootstocks for grafting, lower densities might be desired. If the seedlings are to be transplanted into other beds for additional growth, higher densities (with smaller seedlings) might be more practical. Once the actual density is determined, the necessary rate of sowing can be calculated from certain data obtained from a germination test and from the experiences of the operator at that particular nursery. The following formula is used.

$$W = \frac{A \times S}{D \times P \times G \times L}$$

where W = weight of seeds in pounds to sow per given area.

A = area of seed-bed in square feet.

S = number of living seedlings desired per square foot.

D = average number of seeds per pound at the moisture content at which the seeds are sown.

P = average purity percentage of seeds (expressed as a decimal).

G = average effective germination percentage in the laboratory (expressed as a decimal).

L = average percentage of germinating seeds that will be living trees at the end of the season (expressed as a decimal). This is a correction factor which takes into account the expected losses which experience at that nursery indicates will occur with that species.

Seeds of a particular lot should be thoroughly mixed before planting to be sure that the density in the seed-bed will be uniform. Treatment with a fungicide for control of damping-off is desirable. Small conifer seeds may be pelleted for protection against disease, insects, birds, and rodents. Seeds are planted either by hand or by machine. Depth of planting varies with the size of the seed. In general, a depth of about two to four times the thickness of the seed is a fairly safe estimate, but this also varies with the kind of seed. During early stages of development, seedlings generally must be protected against drying, heat, os cold. In the case of tender plants, glass frames could be placed over the beds, although with most plants a lath shade is sufficient. With some plants, shade should be provided all through the first season, with others, shade is necessary only during the first part of the season. Sprinkling with water to reduce ground temperature during the hot part of the day is useful.

A mulch applied to the seed-bed after planting helps to protect against drying, crusting, and cold, and discourages weed-growth. It is particularly desirable with fall-sown seeds, for seeds which are to remain for long-periods in the seed-bed before germination, or for seeds sown in cold areas. Materials which have been used include sawdust, burlap, pine litter, straw, hay. ground corncobs, wood shavings, sandpaper, canvas, shredded pine cones, sand, manure, and snow. One nurseryman reports the use of old rooting media of sand and peat as a seed-bed covering. The use of polyethylene film as a seed-bed cover has been described. In this case, the seed-bed was sown in the fall, covered lightly with sawdust, well watered, and then covered with a sheet of polythylene film, on top of which burlap was placed. The following spring the film was removed as soon as germination occurred.

During the first year in the nursery, it is important to keep the seedlings growing continuously without any check in development. A continuous moisture supply, cultivation to control weeds, and proper disease and insect control contribute to successful seedling growth. Fertilization with nitrogen is usually necessary, particularly when a mulch has been applied. Decomposition of organic material can produce nitrogen deficiency.

Weed control can be facilitated by careful seed-bed preparation, cultivation, and chemical sprays. Three types of chemical controls are available. *Pre-planting* fumigation with methyl bromide, chloropicrin, DD, Vapam, steam, or the like is effective and also kills disease organisms and nematodes. *Pre-emergence* weed sprays are applied to the soil before crop seeds emerge. *Post-emergence* weed sprays are applied after the crop or nursery seedlings have emerged and are growing. A wide range of selective and nonselective commercial products are available. However, such materials should be used with caution since improper use can cause injury to the young plants. Not only should directions of the manufacturer be followed, but preliminary trials should be made before large-scale use.

The seedlings may remain in the seed-bed for 1 to 3 years, depending upon the kind of plant. Many plants are dug at the end of the first year and placed in *transplant beds* or "lined-out" in the nursery bed for further development. Deciduous plants are planted 4 to 6 in. apart in nursery rows 18 in. to 4 ft apart. Nursery plants in rows are often undercut at the end of 1 year to stimulate subsequent production of a more fibrous root system. Conifers usually go into transplant beds similar to the seed-bed but at wider spacing. Spacing may be several inches apart in rows 6 to 8 in. apart. Other cases involve field planting directly from the nursery.

Seedlings produced in a nursery are often designated by numbers to indicate the length of time in a seed-bed and the length of time in a transplant bed. For instance, a designation of 1-2 means a seeding grown 1 year in a seed-bed and 2 years in a transplant bed or field. Similarly, a designation of 2-0 means a seedling produced in 2 years in a seed-bed and none in a transplant bed.

Direct Nursery Planting

Seeds of many deciduous trees are planted directly in nursery rows rather than in special seed-beds. This procedure is used to produce root-stocks for deciduous fruit and nut varieties. Where plants are to be budded or grafted in place, the width between rows is about 4 ft and the seeds are planted 3 to 4 in. apart in the row. If the seeds show poor germinability, the seeds must be planted closer together to get the desired stand of seedlings. Large seed (walnut) are usually planted 4 to 6 in. deep, medium-sized seed (apricot, almond, peach, and pecan) about 3 in. and small (myrobalan plum) about 1½ in. This may vary with soil type. Plants to be grown as seedlings without budding could be spaced at closer intervals and in rows closer together. If germination is low, and a poor stand results, the surviving trees may, owing to the wide specing, grow too large to be suitable for budding.

Usually, no shading is required by deciduous tree plants grown in the nursery row, although other cultural operations are similar to seed-bed culture. Plants generally remain in place in the nursery row until they are to be transplanted to their permanent location.

A variation of this procedure is sometimes followed with species that germinate well but require some time to grow to budding size. Seeds are either planted close together to get a stand of 10 to 15 seedlings per ft, or planted in a narrow band about 4 in. wide in the nursery row. The seedlings remain relatively small during the first season, after which they are lifted and lined-out in a nursery row. Cherry, apple, or pear seedlings are often produced in this manner.

6

Physiology of Asexual Propagation

Asexual propagation involves reproduction from vegetative parts of plants and is possible because the vegetative organs of many plants have the capacity for regeneration. Stem cuttings have the ability to form adventitious roots. Root cuttings can regenerate a new shoot system. Leaves can regenerate new roots and new shoots. A stem and a root (or two stems) can be grafted together to form a continuous vascular connection when properly combined as a graft.

New plants can start from a single cell. Cells of tobacco pith and carrot root in aseptic cultures have regenerated entire plants identical with the one from which the original cultures were made. Any living cell of the plant appears to have all the genetic information needed to regenerate the complete organism.

Reasons for Using Asexual Propagation

Asexual propagation reproduces *clones*. Such propagation involves mitotic cell division in which there is duplication (usually) of the complete chromosome system and associated cytoplasm from the parent cell to the two daughter cells. Consequently plants propagated vegetatively reproduce all the genetic information of the parent plant. This is why the unique characteristics of any single plant are perpetuated in the establishment of a *clone*. The asexual process is, particularly important in horticulture because the genetic makeup (genotype) of most of the valuable fruit and ornamental cultivars is highly heterozygous, and the unique characteristics of such plants are immediately lost if they are propagated by seed.

Asexual propagation is necessary to grow cultivars that produce no viable seeds, such as some bananas, figs, oranges, and grapes.

Propagation of some species may be easier, more rapid, and more economical by vegetative methods than by seed. Cotoneaster seed has complex dormancy conditions ; leafy stem cuttings, on the other hand, root rapidly and in high percentages. Seedlings of some species grow more slowly than rooted cuttings.

Some plants grown from seed have an extended *juvenile* period and during this time the plant may not only fail to produce flowers and fruit, but may exhibit other undesirable morphological features (*e.g.*, thorniness) that are not present when propagation is by material in the adult stage. On the other hand, it may be desirable to maintain this juvenile stage indefinitely so as to facilitate propagation of difficult-to-root cuttings.

The Clone

Many fruit and ornamental cultivars are groups of plants propagated vegetatively, starting with an individual plant, usually one grown from seed, or a part of a plant, *e.g.*, a *bud sport*. Such a group of plants taken collectively has been given the name *clone*. A *clone* can be defined as "genetically uniform material derived from a single individual and propagated exclusively by vegetative means, such as cuttings, divisions, or grafts". A discovery with any individual member of the clone, *e.g.*, a method of propagation, a certain cultural practice, a method of disease control, or a cross-pollination requirement, applies in the same manner to all other members of that clone.

When a member plant of a certain clone requires cross-pollination to set fruit, it must be pollinated by a plant of a different clone, rather than by a different plant of the same clone, which in reality is just another part of the same plant.

Many clones of horticultural interest have been discovered and perpetuated by man. The 'Bartlett' pear clone originated from seed in England about 1770 and has been maintained asexually ever since. The 'Winesap' apple clone originated similarly some 200 years ago. The original seedling tree has probably been dead for over 100 years. Buds taken from it and budded to rootstocks produced new trees having tops of the same genetic constitution as the original seedling tree. By continuing this process of asexual propagation, thousands upon thousands of trees have since been grown whose budded tops, collectively, make up the 'Winesap' apple clone.

Clones exist in nature, reproducing naturally by such structures as bulbs, rhizomes, runners, stolons and tip layers. The species *Lilium tigrinum* is said to be a single clone. Apomixis in some species of Rosaceae, Gramineae, and Compositeae makes possible the reproduction of some clones naturally by seeds. A clone can perpetuate itself successfully in nature, sometimes better than can seed-propagated plants, as long as the environment remains reasonably constant. If the environment changes drastically, however, a clonally reproduced species will be at a disadvantage because it has no opportunity to evolve forms better adapted to the new environment. Similarly, cultivated clones have a disadvantage in that an adverse situation, such as disease or insect attack, can affect all members of the clone equally and may even destroy it.

The concept of the clone, however, does not mean that all individual members are necessarily identical in all characteristics. The actual appearance and behaviour of a plant, *i.e.*, its *phenotype*, results from the interaction of its genes (*genotype*) with the *environment* in which the plant is growing. Consequently, within a given clone the appearance of the plants, or of fruits and flowers on different plants, may vary some what because of climate, soil, disease, and the like. In many years

'Bartlett' pear trees grown in California produce round, apple-shaped fruit, whereas 'Bartlett' pears grown in Washington and Oregon are relatively long and narrow; this difference is believed to be due to different climatic factors. Variation in fruit shape within any one area and even within an orchard or tree may also be related to the presence, the absence, or the number of seeds within the fruit. Some plants produce leaves of different appearance when growing in shade than when growing in the sun. Some water plants produce leaves quite different in appearance on the part of the plant submerged than on the part in the air. Within a single orchard, fruit trees of the same clone often differ to some extent because of differences in soil, water availability, rootstock, or competition from surrounding plants. Although the environment can modify the growth and appearance of individual members of a clone, such changes are not permanent, since the genotype of the plants is unaffected by the environmental modifications. Further in this chapter several causes of more or less permanent changes within clones are discussed.

The life of the clone is theoretically unlimited. An early belief, that has long prevailed, is that a clone deteriorates with age and can only be rejuvenated by seed propagation. Evidence now shows that modification, sometimes leading to deterioration, can indeed occur in particular clones, and must be guarded against in propagation. The most significant factor seems to be virus infection. In fact, virtually any clone that is grown for any period of time is likely to become infected sooner or later, its success being determined by its ability to tolerate the virus. Genetic changes (mutations) may occur in the clone which, although not always degenerative in a strict sense, can produce off-type individuals that will reduce the value of the clone. An unfavourable environment may lead to progressive deterioration of a clone. For example, lack of sufficient winter chilling for certain strawberry cultivars, or an insufficiently long post-bloom vegetative period for some bulbous plants may lead to gradual decline in vigor and productivity. If a clone is maintained in a proper environment, and procedures are carried out whereby viruses, other pathogens, and off-type mutants are eliminated, a clone could be perpetuated indefinitely. Such procedures, described later in this chapter, are an important aspect of plant propagation.

Another belief, proposed by some people, is that permanent genetic changes in the clone can be induced by the environment or by the influence of one genetically different plant on another if the two are grafted together. Although some aspects of the latter theory have not perhaps been examined critically, the prevailing belief among most scientists is that such permanent genetic changes in clonal material do not occur.

Changes in Clones Associated with Age

The growth cycle of a seedling plant involves a change from a *juvenile* phase to an *adult* phase. The two phases may be distinguished by distinctive morphological and physiological differences. The juvenile-to-adult change may be referred to as *ontogenetic* because more or less permanent changes occur within the apical meristem as it grows, but the basic genetic information in the cells is not altered. New seedlings revert to the juvenile phase and undergo the same transition to the adult phase as did the parent. Evidence that no permanent genetic change is involved is found with nucellar seedlings, which are asexually produced by apomixis and reproduce genetically the same parent plant but only after having gone through the transitional juvenile phase. The juvenile-to-adult change is not the same as the vegetative to-reproductive change, although the two kinds of changes may be correlated. Likewise, the development of senescence is different from either of these phenomena.

The juvenile and adult phases may differ distinctly in appearance or they may show transitional development from one phase to the other. Leaf shape is a common means of identifying phase changes such as illustrated in *Acacia* and *Eucalyptus*. The juvenile phase of many plants—for example, citrus, pear, apple and honey locust—is characterized also by nonflowering, excessive vigor, and thorniness ; the adult phase by flowering and fruiting, reduced vigor, and thornlessness. The juvenile phase of a number of plants, of which English ivy (*Hed-*

era helix) is a classic example, is a trailing vine with alternate palmate leaves. The mature, flowering form is an erect or semi-erect shrub with entire, ovate leaves, produced oppositely around the stem. In certain conifers, such as junipers, the juvenile forms produce needle-shaped leaves, but the adult produces scale-like leaves.

Cuttings taken from juvenile parts of many perennial plants can reproduce adventitious shoots and roots ensily, whereas cuttings taken from the mature phase of the same plant are much less, if at all, capable of forming adventitious roots or shoots.

The ontogenetic change from juvenile to adult as the seedling grows older occurs in the somatic (vegetative) cells and results in differences in the apical meristem in separate parts of the plant at various stages of development. Prevailing evidence suggest that a plant must attain a certain size before the adult phase appears. Consequently the growing points produced in different parts of an individual plant where such ontogenetic changes are taking place may differ considerably in their ontogenetic age. This explains why a portion of an individual plant may be in the juvenile phase and another portion in the adult phase.

The phenomenon that different growing points at different parts of a plant may perpetuate particular phases (juvenile or mature) it used in propagation has been called *topophysis*. For example, buds taken from the lower, juvenile portion of seedling trees of pear, apple, citrus, or honey locust, if used in budding, will produce nursery trees that are vigorous, thorny, and slow to flower ; buds from the upper portion of the same tree can be used to produce nursery trees that are less vigorous, smooth-barked, and thornless, and that flower quickly. By proper selection of propagation material from a seedling plant, it is possible to propagate leaf persistent (juvenile) or deciduous (adult) beech trees, or to propagate fungus (*Keithia thusima*)-resistant (juvenile) or fungus-susceptible (adult) forms of *Thuja plicata*. By such selection of propagating material, the juvenile forms of many conifers can be maintained almost indefinitely ; this can also be done in certain vines, such as *Hedera*. Another type of topophysis occurring in some plants in that the direction of shoot growth produced by a cutting may differ, depending upon whether the cutting was taken from an upright or a lateral branch. For example, cuttings of *Araucaria*, or of coffee, taken from lateral shoots will continue to grow in a horizontal direction, whereas cuttings made from upright shoots of the same plant will develop into upright plants.

Juvenile-to-adult phase changes are important in the initial establishment of a clone from a seedling plant. Various horticultural techniques are designed to speed this transition. To attain the adult phase in the shortest time, selection of propagation wood from the upper, onto-genetically older part of the plant is desirable. In most well-established clones, however, considerable vegetative propagation has already taken place, a flowering and fruiting pattern has become set, and the adult phase has been attained. Consequently plants propagated from material taken from different parts of most grafted or rooted plants do not exhibit such differences. However, sometimes juvenile characteristics can persist for several vegetative generations.

In propagation by rooting cuttings it may be desirable to preserve the juvenile phase of the clone. This can be done to some extent by selection of propagation material, since juvenility tends to persist in the lower part of the plant, particularly in the crown and roots. Likewise, adventitious buds from roots or from heavily cut-back stems, or from "sphaeroblasts" (wart-like growths containing meristematic and conductive tissue) also tend to produce juvenile growth. The success in propagation by stooling is believed to be due to the perpetuation of the juvenile form.

Budsports

A branch which shows changes in one or more inheritable characters that can be perpetuated by asexual means in termed a *bud sport*. Bud sports can originate by any of the somatic mutations or chromosomal changes mentioned earlier. Many of these sports are chimeras and are discovered when a lateral shoot develops from a mutated area. In addition, a bud sport may result when there is a rearrangement of tissues of a chimera such that a shoot may arise from a deeper layer, either from the displacement of epidermal cells with cells of the L-II layer, or by formation of adventitious buds.

Somatic variants such as these are starting points of new clones and, if introduced as cultivars, are given names. Numerous bud sports have been discovered, and a great many more have undoubtedly gone undetected. Some have become important new clones in both fruit and ornamental species. In Florida, pink-fleshed grapefruits were found on a single branch of a tree in a grove of thousands of trees producing white-fleshed grapefruits : an obvious mutation. The seedless 'Washington Navel' orange probably arose as bud mutation of the Brazilian orange, 'Laranja Selecta'. The first citrus fruit to be patented originated in 1929 as a mutation : the red-fleshed 'Ruby' grapefruit. Mutations in apples are common, resulting in changes in fruit colour, size, and shape.

Polyploidy may give rise to "giant" sports in certain plants. Individual tetraploid grapevines occur in vineyards, growers referring to such vines with prodigious growth habits and poor fruitfulness as "bulls," "males," or "giants". Chromosome counts have shown them to be often only partial mutations, (periclinal chimeras) consisting of a single outer layer of diploid cells surrounding the mutated tetraploid inner tissues. Such giant sport canes in the grape are observed to arise at pruning wounds or near areas where a bud or shoot has been injured or killed. These mutant shoots seem to arise from deeply embedded dormant bud initials.

Spontaneous polyploidy also occurs in apples, producing giant sports. Fruits of these sports are usually larger, flatter, and more irregular in shape than those from the parent plant. Trees bearing such fruits are vigorous, with thick twigs, often having wide-angled crotches, which gives them a low, flat shape.

Certain of such giant apple sports are diploid-tetraploid periclinal chimeras, with a $2n$ layer (34 chromosomes per cell), one, two, or three cell layers in thickness, over a $4n$ tetraploid (68 chromosomes per cell) interior. A certain 'Delicious' apple giant sport was found to be tetraploid in all internal tissues, but diploid in the epidermis.

Many (possibly most) mutations produce sports inferior to the parent plant in one or more characteristics. If detected by the propagator, such sports are discarded. Sometimes, as in citrus, these may be propagated unknowingly and cause serious economic loss.

Graft Chimeras

While most chimeras occur naturally, they can also be established artificially by grafting. If scions are cut back severely, almost to the stock, adventitious buds may sometimes arise from the callus in the region of the graft union. Occasionally, a bud may develop which contains tissues of both stock and scion. The resulting structure is a chimera, in which the cells of the two graft components remain distinct regardless of how intermingled they become.

Winkler artificially produced in Germany many graft chimeras, using species in the Solanaceae family. He prepared wedge or saddle grafts of tomato (*Lycopersicon esculentum*) on black nightshade (*Solanum nigrum*) and black nightshade on tomato, both of which are easily grafted and produce callus readily. Winkler's grafts were made. Following union, the graft was cut transversely at the junction, exposing tissues of both components. The cut surface soon became covered with a pad of callus derived from both the tomato and nightshade. From this callus, adventitious buds formed which developed into shoots. Most of these shoots had the appearance of either pure tomato or pure nightshade. Occasionally, however, shoots arose and grew normally which were different from either component. Some showed mixed characters, with half the shoot resembling tomato and the other half the nightshade. Winkler gave it the name "chimera," after the fabulous mythological monster that was part lion and part dragon. Graft chimeras in the periclinal form also developed. One was found to consist mostly of nightshade tissue covered with a single layer of tomato epidermal cells. It bore fruit like that of the true nightshade and its seeds produced nightshade plants.

Graft chimeras among woody perennials have long been known to exist, and some have been perpetuated indefinitely by asexual methods. Probably the earliest known "authentic" chimera is one called the "bizzarria" orange. It is described as bearing a fruit which is citron on one side and orange on the other. It originated supposedly in 1644 in a garden in Florence, Italy, from a grafting operation in which a scion of sour orange (*Citrus aurantium*) was grafted on a stock of citron (*C. medica*). According to the story, the original scion failed to grow, but in the swollen callus around the graft a shoot developed bearing the unusual "bizzarria" orange. It was propagated asexually, and plants were widely distributed, leading to much controversy as to the true nature of its origin and composition. Scientific articles discussing the "bizzarria" orange were still being published 250 years after its origin, agreeing that it probably was a periclinal chimera.

Grafting medlar (*Mespilus germanica*) on a stock of the hawthorn (*Crataegus monogyna*) has led to several recorded instances of chimeras. Near Metz, France, in 1899, two adventitious shoots were found arising at the graft union of an old medlar tree. They had a different appearance from either the medlar or the hawthorn and were different from each other. They were named *Crataegomespilus asnieresi* and *Crataegomespilus dardari* and were regarded for many years as true "*graft hybrids,*" believed to have resulted from a fusion of the nuclei of vegetative cells. Both of these types, sometimes called haw-medlars, have been maintained vegetatively, grafted on the hawthorn and widely distributed to botanical gardens. It is likely, however, that these two forms are periclinal chimeras.

There have been several strong proponents of the graft hybrid hypothesis. Winkler made his Solanaceae graft chimeras in the hope of gaining evidence for the graft hybrid idea. He believed that some of these unusual plant forms were true graft hybrids involving fusion of vegetative cell nuclei. Daniel is probably the most conspicuous supporter of this hypothesis and has written extensively on the subject although many of his views are not generally accepted.

7

Propagation by Cuttings

In propagation by cuttings, a portion of a stem, root, or leaf is cut from the parent plant, after which this plant part is placed under certain favourable environmental conditions and induced to form roots and shoots, thus producing a new independent plant which, in most cases, is identical with the parent plant.

The Importance and Advantages of Propagation by Cuttings

This is the most important method of propagating ornamental shrubs—deciduous species as well as the broad-and narrow-leaved types of evergreens. Cuttings are also used widely in commercial greenhouse propagation of many florists' crops and are commonly used in propagating several fruit species.

For species that can be easily propagated by cuttings, this method has numerous advantages. Many new plants can be started in a limited space from a few stock plants. It is inexpensive, rapid, and simple, and does not require the special techniques necessary in grafting or budding. There is no problem of compatibility with rootstocks or of poor graft unions. Greater uniformity is obtained by absence of the variation which sometimes appears owing to the variable seedling rootstocks of grafted plants. The parent plant is usually reproduced exactly with no genetic change.

It is not always desirable, however, to produce plants on their own roots by cuttings even if it is possible to do so. It is often advantageous or necessary to use a rootstock resistant to some adverse soil condition or soilborne organism, or to utilize available dwarfing or invigorating rootstocks.

Types of Cuttings

Cuttings are almost always made from the vegetative portions of the plant, such as stems, modified stems (rhizomes, tubers, corms and bulbs), leaves, or roots. The reproductive parts of the plant are not ordinarily used, although such parts as the ovary, pedicel and petals, have been reported to form roots.

Several types of cuttings can be made ; these are classified according to the part of the plant from which they are obtained :

- Stem cuttings
  - Hardwood
    - Deciduous
    - Narrow-leaved evergreen
  - Semi-hardwood
  - Softwood
  - Herbaceous
- Leaf cuttings
- Leaf bud cuttings
- Root cuttings

Many plants can be propagated by several of these different types of cuttings with satisfactory results. The type used would depend upon the individual circumstances, the least expensive and easiest usually being selected.

If the particular plant being propagated will root well by hardwood stem cuttings in an outdoor nursery, this method is ordinarily used, because of its simplicity and low cost. Root cuttings of some species are also satisfactory, but cutting material may be difficult to obtain in large quantities. For species more difficult to propagate, it is necessary to resort to the more expensive and elaborate facilities required for rooting the leafy types of cuttings.

In selecting cutting material it is important to use stock plants that are free from diseases, moderately vigorous and productive, and of known identity. Stock plants that are diseased or injured by frost or drought, that have been defoliated by insects or diseases, that have been stunted by excessive fruiting, or that have made rank, overly vigorous growth should be avoided.

A commendable practice for the propagator is the establishment of stock blocks as a source of propagating material, where uniform, true-to-type, pathogen-free mother plants can be maintained and held under the proper nutritive condition for the best rooting of cuttings taken from them.

Stem Cuttings

This is the, most important type of cutting and can be divided into four groups, according to the nature of the wood used in making the cuttings : *hardwood, semi-hardwood, softwood,* and *herbaceous.* In propagation by stem cuttings, segments of shoots containing lateral or terminal buds are obtained with the expectation that under the proper conditions adventitious roots will develop and thus produce independent plants.

The type of wood, the stage of growth used in making the cuttings, the time of year in which the cuttings are taken, and several other factors can be very important in securing satisfactory rooting of some plants.

Hardwood Cuttings (Deciduous Species)

This is one of the least expansive and easiest methods of vegetative propagation. Hardwood cuttings are easy to prepare, are not readily perishable, may be shipped safely over long distances if necessary, and require no special equipment during rooting.

The cuttings are prepared during the dormant season—late fall, winter, or early spring—from wood of the previous reason's growth, although with a few species, such as the fig or olive, two-year-old or older wood is used. Hardwood cuttings are most often used in propagation of deciduous woody plants, although some broad-leaved evergreens, such as the olive can be propagated by leafless hardwood cuttings. Many deciduous ornamental shrubs are started readily by this type of cutting. Some common ones are privet, forsythia, wisteria, honeysuckle, and spiraea. Rose rootstocks, such as *Rosa multiflora* are propagated in great quantities by hardwood cuttings. A few fruit species are propagated commercially by this method—for example fig, quince, olive mulberry, grape, currant, gooseberry, pomegranate, and some plums.

The propagating material for hardwood cuttings should be taken from healthy, vigorous stock plants growing in full sunlight. The wood selected should not be from extremely rank growth with abnormally long internodes, or from small, weakly growing interior shoots. Wood of moderate size and vigor is the most desirable. The cuttings should have an ample supply of stored foods to nourish the developing roots and shoots until the new plant becomes self-sustaining.

Hardwood cuttings vary considerably in length—from 4 to 30 in. Long cuttings, when they are to be used as rootstocks for fruit trees, permit the insertion of the varietal bud into the original cutting following rooting, rather than into a smaller new shoot arising from the original cutting.

At least two nodes are included in the cutting, the basal cut is usually just below a node and the top cut ½ to 1 in. above a node. However, in preparing stem cuttings of plants with short internodes, little attention is ordinarily given to the position of the basal cut, especially when quantities of cuttings are prepared and cut to length, many at a time, by a band saw or paper cutter.

The diameter of the cuttings may range from ¼ in. to 1 or even 2 in., depending upon the species. The "mallet," the "heel", and the straight cutting are three different cuttings. The mallet includes a short section of stem of the older wood, whereas the heel cutting includes only a small piece of the older wood. The straight cutting, not including any of the older wood, is most commonly used, giving satisfactory results in most instances.

Where it is difficult to distinguish between the top and base of the cuttings, it is advisable to make all the basal cuts at a slant rather than at right angles. In large-scale operations, bundles of cutting material are cut to the desired lengths by band saws or other types of mechanical cutters rather than individually by hand.

Cuttings of the first (*Abies sp.*) and pines (*Pinus sp.*) are very difficult to root. In addition, there is considerable variability among the different species in these genera in regard to the ease of rooting of cuttings. As with many other plants, cuttings taken from young seedling stock plants root much more readily than those taken from older trees. Treatments with root-promoting substances particularly indolebutyric acid at relatively high concentrations, are usually beneficial in increasing the speed of rooting and the percentage of cuttings rooted, and in obtaining heavier root systems.

Narrow-leaved evergreen cuttings are best taken between late fall and late winter. Rapid handling of the cuttings after the material is taken from the stock plants is important. The cuttings are best rooted in a greenhouse with relatively high light intensity and under conditions of high humidity or very light misting but without heavy wetting of the leaves. A bottom heat temperature of 75° to 80° F (24° to 26.5° C) has given good results. Dipping the cuttings into a fungicide helps prevent disease attacks. Sand alone has given good results as a rooting medium, as has a 1 : 1 mixture of perlite and peat moss.

Some individual cuttings take longer to root than others. The slower-rooting ones can be restuck in the rooting medium, and often will root eventually.

The type of wood to use in making the cuttings varies considerably with the particular species being rooted. As shown in Fig. 7.1, the cuttings are made 4 to 8 in. long with all the leaves removed from the lower half of the cutting. Mature terminal shoots of the previous season's growth are usually used. In some instances, such as with *Juniperus chinensis pfitzeriana,* older and heavier wood can also be used, thus resulting in a larger plant when it is rooted. On the other hand, some nurserymen use small tip cuttings, 2 to 3 in. long, placed very close together in a flat for rooting. In some species, such as *Juniperus excelsa*, older growth taken from the

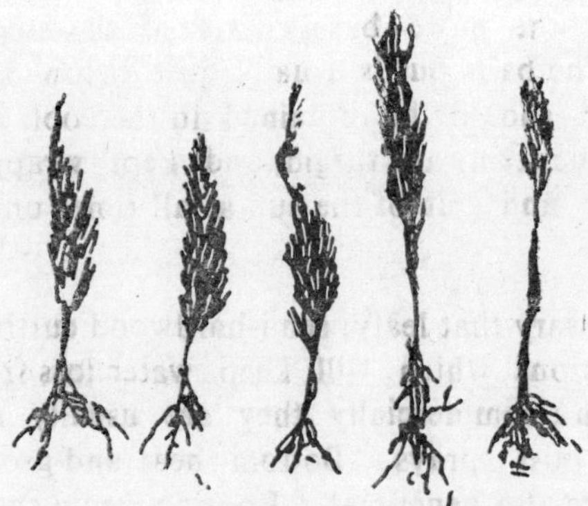

Fig. 7.1. Narrow-leaved evergreen cuttings. These rooted juniper cuttings are about 5 in. long.

sides and lower portion of the stock plant roots better than the more succulent tips. Cuttings of *Taxus* root best if they are taken with a piece of old wood at the base of the cutting. Such cuttings are also less subject to fungus attacks. In certain of the narrow-leaved evergreen species, some type of basal wounding is sometimes beneficial in inducing rooting.

Semi-Hardwood Cuttings

Cuttings of this type are usually made from woody, broad-leaved evergreen species, but leafy summer cuttings taken from partially matured wood of deciduous plants could also be considered as semi-hard-wood. Cuttings of broad-leaved evergreen species are generally taken during the summer months from new shoots just after a flush of growth has taken place and the wood is partially matured. Many ornamental shrubs, such as camellia, pittosporum, euonymus, the evergreen azaleas, and holly are commonly propagated by semi-hardwood cuttings. A few fruit species, such as citrus and olive, can, also be started in this manner.

The cuttings are made 3 to 6 in. long with leaves retained at the upper end but removed from the lower end. If the leaves are very large, they should be reduced in size to lower the water loss and to allow closer spacing in the cutting bed. The terminal ends of the shoots are often used in making the cuttings, but the more basal parts of the stem will usually root also. The basal cut is usually just below a node. The cutting wood should be obtained in the cool, early morning hours when the stems are turgid, and kept wrapped in clean moist burlap and out of the sun at all times until the cutting are made.

It is necessary that leafy, semi-hardwood cuttings be rooted under conditions which will keep water loss from the leaves at a minimum ; commercially they are usually rooted under intermittent mist sprays. Bottom heat and growth regulator treatments are also beneficial. Rooting media such as a 1 : 1 mixture of perlite and peat moss of perlite and vermiculite give satisfactory results.

Softwood (Greenwood) Cuttings

Cuttings prepared from the soft, succulent, new spring growth of deciduous or evergreen species may properly be classed as softwood cuttings. Many ornamental woody shrubs can be started by softwood cuttings. Typical examples are the hybrid French lilacs, forsythia, magnolia, weigela, and spiraea. Some deciduous ornamental trees such as the maples can also be started in this manner. Although fruit tree species are not commonly propagated by softwood cuttings, apple, peach, pear, plum, apricot, and cherry will root, especially under mist.

Softwood cuttings generally root easier and quicker than the other types but require more attention and equipment.

This type of cutting is always made with leaves attached. They must, consequently, be handled carefully to prevent drying, and be rooted under conditions which will avoid excessive water loss from the leaves. Temperature should be maintained during rooting at 75° to 80° F (23° to 27° C) at the base and 70° F (21° C) at the leaves for most species. Softwood cuttings produce roots in 2 to 4 or 5 weeks in most cases. In general, they respond well to treatments with root-promoting substances.

It is important in making softwood cuttings to obtain the proper type of cutting wood from the stock plant. This will vary greatly, however, with the species being propagated. Extremely fast-growing, soft, tender shoots generally are not desirable, as they are likely to deteriorate before rooting. At the other extreme, old, woody stems are slow to root in most cases. The best cutting material has some degree of flexibility, but is mature enough to break when bent sharply. Weak, thin, interior shoots should be avoided as well as vigorous, abnormally thick, or heavy ones. Average growth from portions of the plant in full light is the most desirable to use. Some of the best cutting material is the lateral or side branches of the stock plant. Heading back the main shoots will usually force out numerous lateral shoots from which cuttings can be made. Softwood cuttings ordinarily are 3 to 5 in. long with two or more nodes. The basal cut is usually made just below a node. The leaves on the lower portion of the cutting are removed, but those at the upper portion are retained. If the upper leaves are very large, they should be reduced in size to lower the transpiration rate and to use less space in the propagating bed, although for the best rooting it is desirable to retain the maximum leaf area that is possible without wilting. All flower buds should be removed. In some large nurseries where quantities of cuttings are prepared, bundles of cutting material are rapidly chopped into uniform lengths by paper cutters.

The cutting material is best gathered in the early part of the day and should be kept moist, cool, and turgid at all times by wrapping in damp, clean burlap or placing in large polyethylene bags (kept out of the sun). Laying the cutting material or prepared cuttings in the sun for even a few minutes will cause serious damage. Soaking the cutting material or cuttings in water to keep them fresh is undesirable.

Herbaceous Cuttings

This type of leafy cutting is made from such succulent, herbaceous greenhouse plants as geraniums, chrysanthemums, coleus, and carnations. They are 3 to 5 in. long with leaves retained at the upper end. Most florists' crops are propagated by herbaceous cuttings. They are rooted under the same conditions as softwood cuttings, requiring high humidity. Bottom heat is also helpful. Under proper conditions, rooting is rapid and in high percentages. Although root-promoting substances are not required, they are often used to gain uniformity in rooting and development of heavier root systems. Herbaceous cuttings of some plants that exude a sticky sap, such as the geranium, pineapple, or cactus, do better if the basal ends are allowed to dry for a few hours in the air before they are inserted in the rooting medium. This allows the wounded tissues to dry, which tends to prevent the entrance of decay organisms.

Leaf Cuttings

In this type of cutting, the leaf blade or leaf blade and petiole, are utilized in starting a new plant. In most cases, adventitious roots and an adventitious shoot form at the base of the leaf. Rarely does the original leaf becomes a part of the new plant.

One type of propagation by leaf cuttings is illustrated by *Sansevieria*. These leaf pieces are inserted three-fourths of their length into sand, and after a period of time a new plant forms at the base of the leaf piece. The original leaf cutting does not become a part of the new plant. The variegated form of *Sansevieria*, *S. trifasciata laurenti*, is an example of a periclinal chimera which will not reproduce true to type from leaf cuttings ; to retain its characteristics, it must be propagated by division of the original plant.

In starting plants with thick, fleshy leaves, such as *Begonia rex*, by leaf cuttings, the large veins are cut on the under-surface of the mature leaf, which is then laid flat on the surface of the propagating medium. The leaf is pinned or held down in some manner, with the natural upper surface of the leaf exposed. After a period of time under humid conditions, new plants will form at the point where each vein was cut. The old leaf blade will gradually disintegrate.

Another method, sometimes used with the fibrous-rooted begonias, is to cut large, well-matured leaves into triangular sections each containing a piece of a large vein. The thin outer edge of the leaf is discarded. These leaf pieces are then inserted upright in sand with the pointed end down. The new plant develops from the large vein at the base of the leaf piece. The original leaf dies, not becoming a part of the new plant.

The African violet (*Saintpaulia*) is typical of leaf cuttings which can be made of an entire leaf (leaf blade plus petiole), the leaf blade only, or just a portion of the leaf blade. The new plant forms at the base of the petiole or midrib of the leaf blade.

Leaf cuttings should be rooted under the same conditions of high humidity that are used for softwood or herbaceous cuttings. Root-promoting chemicals are usually helpful.

Leaf-Bud Cuttings

This type of cutting consists of a leaf blade, petiole, and a short piece of the stem with the attached axillary bud.

Such cuttings are of particular value for species that are able to initiate roots but not shoots from detached leaves. In such cases, the axillary bud at the base of the petiole provides for the essential shoot formation. A number of plant species such as the black raspberry (*Rubus occidentalis*), blackberry, boysenberry, lemon, camellia, and rhododendron can be readily started by leaf-bud cuttings, as well as many tropical shrubs and most herbaceous green-house plants usually started by stem cuttings. Red raspberries (*Rubus idaeus*) apparently do not reproduce by leaf-bud cuttings.

This method is particularly valuable when propagating material is scarce, because it will produce at least twice as many new plants from the same amount of stock material as can be started by stem cuttings. Each node can be used as a cutting. Leaf-bud cuttings should only be made from material having well-developed buds and healthy, actively growing leaves.

Treatment of the cut surfaces with one of the root-promoting substances may stimulate root production. The cuttings should be inserted in the rooting medium with the bud as soon as the plants become well established with good root development they can be transplanted to their permanent location.

Plants with Large Roots, Propagated Out-of-Doors

Root cuttings of this type are made 2 to 6 in. long. They are tied in bundles, care being used to keep the same ends together to avoid planting upside down later. The cuttings are then packed in boxes of damp sand, sawdust, or peat moss for about 3 weeks. During this storage period the cuttings should be kept cool, about 40° F (4.5° C). They should be planted 2 to 3 in. apart in a well-prepared nursery soil with the tops of the cuttings level with, or just below, the top of the soil.

Table 7.1. lists many of the species which can be propagated by root cuttings.

ROOT PROMOTING SUBSTANCES

Herbaceous Plants

Cuttings of most plants of this type have shown a definite increase in rooting when treated with growth regulators. The chief effect has been to speed the rate of rooting and to produce heavier clumps of roots. Plants which have shown good response include chrysanthemums, geraniums, carnations, begonias, poinsettias, English ivy, and African violets. Cuttings of most varieties of such herbaceous plants usually root so easily under proper care without growth-regulator treatments, however, that it is debatable whether their use is justified.

Table 7.1. Some Species That Are Commonly Propagated by Root Cuttings.

Aesculus parviflora (bottle-brush buckeye)	*Populus alba* (white poplar)
Ailanthus altissima (tree-of-heaven)	*Populus tremula* (European aspen)
Albizia julibrissin (silk tree)	*Populus tremuloides* (quaking aspen)
Amelanchier stolonifera (dwarf june-berry)	*Prunus glandulosa* (dwarf flowering almond)
Aralia spinosa (devil's walking stick)	*Pyrus calleryana* (oriental pear)
Campsis radicans (trumpt vine)	*Rhus copallina* (shining sumac)
Celastrus scandens (American bittersweet)	*Rhus glabra* (smooth sumac)
Chaenomeles japonica (Japanese flowering quince)	*Rhus typhina* (staghorn sumac)
Clerodendrum trichotomum (glorybower)	*Rhus typhina* (staghorn sumac)
Comptonia peregrine (sweet fern)	*Robinia hispida* (rose acacia)
Hydrangea quercifolia (oak leaf hydrangea)	*Robinia pseudoacacia* (black locust)
Malus sp. (apple, flowering crabapple)	*Rosa blanda* (rose)
Morus sp. (mulberry)	*Rosa nitida* (rose)
Myrica pennsylvanica (bayberry)	*Rosa virginiana* (rose)
Papaver orientale (oriental poppy)	*Rubus* sp. (blackberry, raspberry)
Phlox sp. (phlox)	*Sassafras albidum* (sassafras)
Plumbago sp. (leadwort)	*Sophora japonica* (Japanese pagoda tree)
	Syringa vulgaris (lilac)
	Ulmus carpinifolia (smooth-leaved elm)
	Wisteria sp. (wisteria)
	Xanthorhiza simplicissima (shrub yellow root)

Broad-Leaved Evergreens

Propagation of this type of plant, by softwood or semi-hardwood cuttings, is often considerably improved by growth-regulator treatments, provided that the other necessary requirements for rooting are satisfied. The citrus species—orange, lemon, and grapefruit—as well as the olive are evergreen tree fruits that respond well to treatment. A great many ornamentals in this group have also given good response, including camellia, holly, rhododendron, box, daphne, euonymus, gardenia, oleander, escallonia, pittosporum, and pyracantha.

Narrow-Leaved Evergreens

For most species, if the cuttings are taken at the right time of year, and the proper material and concentration used, improved rooting can be expected to result from the use of root-promoting chemicals. Species found to respond to treatments include fir, juniper, spruce, pine, yew, aborvitae, and hemlock.

Deciduous Plants

Cuttings may be made of this type of plant either as hardwood cuttings during the dormant season or as softwood cuttings during the active growing period. Experience has shown a response to growth-regulator treatments with both types of cuttings. Plants of this type that have responded to treatments include the apple, apricot, pear, peach, plum, cherry, walnut, filbert, maple, birch, beech, and mulberry.

Environmental Conditions for Rooting Leafy Cuttings

For the successful rooting of leafy cuttings, the essential environmental requirements are proper temperature (65° to 75° F ; 18° to 27° C), an atmosphere conductive to low water loss from the leaves ; ample light ; and a clean, moist, well-aerated, and well-drained rooting medium. There are many possible types of equipment that are satisfactory for providing these conditions—from a simple glass jar placed over a few cuttings stuck in sand to elaborate greenhouse benches with automatic mist control and automatically controlled electric soil cables below the cuttings. Probably one of the simplest devices, which is entirely satisfactory for rooting a limited number of cuttings, is an ordinary wooden box half-filled with the rooting medium, with a pane of glass or a sheet of polyethylene film over the top. If this is placed in a heated room in the winter near a window, cuttings of many species can be rooted in it. Such a box can also be used successfully in the summer for rooting cuttings of some plants, if placed out-of-doors in the shade. Cuttings of many plants can be rooted successfully with only artificial light, if placed under an ordinary large fluorescent lamp fixture.

In commercial operations, large-scale rooting of leafy cuttings is done in hotbeds, cold frames, or beds in greenhouses or plastic houses. In these structures, proper environmental conditions can be established and maintained.

Sanitation

Most commercial propagators recognize the value of maintaining strict sanitary procedures during all stages of making and rooting leafy cuttings. It is much easier to prevent attacks of disease organisms than to try to stop them. Losses could be considerable from a disease attack where hundreds of thousands of cuttings might become involved.

During the preparation of facilities for rooting the cuttings and while making and inserting them in the rooting medium, the following procedures, if observed closely, will aid considerably in preventing losses from attacks by noxious organisms.

Propagating benches in the greenhouse should be washed thoroughly with water and sprayed with a copper naphthenate solution (1 part to 5 parts paint thinner). (*Caution* : Creosote and certain phenolic materials used as wood preservatives give off fumes which are toxic to plants, and should not be used.) Flats for rooting the cuttings should be thoroughly washed, filled with the rooting medium, and then sterilized by steam or methyl bromide. The work benches where the cuttings are to be prepared should be washed thoroughly with water and sprayed with a Clorox (1 part to 4 parts water) solution or a formaldehyde (1 qt 38 per cent formaldehyde to 5 gal. of water) solution. Tools used in preparing the cuttings should be dipped in such solutions several times daily.

The cutting material itself should be free of insects and disease organisms. As the cutting material is gathered, it can be placed on and wrapped in large sheets of clean, washed black polyethylene. When full these are protected from the sun and brought into the propagating room as soon as possible. The cutting material can be placed on a large raised wire rack and washed thoroughly with water. Mist nozzles should be placed over this rack to keep the material fresh while the cuttings are being prepared. After the cuttings are made, they may be soaked for about 10 minutes in a mild fungicide, such as Morsodren (1 tsp per 5 gal. of water). Following this, the cuttings are allowed to drain in a wire rack and are then treated with a root-promoting compound just before they are stuck into the propagating medium.

Efforts to conduct propagating operations under clean, sterilized conditions are of no avail, however, unless all components are included—the cuttings themselves, the flats, the rooting medium, the area where the cuttings are prepared, the tools used in making the cuttings, and the rooting bench. Keep the watering house nozzle off the floor where it could pick up and later distribute harmful organisms.

Preparing the Rooting Frame and Inserting the Cuttings

The frames or benches should preferably be raised or, if on the ground, equipped with drainage tile, so there is never any question of perfect drainage of excess water.

The frames or flats should be deep enough so that about 4 in. of rooting medium can be used. The depth should be enough so that an average-length cutting—3 to 5 in.—can be inserted up to half its total length, with the end of the cutting still an inch or more above the bottom of the frame. The rooting medium should be watered thoroughly before the cuttings are inserted, which should be as soon as possible after they are prepared. It is very important that the cuttings be protected from drying at all stages during their preparation and insertion.

After a section of the rooting bench or a flat is filled with cuttings, it should be well watered to settle the rooting medium around the cuttings.

Care of Cuttings During Rooting

Hardwood stem cuttings or root cuttings started out-of-doors in the nursery require only the usual care given to other crop plants, such as adequate soil moisture, freedom from weed competition, and insect and disease control. The best results with most species are likely to be obtained if the nursery is established in full sun where shading and root competition from large trees or shrubbery do not occur.

Leafy softwood or semi-hardwood stem cuttings and leaf-bud or leaf cuttings being rooted under high humidity require close attention throughout the rooting period. The temperature should be controlled carefully. The cuttings must not be allowed to show wilting for any length of time. Glass-covered frames, exposed even for a few hours to strong sunlight, will build up excessively high and injurious temperatures, owing to the heat accumulating under the glass. Such equipment should always be protected by cloth screens, whitewash on the glass, or some other method of reducing the light intensity.

If bottom heat is provided, thermometers should be inserted in the rooting medium to the level of the base of the cutting and checked at frequent intervals, especially at first. A temperature of 65° to 75° F (18° to 24° C) is desirable. Excessively high temperatures in the rooting medium even for a short time are likely to result in death of the cuttings.

It is very important to maintain humidity conditions as high as possible in rooting leafy cuttings to reduce water loss from the leaves to a minimum. Syringing the leaves with a spray nozzle at frequent intervals is a common practice especially during hot weather. Although more time consuming, several light sprinklings with water each day are better than heavy soakings at less frequent intervals. A nozzle should be used that breaks the water into a fine spray. A drop of the humidity to a low level with a consequent pronounced wilting of the cuttings, if prolonged for any length of time, may so injure the cuttings that rooting will not occur, even though high-humidity conditions are subsequently resumed. However, most commercial nursery operations use intermittent-mist propagating beds to overcome such problems.

Although high humidity in the propagating frame is important while roots are forming, it is also necessary that adequate drainage be provided so that excess water can escape and not cause the rooting medium to become soggy and waterlogged. When peat or sphagnum moss is used as a component of the rooting medium, it is especially important to see that it does not become excessively wet.

It is also necessary to maintain sanitary conditions in the propagating frame. Leaves that drop should be removed promptly, as well as any obviously dead cuttings. Organisms find ideal conditions in a humid, closed propagating frame with low light intensity and, if not controlled, can destroy thousands of cuttings overnight.

Disease problems under mist propagation conditions have not been serious. Probably the frequent washing of the leaves by the aerated water removes spores before they are able to germinate. The greater light intensity and air movement also decrease the incidence of diseases.

If insects, such as red spiders, aphids, or mealy bugs, appear on the leaves of the cuttings, immediate control measures are necessary.

HANDLING CUTTINGS AFTER ROOTING

Hardwood Cuttings

Rooted hardwood cuttings in the nursery row are usually dug during the dormant season after the leaves have dropped. With fast growing species, the cuttings may be sufficiently large to dig after one season's growth. Slower growing species may require 2 or even 3 years to become large enough to transplant.

The digging should take place on cool, cloudy days, when there is no wind. If possible, digging should not be done when the soil is wet, especially if it has a high clay content.

Most of the soil should drop readily from the roots after the plants are removed. After the plants are dug, they should be quickly heeled-in in a convenient location or else replanted immediately in their permanent location. "Heeling-in" is to place dug, barerooted deciduous nursery plants close together in trenches with the roots well covered. This is a temporary provision for holding the young plants until they can be set out in their permanent location.

Commercial nurseries often store quantities of deciduous plants for several months in cool, dark rooms with the roots protected by damp wood shavings, shingle tow, or some similar material. Nursery stock to be kept for extended periods should be held under refrigerated conditions at 32° to 35° F (0° to 2° C).

If only a few plants are to be removed from the nursery row, they can be dug with shovels, but in large-scale nursery operations some type of mechanical digger is generally used. This digger "undercuts" the plants. A sharp U-shaped blade travels 1 to 2 ft below the soil surface under the nursery row, cutting through the roots. Sometimes a horizontal "lifting" blade is attached to, and travels behind, the cutting blade. This slightly lifts the plants out of the soil, making them easy to pull by hand.

Unless very small, plants of evergreen species usually cannot be successfully handled bare-root, as is done with dormant and leafless deciduous plants. The presence of leaves on evergreen plants requires the continuous contact of the roots with soil. Therefore, large, salable plants of broad-or narrow-leaved evergreens, and occasionally deciduous plants, are either grown in containers or dug and sold "balled and burlapped." By the latter method, the plants are removed from the soil by carefully digging a trench around each individual. The soil mass around the roots is sometimes tapered at top and bottom, resulting in a half of soil in which the roots are embedded. It is important that the soil be at the proper moisture level—not too wet and not too dry—otherwise it will fall apart. The ball is tipped gently onto a large square of burlap, which is then pulled tightly and sewed in place with heavy twine. If properly done, this adequately holds the soil to the roots, and the plant can then be moved safely of considerable distances and replanted successfully.

To properly prepare field-grown nursery stock for transplanting, either by hand "ball and burlapping" or by machine digging prior root pruning is necessary. This is to produce a compact, fibrous root system and should be started when the young plants (*liners*) are first set in the field. Long or curled and twisted roots should be cut back to a few inches from the crown of the liner. Root pruning the second year should also be done, either by a tractor-powered U-blade cutter for large-scale operations, or with a sharp shovel where only a few plants are involved. This procedure helps confine the roots to the soil mass which will be taken with the plant during the digging and balling operation.

The production of balled and burlapped nursery stock is gradually being replaced, especially in areas with mild winters, by the production of plants in containers primarily because of the lower labour costs involved and the greater opportunities for mechanization.

Softwood, Herbaceous, Semi-Hardwood, Leaf-Bud, Leaf Cuttings

Cuttings such as these, rooted with leaves attached and under conditions of high humidity, require considerable care in being removed from the rooting medium. After rooting has started, the humidity should be lowered and ventilation of the bed provided. They should be dug as soon as a substantial root system with secondary roots has formed. Many propagators have experienced rapid rooting of cuttings only to have them die when they are dug and potted. Sometimes this trouble may be overcome by leaving the cuttings in the rooting bed longer, until after the first-formed primary roots have branched to develop a dense, fibrous secondary root system to which the rooting medium clings in a ball.

Most of the rooting media used contain little or no mineral nutrients available to the plants, hence the cuttings are dependent upon the nutrients already stored in the stems and leaves at the time the cuttings were made. This is generally sufficient to maintain the cuttings during the period of root formation. With some species, however, it may be helpful to water the cuttings with a nutrient solution about 10 days before they are to be removed, especially if digging has been delayed until a secondary root system has developed.

If for any reason the cuttings are to be kept for a prolonged period in the rooting medium, they should be watered several times with a nutrient solution.

In digging, the cuttings should be lifted gently from the rooting medium with a trowel or some similar device, taking care not to break off the roots. It is desirable for the cuttings to be lifted out with a mass of the rooting medium still adhering to the roots. This can be done if some material such as peat moss or vermiculite is included in the rooting medium. The cuttings are ready for potting when most of the roots are 1 to 2 in. long.

The rooted cuttings should be watered thoroughly shortly after potting. It is very important that potted cuttings be moved gradually from the protected conditions under which they were rooted (high humidity, low light intensity) to out-of-door conditions (low humidity, high light intensity, and wind). It is often best to leave the cuttings for several days after they have been potted under the same conditions in which they were rooted. If they were rooted in mist, they must be given especially close attention and moved gradually to the dryer atmosphere. Before placing the rooted cuttings in full sun, they should be hardened-off for 1 or 2 weeks in a cold frame or lathhouse or under some partial protection from the sun.

Cold Storage of Rooted and Unrooted Leafy Cuttings

Sometimes it may be convenient to take cuttings at certain times, such as when nursery plants are being sheared and shaped, for later rooting. This was done experimentally with the Kurume type of azaleas overwintered in a cold greenhouse. Softwood cuttings were taken in the spring and held in polyethylene bags at temperatures from 31° to 40° F (-0.5° to 4.5° C) for up to 10 weeks, with subsequent rooting equal to that from unstored cuttings. Unrooted carnation cuttings have long been stored commercially for subsequent rooting. In test on the effects of storage on subsequent performance of the plants, it was found that cuttings rooted after storage at 33° F (0.5° C) gave better results than those stored after rooting.

In some situations cuttings may be rooted in late summer or early fall for planting outdoors the following spring. Several studies have shown that it is possible to hold rooted cuttings of some species in polyethylene bags for prolonged periods under cold storage temperatures of 35° to 40° F (1.5° to 4.5° C). In one test, rooted cuttings of certain plants were safely stored for about 5 months at temperatures of 34° to 39° F (1° to 4° C) when placed in polyethylene bags. Packing material around the roots was of no advantage. Other studies on cold storage of rooted cuttings of 31 different species fos 6 months at two temperatures (32° and 40° F) showed better survival with many of the species at 32° than at 40°, although no difference was noted with other species. With some species, dusting prior to storage with 5 per cent Captan increased survival, but in others this made no difference. Apparently some species survive cold-storage treatments much better than others.

8

Physiology of Grafting and Budding

Grafting is the art of joining parts of plants together in such a manner that they will unite and continue their growth as one plant. The part of the graft combination which is to become the upper portion or top of the new plant is termed the *scion* (*cion*), and the part which is to become the lower portion or root is termed the *rootstock*, or *understock*, or sometimes the *stock*. All methods of joining plants are properly termed *grafting*, but when the scion part is a small piece of bark (and sometimes wood) containing a single bud, the operation is termed *budding*. Most of the situations discussed in this chapter affecting grafting would therefore apply equally well to budding, although the techniques involved are somewhat different.

Natural Grafting

One occasionally sees branches of trees that have become grafted together naturally following a long period of being pressed together without disturbance. The English Ivy (*Hedera helix*) forms such grafts, and detailed studies have been made of translocation in natural grafts of this species. Not so obvious but of much greater occurrence, particularly in forest stands, is the natural grafting of roots. Such grafts are most common between roots of the same tree or between roots of trees of the same species. Grafts between roots of trees of different species are rare. In the forests living stumps sometimes occur, kept alive because their roots have become grafted to those of nearly intact, living trees. The anatomy of natural grafting of aerial roots has been studied ; the initial contact is established by the formation and fusion of epidermal hairs. Such natural grafting provides a means of virus transmission from an infected tree to its neighbours. This can be important in closely set orchard and nursery plantings of fruit trees, where numerous root grafts could occur and result in the slow spread of the virus throughout the planting. Natural root grafting is a potential source of error in virus indexing procedures where virus-free and virus-infected trees are grown in close proximity. The fact that natural grafting does take place throws some doubt on the assumption usually held that each tree is an independent, discrete organism.

Formation of the Graft Union

A number of detailed studies have been made of the healing of graft unions, mostly with woody plants.

Briefly, the usual sequence of events in the healing of a normal graft union is as follows :

(*a*) Freshly cut scion tissue capable of meristematic activity is brought into secure, intimate contact with similar freshly cut stock tissue in such a manner that the cambial regions of both are in close proximity. Temperature and humidity conditions must be such as to promote activity in the newly exposed, and surrounding, cells.

(*b*) The outer exposed layers of cells in the cambial region of both scion and stock produce parenchyma cells which soon intermingle and interlock ; this is called *callus tissue.*

(*c*) Certain cells of this newly formed callus which are in line with the cambium layer of the intact scion and stock differentiate into new cambium cells.

(*d*) These new cambium cells produce new vascular tissue, xylem toward the inside and phloem toward the outside, thus establishing vascular connection between the scion and stock, a requisite of a successful graft union.

The healing of a graft union can be considered as the healing of a wound. Such injury to tissue as would occur if the end of a branch were split longitudinally would heal quickly if the split pieces were bound tightly together. New parenchyma cells would be produced by abundant proliferation of cells of the cambium region of both pieces, forming callus tissue. Some of the interlocking parenchyma cells differentiate into cambium cells, which subsequently produce xylem and phloem.

If between the two split pieces, one interposed a third, detached, piece which had been cut so that a large number of its celis in the cambial region could be placed in intimate contact with cells of the cambial region of the two split pieces, proliferation of cells from all cambial areas would soon result in complete healing, with the foreign, detached piece joined completely between the two original split pieces. A graft union is essentially a healed wound, with an additional, foreign, piece of tissue incorporated into the healed wound.

This added piece of tissue, the scion, will not resume its growth successfully, however, unless vascular connection has been established so that it may obtain water and mineral nutrients. In addition, the scion must have a terminal meristematic region—a bud—to resume shoot growth.

In the healing of a graft union, the parts of the graft that are originally prepared and placed in close contact do not themselves move about or grow together. The union is accomplished entirely by cells which develop *after* the actual grafting operation has been made.

In addition, it should be stressed that in a graft union *there is no intermingling of cell contents.* Cells produced by the stock and by the scion each maintain their own distinct identity.

Considering in more detail the steps involved in the healing of a graft union, we may say that the first one listed below is a preliminary step, but nevertheless, it is essential and one over which the propagator has control.

(*a*) *Establishment of intimate contact of a considerable amount of the cambial region of both stock and scion under favourable environmental conditions.*

Temperature conditions that will cause high cell activity are necessary. Usually, temperatures from 55° to 90° F (12.8° to 32° C), depending upon the species, are conductive to rapid cell growth. The grafting operation should thus take place at a time of year when such favourable temperatures can be expected and when the plant tissues, especially the cambium, are in a naturally active state. These conditions generally occur during the early spring months.

The new callus tissue arising from the cambial region is composed of thin-walled, turgid cells which can easily become desiccated and die. It is important for the production of these parenchyma cells that the humidity in the vicinity of the cambial region of the graft union be kept at a high level. This explains the necessity of thoroughly waxing the graft union or using some other method to maintain a high degree of tissue hydration.

It is important, too, that the region of the graft union be kept as free as possible from pathogenic organisms. The thin-walled parenchyma cells, at relatively high humidity and temperature, will provide favourable conditions for growth of certain fungi and bacteria which are exceedingly detrimental to the successful healing of the graft union. Prompt waxing of the graft helps prevent disease infection.

It is essential that the two original graft components be held together firmly by some means, such as wrapping, tying, or nailing, or better yet, by wedging (as in the cleft or saw-kerf grafts) so that the parts will not move about and dislodge the interlocking parenchyma cells after proliferation has started.

The statement is often made that for successful grafting the cambium layers of stock and scion must be "matched." Although this is desirable, it is unlikely that complete matching of the two cambium layers is, or ever can be, attained. In fact, it is only necessary that the cambial regions be close enough together so that the parenchyma cells from both stock and scion produced in this region can become interlocked. It is in the region of the cambium that the essential callus production in the highest. Two badly matched cambial layers may delay union or, if extremely mismatched, prevent union. In studies of grafting monocotyledonous plants, it was found that a cambium layer is not required for a successful graft union but that any meristematic tissue could be utilized for this purpose and is capable of forming a union between stock and scion.

(b) Production and interlocking of parenchyma cells (callus tissue) by both stock and scion.

During the grafting operation the cells cut and damaged by the grafting knife turn brown and die. Underneath these dead cells new parenchyma cells arise from both stock and scion, coming from the parenchyma of the phloem rays and the immature parts of the xylem. The actual cambial layer itself seems to take little or no part in this first development of the callus. In grafting scions on established stocks, the stock produces most of the callus. These parenchyma cells, comprising the spongy callus tissue, fill the space between the two original components of the graft (the scion and the stock), becoming intimately interlocked and providing some mechanical support, as well as allowing for some passage of water and nutrients from the stock into the scion. For a time, between the callus arising from the stock and that arising from the scion, there is a more or less continuous brown line consisting of the above-mentioned dead and crushed cells remaining from the grafting cuts. This line of cells is gradually resorbed, however, and disappears. At the final stage of healing the cells of the outer layer of callus become suberized. Exposed dead cells (tracheids, vessels) are sealed off with a deposit of gum.

(*c*) *Production of a new cambium across the callus "bridge."* At the edges of the newly formed callus mass, parenchyma cells which are touching the cambial cells of the stock and scion will differentiate into new cambium cells. This cambial formation in the callus mass proceeds farther and farther inward away from the original stock and scion cambium, and on through the callus bridge, until a continuous cambial connection between stock and scion has finally formed.

(*d*) *Formation of new xylem and phloem from the new vascular cambium in the callus bridge.*

The newly formed cambial sheath in the callus bridge begins typical cambial activity, laying down new xylem and phloem, along with the original vascular cambium of the stock and scion, continuing this throughout the life of the plant.

In the formation of new vascular tissues following cambial continuity, it appears that the type of cells formed by the cambium may be influenced by the cells of the stock adjacent to the cambium. For example, xylem ray cells are formed where the cambium is in contact with xylem rays of the stock, and xylem elements where they were in contact with xylem elements.

The new xylem tissue originates from the scion rather than from the stock. This is shown by "ring grafting" (where a ring of bark from a young tree is removed and replaced by a ring of bark from another tree). Researchers found that using bark rings of the 'Scugog' apple variety, which has purple xylem, subsequent xylem growth of the tree following grafting was entirely purple in colour just under the 'Scugog' ring of bark, whereas in the remainder of the tree the xylem remained white.

This production of new xylem and phloem then establishes vascular connection between the scion and the stock. Under high transpiration rates this must occur before much new shoot growth takes place from buds on the scion ; otherwise, the enlarging leaf surfaces on the scion shoots will have little or no water supply to offset that lost by transpiration, and the scion will quickly become desiccated and die. Under some conditions, however, where vascular connections fail to occur, enough translocation can take place between the parenchyma cells of the callus to permit survival. In grafts of the monocot, *Vanilla* orchid, scions survived and grew for two years with only parenchyma unions.

A somewhat different developmental sequence occurs in tobacco and cotton. Here, xylem tracheary elements or phloem sieve tubes, or both, form directly by differentiation of callus into these vascular elements. A cambium layer subsequently differentiates between the two vascular elements. Apparently parenchyma cells, which make up the callus, can differentiate into tracheid-like elements with relative ease. Buds, as would occur on the scions of grafts, seem to be effective in inducing processes of dedifferentiation and subsequent redifferentiation in the tissue on which they are grafted. This bud influence has been shown by inserting a bud into a piece of *Cichorium intybus* root consisting only of old vascular parenchyma, and observing that under the influence of the bud, the old parenchyma cells dedifferentiate and redifferentiate into groups of conducting elements.

The Healing Process in T-Budding

In T-budding, the bud piece usually consists of the periderm, cortex, phloem, and often some xylem tissue to which is attached a lateral bud. In budding, this bark piece is laid against the exposed xylem of the stock.

Detailed studies of the healing process in T-budding have been made for the rose, citrus and apple.

When the bud piece is removed from the budstick, and when the flaps of bark on either side of the "T" incision on the stock are raised, the cambium and newly formed cambial derivatives in these tissues are usually destroyed, owing to their very tender, succulent nature.

In the apple when the bark is lifted in the budding operation the separation occurs in the young, undifferentiated xylem.

The entire cambial zone remains attached to the inside of the bark flaps. After the bud shield is inserted, callus is produced, which surrounds the bud shield and holds it in place. The callus originates almost entirely from the rootstock tissue, mainly from the exposed surface of the xylem cylinder. Very little callus is produced from the sides of the bud shield. Callus starts development 2 days after budding and continues rapidly for 2 to 3 weeks until all internal air pockets are filled. Following this, a continuous cambium is established between the bud and the rootstock. The callus then begins to lignify, and isolated tracheary elements appear. Lignification of the callus is completed about 12 weeks after budding.

In the rose, about 3 days after budding the terminal cells of the broken xylem rays and adjacent cambial derivatives on the exposed surface of the stock begin to enlarge and divide, leading to the production of callus strands. In the same manner, callus strands develop from terminal cells of broken phloem rays and adjacent young secondary phloem cells on the cut surface of the inner side of the bud piece. Within 14 days the space between the stock and the bud piece is completely filled with callus, which has developed mainly from the proliferating immature secondary xylem of the stock and the immature secondary phloem of the bud piece. During the second week, short areas of cambium cells appear in this newly developed callus tissue. By the tenth day, a completed band of cambium tissue extends over the face of the stock and is joined to the uninjured cambium on either side of the bud piece.

After cambial continuity is completed, continuity of vascular tissues soon becomes established between the bud and stock. In T-budding, then, the primary union is between the surface of the phloem on the inner face of the shield and the meristematic xylem surface of the stock. A secondary type of union may occur, however, at the edges of the shield piece, as it would in chip budding.

The various stages in the healing process of the union in T-budding of citrus have been determined as follows:

	Stage of Development	*Approximate Time after Budding*
1.	First cell division	24 hours
2.	First callus bridge	5 days
3.	Differentiation :	
	(*a*) in the callus of the bark flaps	10 days
	(*b*) in the callus of the shield	15 days
4.	First occurrence of xylem tracheids :	
	(*a*) in the callus of the bark flaps	15 days
	(*b*) in the callus of the shield	20 days
5.	Lignification of the callus completed :	
	(*a*) in the bark flaps	25 to 30 days
	(*b*) under the shield	30 to 45 days
6.	First occurrence of meristematic layers in the callus between shield and bark flaps	15 days

Factors Influencing the Healing of the Graft or Bud Union

As anyone experienced in grafting or budding knows, the results obtained in grafting are often inconsistent, an excellent percentage of "takes" occurring in some operations, whereas in others the results are very discouraging. There are a number of factors which influence the healing of graft unions.

Incompatibility

One of the symptoms of incompatibility in grafts between distantly related plants is a complete lack, or a very low percentage, of successful unions. Grafts between some plants known to be incompatible, however, will initially make a satisfactory union, even though the combination eventually fails.

Kind of Plants

Some plants are much more difficult to graft than others even when no incompatibility is involved. Difficult ones, for example, are the hickories, oaks, and beeches. Nevertheless, such plants, once successfully grafted, grow very well with a perfect graft union. In topgrafting apples and pears, even the simplest techniques usually give a good percentage of successful unions, but in topgrafting certain of the stone fruits, such as peaches and apricots, much more care and attention to details are necessary. Strangely enough, topgrafting peaches to some other compatible species, such as plums or almonds, is more successful than reworking them back to peaches. Many times one method of grafting will give better results than another, or budding may be more successful than grafting, or vice versa. For example, in topworking the native black walnut (*Juglans hindsii*) to the Persian walnut (*Juglans regia*) in California, the bark graft method is more successful than the cleft graft.

Some easily grafted plants, such as the apple, form a "wound gum" plugging exposed xylem elements after the grafting operation, thus preventing excessive dessication and the death of tissues. Other plants, such as the walnut, in which graft unions heal with difficulty, form such "wound gum" very slowly, and in these desiccation and death of tissues in the area of the graft union may be extensive.

Some species, such as the Muscadine grape (*Vitis rotundifolia*), mango (*Mangifera indica*), and *Camellia reticulata*, are so difficult to propagate by the usual grafting or budding methods that "approach grafting," in which both partners of the graft are maintained for a time on their own roots, is often used. This variation among plant species and varieties in their grafting ability is probably related to their production of callus, which is essential for a successful graft union. *Camellia reticulata*, for example, is difficult to graft and is a very poor callus producer.

Temperature and Moisture Conditions During and Following Grafting

There are certain environmental requirements which must be met for callus tissue to develop.

Temperature has a pronounced effect on the production of callus tissue (Fig. 8.1). In apple grafts little, if any, callus is formed below 32° F (0° C) or above about 104° F (40° C). Even around 40° F (4° C), callus development is slow and meager,

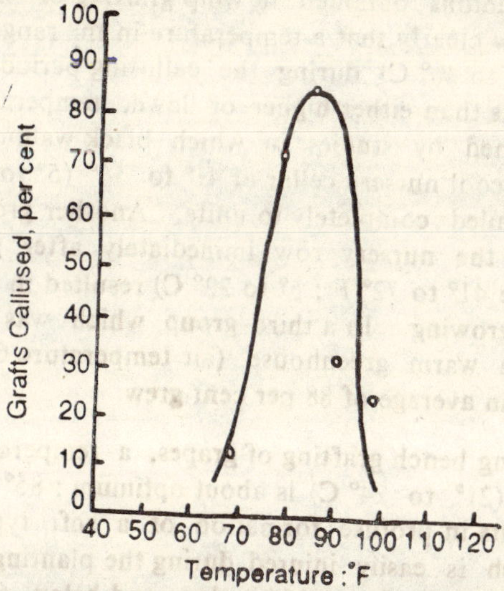

Fig. 8.1. Influence of temperature on the callusing of walnut (*Juglans*) grafts. Callus formation is essential for the healing of the graft union. Maintaining an optimum temperature following grafting is very important for successful healing of walnut grafts. Adapted from data of Sitton (149).

and at 90° F (32° C) and higher, callus production is retarded, with cell injury becoming more apparent as the temperature increases, until death of the cells occurs at 104° F. Between 40° and 90° F, however the rate of callus formation increases directly with the temperature. In such operations as bench grafting, callusing may be allowed to proceed slowly for several months by storing the grafts at relatively low temperatures (45° to 50° F ; 7° to 10° C), or if rapid callusing is desired, they may be kept at higher temperatures for a shorter time. Since excessive callusing of root graft unions may result in the formation of undersirable "callus knots" (sometimes mistaken by plant inspectors for crown gall tumors), it can be controlled by storing well-callused grafts at reduced temperatures to prevent further callus development.

Temperature during the healing period following grafting was found to have a pronounced effect upon the degree of successful unions obtained in whip grafting of black walnuts. Studies show clearly that a temperature in the range of 77° to 86° F (25° to 30° C) during the callusing period gives much better results than either higher or lower temperatures. This was confirmed by studies in which black walnut root grafts placed in a cool nursery cellar at 41° to 56° (5° to 13° C) for callusing failed completely to unite. Another group of grafts planted in the nursery row immediately after grafting (air temperature 41° to 72° F ; 5° to 22° C) resulted in only about 3 per cent growing. In a third group which was allowed to callus in a warm greenhouse (air temperature 65° to 80°, 8° to 27° C), an average of 88 per cent grew.

Following bench grafting of grapes, a temperature of 70° to 75° F (21° to 24° C) is about optimum ; 85° F (29° C) or higher results in profuse formation of a soft type of callus tissue which is easily injured during the planting operations. Below 70° F, callus formation is slow, and below 60° F (5° C), it almost ceases.

Grafting operations performed late in the spring when excessively high temperatures may occur, often result in failure. Tests of walnut topgrafting in California during very hot weather in May showed that whitewashing the area of the completed graft union definitely promoted healing of the union. Whitewash would reflect a considerable portion of the radiant energy of the sun rather than allow it to be absorbed, thus resulting in lower bark temperatures. In addition, in these tests, scions placed on the north and east sides of the stub survived much better than those on the south and west. Undoubtedly this was due to the lower temperatures resulting from their shaded position.

Since the parenchyma cells comprising the important callus tissue are thin-walled and tender, with no provision for resisting desiccation, it is obvious that if they are exposed to drying air for very long, they will be killed. This was found to be the case in studies of the effect of humidity on healing of apple grafts. Air moisture levels below the saturation point inhibited callus formation, the rate of desiccation of the cells increasing as the humidity dropped. In fact, the presence of a film of water against the callusing surface was much more conductive to abundant callus formation than just maintaining the air at 100 per cent relative humidity. Unless a completed graft union is kept by some means at a very high humidity level, the chances of successful healing are rather remote. With most plants, thorough waxing of the graft union, which retains the natural moisture of the tissues, is all that is necessary.

Often root grafts are not waxed but stored in a moist packing material during the callusing period. Damp peat moss is an excellent material for callusing, because it provides proper moisture and aeration.

It has been shown that oxygen is necessary at the graft union for the production of callus tissue. This would be expected, since such rapid cell division and growth is accompanied by relatively high respiration, which requires oxygen. For some plants, a lower percentage of oxygen than is found naturally in air is sufficient, but for others, healing of the graft union is better if the union is left unwaxed and placed in an enclosure of water-saturated air. This may indicate that the latter plants have a high oxygen requirement for callus formation. Waxing restricts air movement to such an extent that oxygen may become limiting, and callus tissue fails to form. This situation apparently exists in grafting the grape, in which usually the union is not covered with wax or other air excluding material.

Growth Activity of the Stock Plant

Some propagation methods, such as T-budding and bark grafting, dependent upon the bark "slipping," which means the vascular cambium cells are actively dividing, producing young thin-walled cells on each side of the cambium. These newly formed cells readily separate from one another, so the bark "slips." Initiation of such cambial activity in the spring is thought to result from swelling of the buds in warm weather, since shortly afterwards cambial activity can be detected beneath each developing bud, with a wave of cambial activity progressing down the stems and trunk. This stimulus is believed to be due to increased auxin levels originating in the expanding buds.

In budding seedlings in the nursery row in late summer it is important that they have an ample supply of soil moisture just before, during, and following the budding operation. If they should lack water during this period, active growth is checked, cell division in the combium stops, and the chances of getting the buds to unite are lessened.

There is evidence to show that callus proliferation—essential for a successful graft union--occurs most readily at the time of year just before and during "bud break" in the spring, diminishing through the summer and into the winter. Increasing callus proliferation takes place again in late winter, but this is not dependent upon breaking of bud dormancy.

Plants exhibiting strong root pressure (such as the walnut) will show, at certain periods of high growth activity in the spring, excessive sap flow or "bleeding" when cuts are made preparatory to grafting. Grafts made with moisture exudation around the union will not heal, and should be made at some other stage of growth ; such "bleeding" can be overcome by making slanting knife cuts around the tree through the bark and into the xylem to permit the exudation to take place below the graft union. When potted rootstock plants to be grafted (in such species as *Fagus, Betula*, or *Acer*) show excessive root pressure, they should be put in a cool place with reduced watering until the "bleeding" stops.

On the other hand, potted rootstock plants, such as junipers or rhododendrons, when first brought into a warm greenhouse in winter for grafting, are dormant, and grafting done then would be unsuccessful. Grafting should be delayed until the rootstock plants have been held for several weeks at 60 to 65° F (15° to 18° C) and new roots start to form ; then the rootstock plant is physiologically active enough for the union to heal.

When the rootstock plant is physiologically overactive (excessive root pressure and "bleeding"), or underactive (no root growth being made), some form of side graft in which the rootstock top is not at first removed, should be used. In situations in which the rootstock is neither overactive or underactive, one of the many forms of topgrafting, in which the top of the stock is completely removed at the time the graft is made, is likely to be successful.

Propagation Techniques

Even in using the standard grafting or budding methods, there are numerous possible variables which may affect the success of the operation. There are many opinions, often conflicting, about the proper techniques to use.

Sometimes the grafting technique is so poor that only a very small portion of the cambial regions of the stock and scion are brought together. Although healing occurs in this region and growth of the scion may start, when a large leaf area develops and high temperatures and high transpiration rates occur, sufficient movement of water through the limited conducting area cannot take place, and the scion subsequently dies. Other errors in grafting technique, such as poor or delayed waxing, uneven cuts, or use of desiccated scions can, of course, result in grafting failure.

Poor grafting techniques, although they may delay adequate healing for some time—weeks, or even years—do not in themselves cause any permanent incompatibility. Once the union is adequately healed, growth can proceed normally.

Virus Contamination, Insect Pests, and Diseases

Using virus-infected propagating materials in nurseries can reduce bud "take" as well as the vigor of the resulting plant. In stone fruit propagation the use of budwood free of ring spot virus has consistently given improved percentages of "takes" over infected wood.

Topgrafting olives in California is seriously hindered in some years by attacks of the American plum borer (*Euzophera semifuneralis*), which feeds on the soft callus tissue around the graft union, resulting in the death of the scion. (This is easily controlled, however, by painting the completed graft, before waxing, with a 3 oz per gal. solution of 50 per cent DDT in water). In England, nurserymen are often plagued with the red bud borer (*Thomasiniana oculiperda*), which eats the healing callus beneath the bud-shield in newly inserted T-buds.

Sometimes bacteria or fungi gain entrance at the wounds made in preparing the graft or bud unions. For example, it was found that a rash of failures in grafts of *Cornus florida rubra* on common *C. florida* stock was due to the presence of the fungus *Chalaropsis thielavioides.* Chemical control of such infections materially aids in promoting healing of the unions.

In South and Central American, rubber (*Hevea*) trees are propagated by a modification of the patch bud A major cause of budding failures in these countries has been infection of the cut surfaces by a fungus, *Diplodia theobromae*. Control of this infection has been obtained by fungicidal treatments.

In topgrafting mangos in Florida, control of the fungus diseases anthracnose and scab is essential for success. This is done by spraying the rootstock trees and the source of scion-wood regularly with copper fungicides before grafting is attempted.

Possible Relation of Growth Substances to Healing of the Graft Union

Studies of the practical application of growth substances, particularly auxin, to tree wounds or to graft unions and their effect on promoting subsequent healing have not given consistent results ; consequently such materials are not generally used for this purpose. It is known from tissue culture studies, however, that there is a definite relationship between callus production (which is essential for graft healing) and the levels of certain endogenous growth substances, particularly kinetin, and auxin. Fig. 8.2 shows the relationship between auxin (indoleacetic acid) and kinetin levels in the production of callus in tobacco segments maintained under aseptic conditions. Perhaps further studies may show some practical benefit from the use of cambinations of auxin and kinetin in stimulating callus formation and subsequent healing of graft unions, thereby facilitating grafting of combinations considered to be difficult.

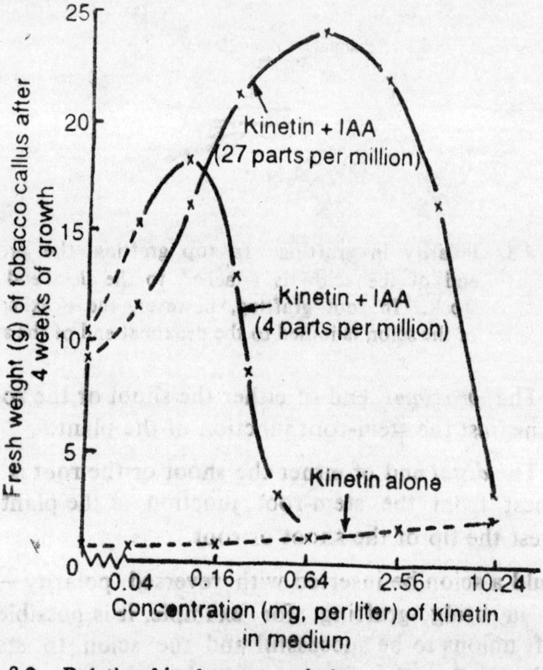

Fig. 8.2. **Relationship between relative concentration of auxin (indoleacetic acid) and kinetin upon callus production. Tobacco callus grown in aseptic culture medium. From original data of Murashige and Skoog.**

Polarity in Grafting

Attention to proper polarity is very important if the graft union is to be permanently successful. In all commercial grafting operations correct polarity is strictly observed.

As a general rule, as shown in Fig. 8.3 in grafting two pieces of stem tissue together, the morphologically proximal end of the scion should be inserted into the morphologically

distal end of the stock. But in grafting a piece of stem tissue on a piece of root, as is done in root grafting, the proximal end of the scion should be inserted into the proximal end of the root piece.

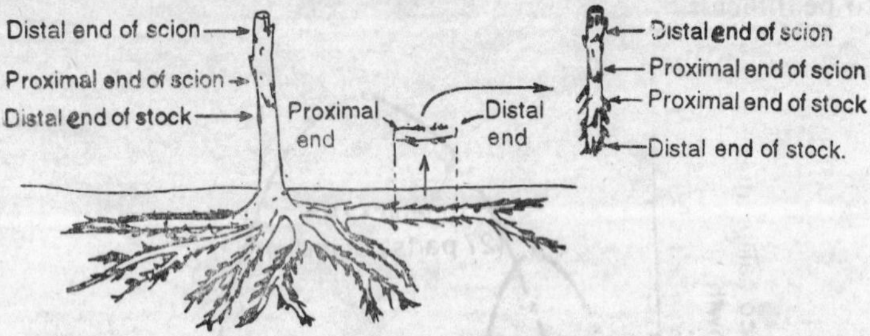

Fig. 8 3. Polarity in grafting. In top grafting, the proximal end of the scion is attached to the distal end of the stock. In root grafting, however, the proximal end of the scion is joined to the proximal end of the stock.

The *proximal* end of either the shoot or the root is that nearest the stem-root junction of the plant.

The *distal* end of either the shoot or the root is that furthest from the stem-root junction of the plant and nearest the tip of the shoot or root.

Should a scion be inserted with reversed polarity—"upside down"—in bridge-grafting, for example, it is possible for the two graft unions to be successful and the scion to stay alive for a time. But, the reversed scion does not increase from its original size, whereas the scion with correct polarity enlarges normally.

In nurse-root grafting, the rootstock may be purposely grafted to the scion with reversed polarity. Union will occur, and the root will supply water and mineral nutrients to the scion. Since the scion is unable to supply necessary organic materials to the rootstock, the stock eventually dies. In this case, the graft union is purposely set well below the ground level, and the scion itself produces adventitious roots which ultimately become the entire root system of the plant.

In T-budding or patch budding, the rule for observance of correct polarity is not as exacting. Buds can be inserted with reversed polarity and still make permanently successful unions. The buds start growing downward, then the shoots recurve and start upward. In the inverted bud piece, the cambium seems capable of continued functioning and growth. However, in the vessels as well as in the fibers formed from cambial activity, there is a twisting configuration which apparently keeps translocation and water conduction oriented as in the original position.

If a complete ring of bark is removed from the trunk of a young tree, inverted, and regrafted into place, the growth of the tree will be markedly retarded. Organic materials that stimulate growth of the roots move downward through the phloem. When they reach the inverted bark piece, their movement is obstructed. Presumably, the lack of these materials retards root growth subsequently dwarfing the entire tree. Such dwarfing is temporary, however ; experiments have shown that within 2 years trees with such inverted bark pieces resume normal growth. This is due to a bridge of normally oriented tissues forming at the vertical seams of the ring grafts.

Limits of Grafting

Since one of the requirements for a successful graft union is the close matching of the callus-producing tissues near the cambium layers, grafting is generally confined to the dicotyle dons in the angiosperms, and to the gymnosperms, the cone-bearing plants. Both have a vascular cambium layer existing as a continuous tissue between the xylem and the phloem. In the monocotyledonous plants of the Angiospermae, which do not have a vascular cambium, grafting is more difficult and the percentage of "takes" much lower than in the dicots. There are cases of successful graft unions between the stem parts of monocots. By making use of the meristematic properties found in the intercalary tissues (located at the base of internodes), successful grafts have been obtained with various grass species as well as the large tropical monocotyledonous vanilla orchid of commerce.

Before a grafting operation is started, it should be determined that the plants to be combined are capable of uniting. There is no definite rule that can be followed exactly which will give this information. *Generally, the more closely the plants to be grafted are related botanically, the better the chances are of the graft union being successful.* This will not hold true consistently, however, since botanical classifications are based on reproductive characteristics, whereas grafting is concerned primarily with the vegetative properties of plants.

Grafting Within a Clone

A scion can be grafted back on the plant from which it came, and a scion from a plant of a given clone can be grafted to any other plant of the same clone. For example, a scion taken from an "Elberta' peach tree could be grafted successfully to any other 'Elberta' peach tree in the world.

Grafting Between Clones Within a Species

Almost always, different clones within a species can be grafted together without difficulty and will produce normal plants.

Grafting Between Species Within a Genus

For plants in different species but in the same genus, the situation is confused. In some cases grafting can be done successfully, in others it cannot. Grafting between most species in the genus *Citrus*, for example, is successfully and widely used commercially. Varieties of the almond (*Prunus amygdalus*), the apricot (*Prunus armeniaca*), the European plum (*Prunus domestica*), and the Japanese plum (*Prunus salicina*) all different species—are grafted commercially on the peach (*Prunus persica*), a still different species, as a rootstock. On the other hand, the almond and the apricot, both in the same genus, cannot be intergrafted successfully. The complexity of the situation is further illustrated by the fact that the 'Beauty' variety of Japanese plum (*Prunus salicina*) makes a good union when grafted on the almond, but another variety of *P. salicina*, 'Santa Rosa,' cannot be successfully grafted on the almond.

Compatibility between species may depend upon the particular clone or seedling used, either for stock or scion. For example, choice of the almond variety (clone) is highly important when grafting to 'Marianna 2624' plum as a rootstock, some varieties being, successful, others not. Likewise, clonal selections from different seedlings of the St. Julien plum (*Prunus insititia*), when used as rootstocks, have resulted in greatly different behaviour of the trees after topworking to the same peach clone. For example, clonal selections A and C produced good peach trees with no signs of incompatibility, while trees of the same peach variety worked on selection B died within 5 years ; peach trees on rootstock selection D died after 1 year.

There are some cases in which a given interspecies graft is successful, but the reciprocal combination is not. For instance, the 'Marianna' plum (*Prunus cerasifera* × *P. munsoniana* ?) on the peach (*Prunus persica*) makes an excellent graft combination, but grafts of the peach on 'Marianna plum either soon die or fail to develop normally. Although many varieties of Japanese plums (*Prunus salicina*) can be successfully grafted on the European plum (*P. domestica*), grafts of most varieties of the European plum on the Japanese are unsuccessful.

Grafting Between Genera Within a Family

When the plants to be grafted together are in different genera but in the same family, the chances of the union being successful become more remote. Cases can be found in which such grafts are highly successful and used commercially, but in most instances such combinations are failures.

Trifoliate orange (*Poncirus trifoliata*) is used commercially as a dwarfing stock for various species in the genus *Citrus*. The quince (*Cydonia oblonga*) has long been used as a dwarfing rootstock for certain varieties of the pear (*Pyrus communis*). The reverse combination, quince on pear, though, is unsuccessful. Intergeneric grafts in the nightshade family, Solanaceae, are quite common. Tomato (*Lycopersicon esculentum*) can be grafted successfully on Jimson weed (*Datura stramonium*), tobacco

(*Nicotiana tabacum*), potato (*Solanum tuberosum*), and black nightshade (*Solanum nigrum*).

Grafting Between Families

Grafting such plants together is usually considered impossible, but there are reported instances in which it has been accomplished. These are with short-lived, herbaceous plants, though, for which the time involved is relatively short. Grafts, with vascular connections between the scion and stock, were successfully made, using white sweet clover, *Melilotus alba* (Leguminosae), as the scion and sunflower. *Helianthus annuus* (Compositae), as the stock. Cleft grafting was used, with the scion inserted into the pith parenchyma of the stock. The scions continued growth with normal vigor for over 5 months. As far as is known, however, there are no instances in which woody perennial plants belonging to different families have been successfully and permanently grafted together.

For any successful grafting operation there are five important requirements :

(*a*) *The stock and scion must be compatible*. They must be capable of uniting. Usually, but not always, plants closely related, such as two apple varieties, can be grafted together. Distantly related plants, such as an oak tree and an apple tree, cannot be grafted together.

(*b*) *The cambial region of the scion must be in intimate contact with that of the stock.* The cut surfaces should be held together tightly by wrapping, nailing, or some other such method. Rapid healing of the graft union is necessary so that the scion may be supplied with water and nutrients from the stock by the time the buds start to open.

(*c*) *The grafting operation must be done at a time when the stock and scion are in the proper physiological stage.* Usually this means that the scion buds are dormant. For deciduous plants, dormant scion wood is often collected during the winter and kept inactive by storing at low temperature. The rootstock plant may be dormant or in active growth, depending upon the grafting method used.

(*d*) *Immediately after the grafting operation is completed, all cut surfaces must be carefully protected from desiccation.* This is done either by covering with grafting wax or by placing the grafts in moist material or in a covered grafting frame.

(*e*) *Proper care must be given the grafts for a period of time after grafting.* Shoots coming from the stock below the graft will often choke out the desired growth from the scion. Or, in some cases, shoots from the scion will grow so vigorously that they break off unless staked ond tied or cut back.

METHODS OF GRAFTING

Whip, or Tongue, Grafting

This method, as shown in Figures 8.4 and 8.5 is particularly useful for grafting relatively small material, 1/4 to 1/2 in. in diameter. It is highly successful if properly done because there is considerable cambial contact. It heals quickly and makes a strong union. Preferably, the scion and stock should be of equal diameter. The scion should contain two or three buds with the graft made in the smooth internode area below the lower bud.

The cuts made at the top of the stock should be exactly the same as those made at the bottom of the scion. First, a long, smooth, sloping cut is made, 1 to 2½ in. long. The longer cuts are made when working with large material. This first cut should be made, preferably with one single stroke of the knife, so as to leave a very smooth surface. To do this, the knife must be sharp. Wavy, uneven cuts will not result in a satisfactory union.

On each of these cut surfaces, a reverse cut is made. It is started downward at a point about one-third of the distance from the tip and should be about one-half the length of the first cut. To obtain a smooth-fitting graft, this second cut should not just split the grain of the wood but should follow along under the first cut, tending to parallel it.

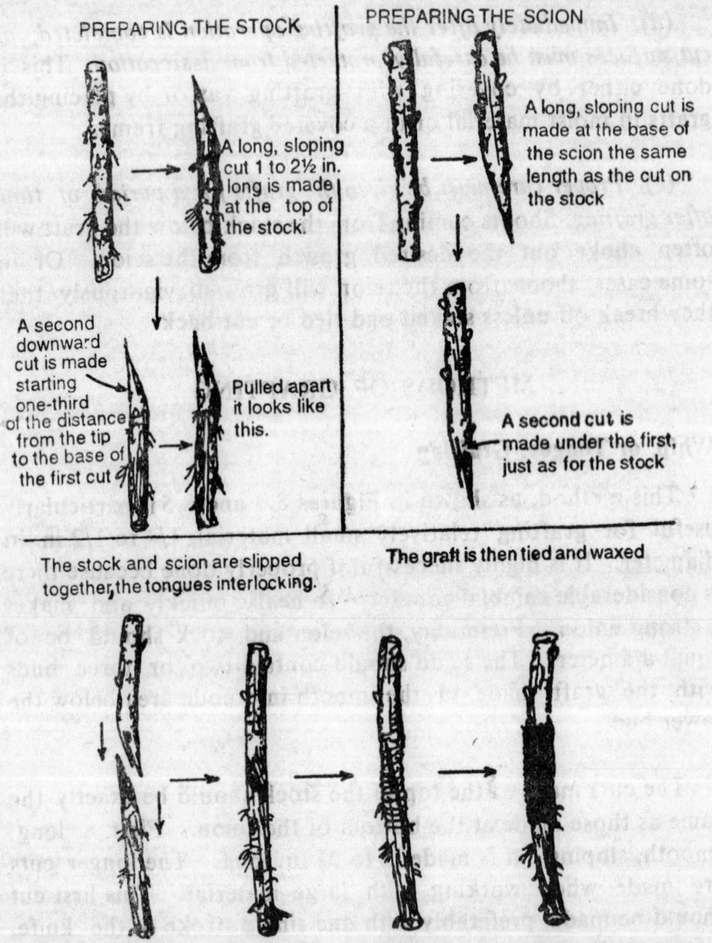

Fig. 8.4. The whip, or tongue, graft. This method is widely used in grafting small plant material and is especially valuable in making root grafts as illustrated here.

The stock and scion are then inserted into each other, with the tongues interlocking. It is extremely important that the cambium layers match along at least one side, preferably along both sides. The lower tip of the scion should not overhang the stock, as there is a likelihood of the formation of large callus knots. In some species, such callus overgrowths

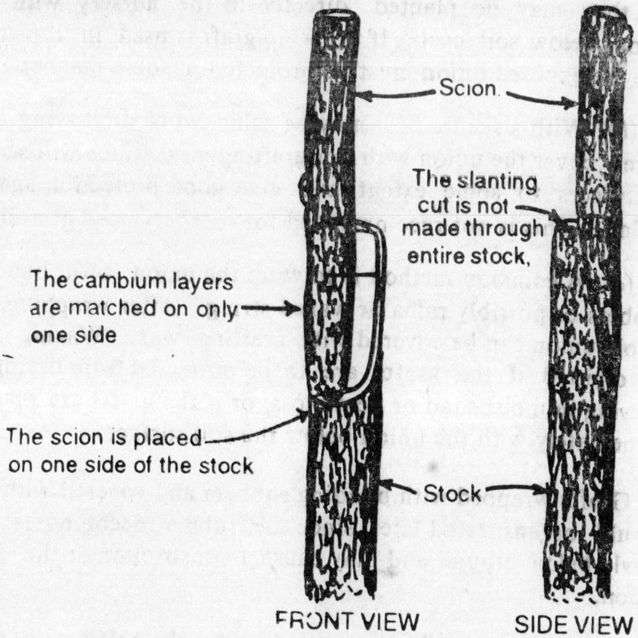

Fig. 8.5. **Method of making a whip, or tongue, graft when the scion is considerably smaller than the stock.**

are often mistaken for crown gall knots, which are caused by bacteria. The use of scions larger than the stock should be avoided for the same reason. If the scion is smaller than the stock, it should be set at one side of the stock so that the cambium layers will be certain to match along that side. If the scion is considerably smaller than the stock, the first cut on the stock consists only of a slice taken off one corner.

After the scion and stock are fitted together, they should be held securely in some manner until the pieces have united. There are a number of possible ways of doing this.

(*a*) If the unions are very well made with a tight, snug fit, it is possible that no additional wrapping or tying is needed, but it is safer to provide some type of wrapping. If not wrapped, the grafts must be protected from drying by burying in moist sand, peat moss, or sawdust until the union has healed.

Or they may be planted directly in the nursery with the union below soil level. If the whip graft is used in topworking, the exposed union must be protected in some manner.

(*b*) With a secure fit it may be sufficient to omit tying and merely cover the union with hot grafting wax, which will secure the pieces to some extent and give good protection against drying. This is not recommended for inexperienced grafters.

(*c*) A common method is to wrap the union with budding rubbers or possibly raffia or waxed string. After wrapping, the whole union can be covered with grafting wax. Waxing may be omitted if the grafts are to be protected from drying by burying in moist sand or peat moss, or if the grafts are planted immediately with the union below the soil surface.

Grafts wrapped with budding rubbers and covered with soil should be inspected later, since the rubber decomposes very slowly below ground and may cause a constriction at the graft union.

(*d*) A practice widely used is to wrap the grafts with some type of adhesive tape. A special nurseryman's tape is available. The tape is drawn tightly around the graft union with the edges slightly overlapping. This holds the parts together very well and prevents drying, thus eliminating the need for waxing. If just one thickness of tape is used, it will decompose sufficiently fast (if the union is below ground) that no constriction of growth will develop. If used above ground, the tape should be cut after the graft has healed. The use of tight wrapping material such as this is especially recommended when difficulty is encountered with the formation of excessive callus.

(*e*) Plastic tapes are available for wrapping grafts. They are used just as adhesive tape, although they are not adhesive. The final turn of the tape is secured by slipping it under the previous turn. This tape has some elasticity. Also, it deteriorates more slowly below ground than above.

The whip graft can also be used in topworking young trees. The stocks in this case would be small, pencil-sized branches, well distributed around the tree. The grafts are usually tied with nurseryman's adhesive tape or with string, the latter covered with a coat of grafting wax. After the parts of the graft have united, the tying material must be cut ; otherwise, the branch may be constricted as growth commences.

Splice Grafting

This method is the same as the whip, or tongue, graft except that the second or "tongue" cut is not made in either the stock or scion. A simple slanting cut of the same length and angle is made in both the stock and the scion. These are placed together and wrapped or tied as described for the whip graft. The splice graft is simple and easy to make. It is particularly useful in grafting plants that have a very pithy stem or which have wood that is not flexible enough to permit a tight fit when a tongue is made as in the whip graft.

Side Grafting

There are numerous variations of the side graft ; three of the most useful are described. As the name suggests, the scion is inserted into the side of the stock, which is generally larger in diameter than the scion.

Stub Graft

This method is useful in grafting branches of trees that are too large for the whip graft yet not large enough for other methods such as the cleft or bark graft. For this type of side graft, the best stocks are branches about 1 in. in diameter. An oblique cut is made into the stock branch with a chisel or heavy knife at an angle of 20 to 30 degrees. The cut should be about 1 in. deep, and at such an angle and depth that when the branch is pulled back the cut will open slightly but will close when the pull is released.

The scion should contain two or three buds and be about 3 in. long and relatively thin. At the basal end of the scion, a wedge about 1 in. long is made. The cuts on both sides of the scion should be very smooth, each made by one single cut with a sharp knife. It is best to insert the scion into the stock at an angle as shown in Figure 8.6 so as to obtain maximum contact of the cambium layers. The grafter inserts the scion into the cut while the upper part of the stock is pulled backward, using care to obtan the best cambium contact. Then the stock is released. The pressure of the stock should grip the scion tightly, making tying unnecessary, but if it is desired, the

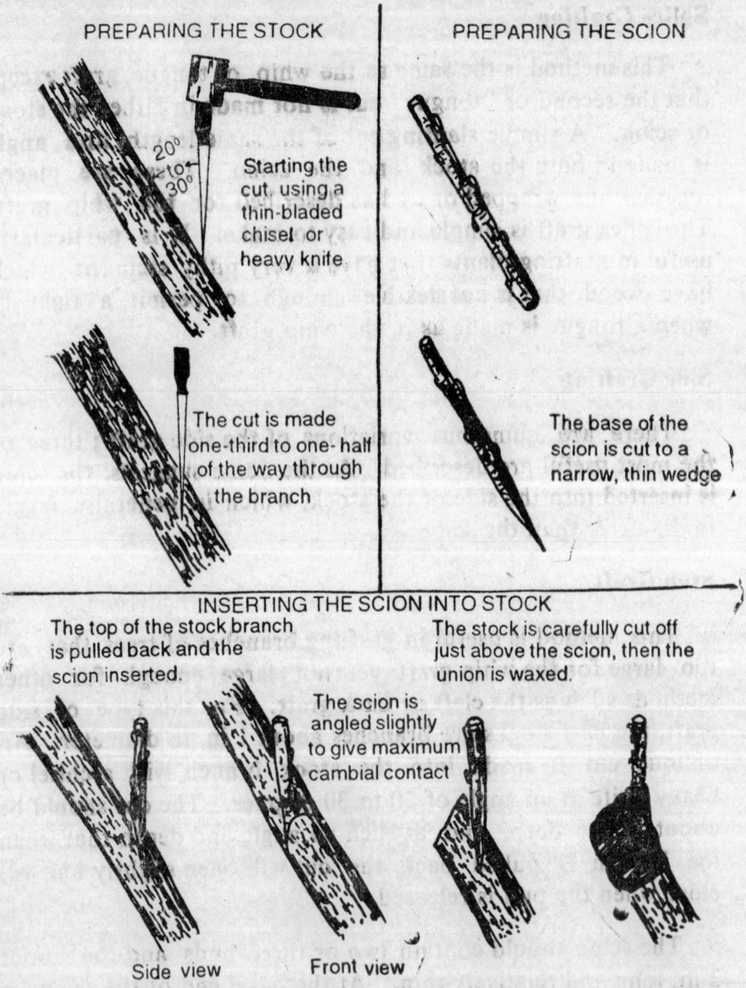

Fig. 8.6. **Steps in preparing the side, or stub, graft. A thinbladed chisel as illustrated here is ideal for making the cut, but a heavy butcher knife could be used satisfactorily.**

scion can be further secured by driving two small flat-headed wire nails (20 gauge, 5/8 in. long) into the stock through the scion. Wrapping the stock and scion at the point of union with nurseryman's tape may also be helpful. After the graft is completed, the stock may be cut off just above the union. This must be very carefully done or the scion may become dislodged. The entire graft union must be thoroughly covered with grafting wax, sealing all openings. The end of the scion should also be covered with wax.

Under some conditions, healing will be more rapid and certain if the stock is left intact and cut off above the scion later. In grafting citrus seedlings by use of side grafting methods, it is a common practice to lop over the top of the stock about 6 in. above the graft. Some time later, after growth starts from the scion, the top is completely removed just above the graft.

This method may be used to provide a new branch at a position in a tree where it is particularly needed. To force the new scion into active growth, it may be necessary to prune back rather severely the top of the stock branch above the graft.

This grafting method is recommended as being particularly useful in the nursery for the purpose of spring grafting young nursery trees which were fall-budded but in which the bud failed to grow.

Side-Tongue Graft

This type of side graft, shown in Fig. 8.7, is useful for small plants, especially some of the broad- and narrow-leaved evergreen species. The stock plant should have a smooth section in the stem just above the crown of the plant. The diameter of the scion should be slightly smaller than that of the stock. The cuts at the base of the scion are made just as for the whip graft. Along a smooth portion of the stem of the stock, a thin piece of bark and wood, the same length as the cut surface of the scion, is completely removed. Then a reverse cut is made downward in the cut on the stock, starting

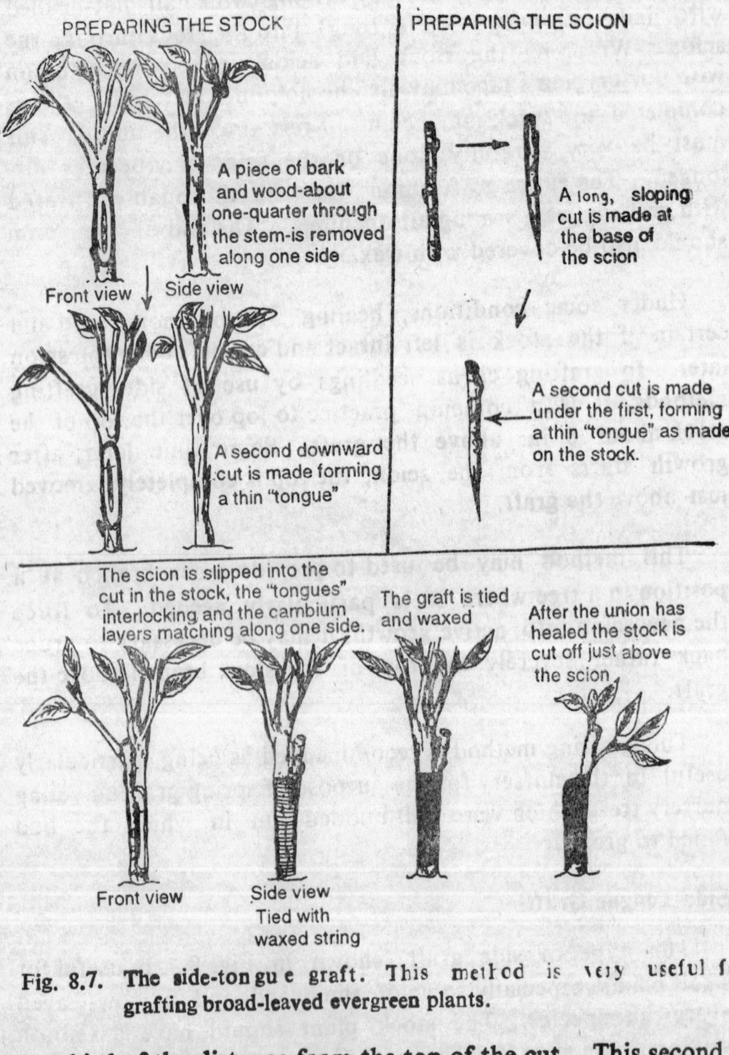

Fig. 8.7. **The side-tongue graft. This method is very useful for grafting broad-leaved evergreen plants.**

one-third of the distance from the top of the cut. This second cut in the stock should be the same length as the reverse cut in the scion. The scion is then inserted into the cut in the stock, the two tongues interlocking, and the cambium layer(s) matching.

The top of the stock is left intact for several weeks until the graft union has healed. Then it may be cut back above

the scion gradually or all at once. This forces the buds on the scion into active growth.

Side-Veneer Graft (Spliced Side Graft)

This variation of side grafting, shown in Fig. 8.8, is widely used, especially for grafting small potted plants, such as seedl-

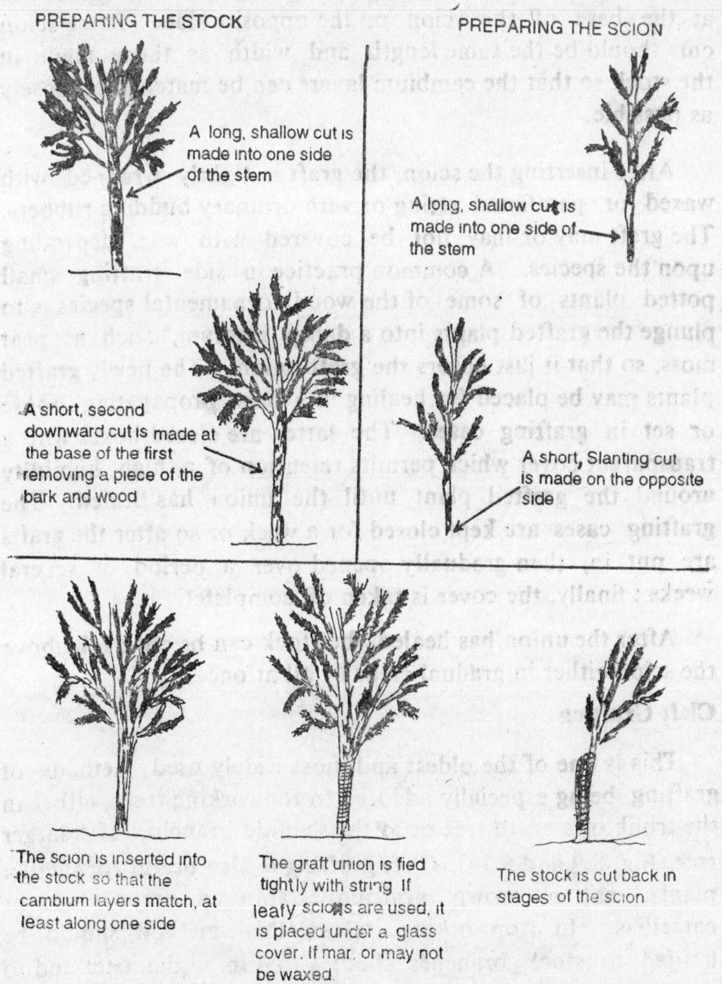

Fig. 8.8. Steps in making the side-veneer graft. This method is widely used in propagating narrow-leaved evergreen species that are difficult to start by cuttings.

ing evergreens. A shallow downward and inward cut from 1 to 1½ in. long is made in a smooth area just above the crown of the stock plant. At the base of this cut, a second short inward and downward cut is made, intersecting the first cut, so as to remove the piece of wood and bark. The scion is prepared with a long cut along one side and a very short one at the base of the scion on the opposite side. These scion cuts should be the same length and width as those made in the stock so that the cambium layers can be matched as closely as possible.

After inserting the scion, the graft is tightly wrapped with waxed or paraffined string or with ordinary budding rubbers. The graft may or may not be covered with wax, depending upon the species. A common practice in side grafting small potted plants of some of the woody ornamental species is to plunge the grafted plants into a damp medium, such as peat moss, so that it just covers the graft union. The newly grafted plants may be placed for healing in a mist propagating house or set in grafting cases. The latter are closed boxes with a transparent cover which permits retention of a high humidity around the grafted plant until the union has healed. The grafting cases are kept closed for a week or so after the grafts are put in, then gradually opened over a period of several weeks ; finally, the cover is taken off completely.

After the union has healed, the stock can be cut back above the scion either in gradual steps or all at once.

Cleft Grafting

This is one of the oldest and most widely used methods of grafting, being especially adapted to topworking trees, either in the trunk of a small tree or in the scaffold branches of a larger tree (Fig. 8.9 and 8.10). Cleft grafting is also useful for smaller plants, as in crown grafting established grape vines or camellias. In topworking trees, this method should be limited to stock branches about 1 to 4 in. in diameter and to species with fairly straight-grained wood which will split evenly. Although cleft grafting can be done any time during the dormant season, the chances for successful healing of the graft union

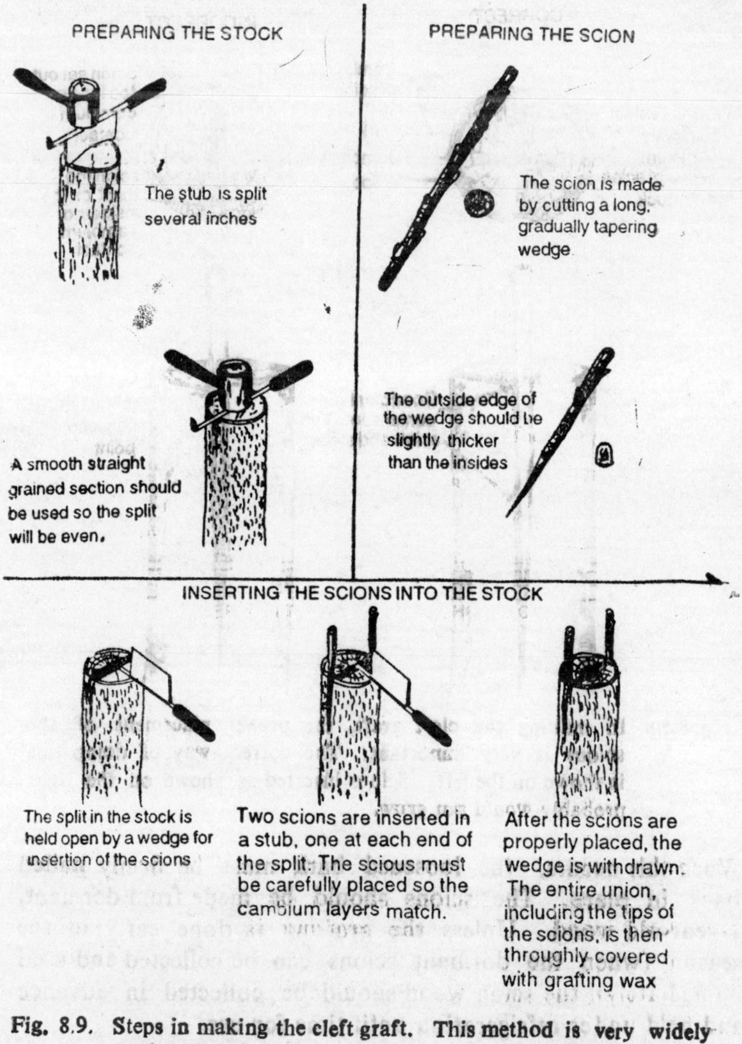

Fig. 8.9. **Steps in making the cleft graft. This method is very widely used and is quite successful if the scions are inserted so that the cambium layers of stock and scion match properly.**

are best if the work is done in early spring just when the buds of the stock are beginning to swell but before active growth has started. If cleft grafting is done after the tree is in active growth, it is likely that the bark of the stock will separate from the wood, causing difficulties in obtaining a good union.

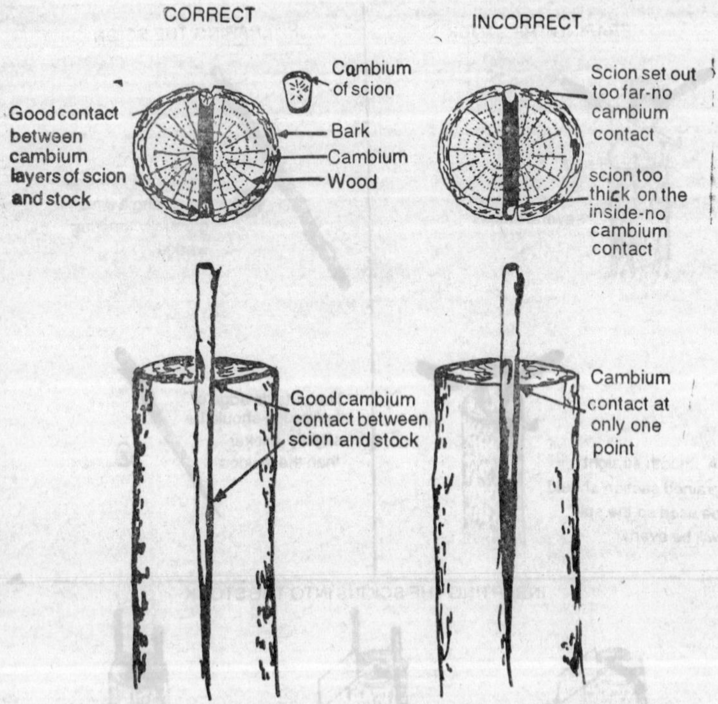

Fig. 8.10. In making the cleft graft, the proper placement of the scions is very important. The correct way of doing this is shown on the left. Scions inserted as shown on the right probably would not grow.

When this occurs, the loosened bark must be firmly nailed back in place. The scions should be made from dormant, 1-year-old wood. Unless the grafting is done early in the season (when the dormant scions can be collected and used immediately), the scion wood should be collected in advance and held under refrigeration until time for use.

In sawing off the branch for this and other topworking methods, the cut should be made at right angles to the main axis of the branch.

In making the cleft graft, a heavy knife, such as a butcher knife, or one of several special cleft grafting tools, is used to make a vertical split for a distance of 2 to 3 in. down the center of the stub to be grafted. This is done by pounding the knife in with a hammer or mallet. It is very important to have the branch sawed off in such a position that the end of the stub which is left is smooth, straight-grained, and free of knots for at least 6 in. Otherwise, when the split is made it may not be straight, or the wood may split one way and the bark another. The split should be in tangential rather than radial direction in relation to the center of the tree. This permits better placement of the scions for their subsequent growth. Sometimes this cleft is made by a longitudinal saw cut rather than by splitting. After a good, straight split is made, a screwdriver, chisel, or the wedge part of the cleft-grafting tool is driven into the top of the split to hold it open.

Two scions are usually inserted, one at each side of the stock where the cambium layer is located, although in large branches, two splits are sometimes made at right angles to each other and four scions inserted. The scions should be 3 or 4 in. long and have two or three buds. The basal end of each scion should be cut into a long, gently sloping wedge— about 2 in. long. It is not necessary that the end of the wedge come to a point. The side of the wedge which is to go to the outer side of the stock should be slightly wider than the inside edge. Thus, when the scion is inserted and the tool is removed, the full pressure of the split stock will come to bear on the scions at the position where the cambium of the stock touches the cambium layer on the outer edge of the scion. Since the bark of the stock is almost always thicker than the bark of the scion, it is usually necessary for the outer surface of the scion to set slightly in from the outer surface of the stock in order to match the cambium layers.

In all types of grafting the scion must be inserted right side up. That is, the points of the buds on the scion should be pointing away from the stock. Failure to observe this means ultimate failure of the graft to grow.

The long, sloping wedge cuts at the base of the scion should be smooth, made by a single cut on each side with a very sharp knife. Both sides of the scion wedge should press firmly against the stock for their entire length. A common mistake in cutting scions for this type of graft is to make the cut on the scion too short with too abrupt a slope, so that the only point of contact is just at the top. Shaving slightly the sides of the split in the stock will often permit a smoother contact.

After the scions are properly made and inserted, the tool is carefully withdrawn, using care not to disturb the scions. They should be held so tightly by the pressure of the stock that they cannot be pulled loose by hand. No further tying or nailing is needed unless very small stock branches have been used, in which case wrapping around the top of the stock tightly with string or waxed cloth can be done to hold the scions in place more securely.

Thorough waxing of the completed graft is essential. The top surface of the stub should be entirely covered, permitting the wax to work into the split in the stock. The sides of the grafted stub should be well covered with wax as far down the stub as the split has gone. The top of the scions should be waxed but not necessarily the bark or buds of the scion. Two or 3 days later all the grafts should be inspected and rewaxed where openings appear. Lack of thorough and complete waxing in this type of graft is almost certain to result in failure.

Saw-Kerf (Notch) Grafting

This can be used in place of the left graft. (see Fig. 8.11). It is especially useful in topworking trees with branches 3 to 4 in. or more in diameter. It does not have the serious disadvantage of a deep split in the stock (as does the cleft graft), which may permit the entrance of decay-producing organisms. Saw-kerf grafting can be done over a long period of time-2 to 3 months—before growth of the stock starts in the spring. If the work is to be performed late in the season, at the time active growth is starting, it is necessary to collect the scion wood earlier and hold it under refrigeration to keep it dormant

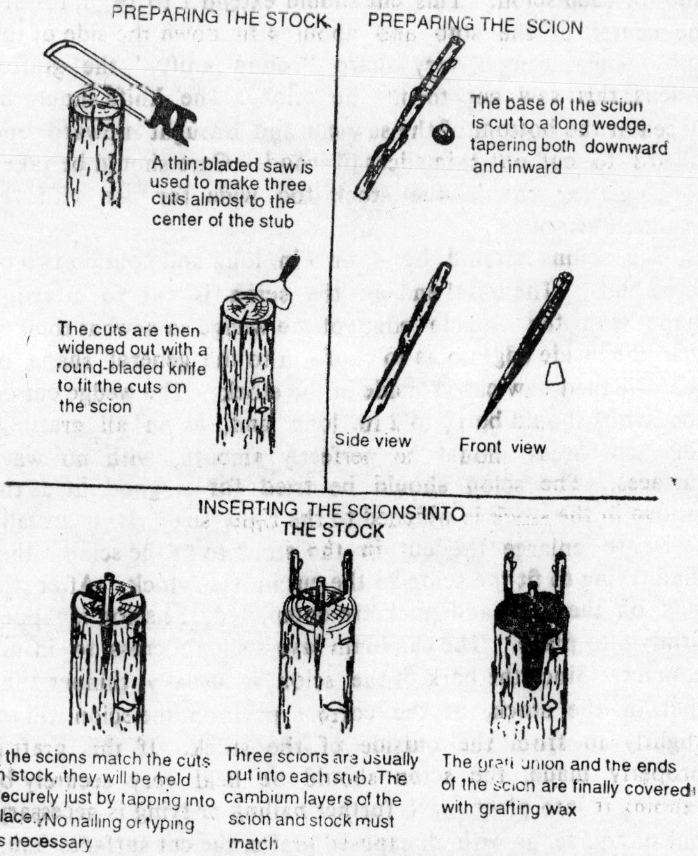

Fig. 8.11. Steps in making the saw kerf (notch) graft.

until the grafting is to be done. Curly-grained stock branches, which could not split evenly for the cleft graft, can be worked readily by the saw-kerf graft. The saw-kerf graft is somewhat more difficult for beginners to perform than most of the other types, but in the hands of experienced workers it can be done rapidly and is highly successful, especially with certain hard-to-graft species, such as the peach.

There are two types of saw-kerf grafts—deep and shallow. Usually, three scions are inserted into each stock branch. A cut with a thin-bladed, fine-toothed saw is made into the stub for each scion. This cut should extend 1 to 1½ in. toward the center of the stub and about 4 in. down the side of the stub. Then, using a very sharp "round knife," the grafter widens this saw cut to fit the scion. The knife should be placed at the bottom of the saw cut and brought upward and inward to cut out thin slices of wood. Care should be taken not to get the cut in the stock too wide for use with the available scions.

The scions should be 4 or 5 in. long and contain two or three buds. The basal end of the scion is cut to a wedge shape with the outside edge of the wedge somewhat thicker than the inside edge so as to conform to the general shape of the widened saw cut as made in the stock. The wedge cut on the scions should be 1½ to 2 in. long and, as in all grafting, the cut areas should be perfectly smooth, with no wavy surfaces. The scion should be tried for a good fit as the groove in the stock is widened to the right size. It is usually easier to enlarge the cut in the stock to fit the scion rather than trying to fit the scion to the cut in the stock. After the cuts on the scion and stock are completed, the scion is tapped firmly into place. The cambium layers should cross to insure contact. Since the bark of the scion is usually thinner than that of the stock, at the correct position the scion will set slightly in from the outside of the stock. If this graft is properly made, the scion should be held very securely by tapping it into place. No further nailing or tying is necessary, but of course, as with all exposed grafts, the cut surfaces must be thoroughly waxed.

The round knife is a useful tool in making this graft. It may be adapted from a leather-worker's knife or a cook's mincing knife. The edge of the knife should be sharpened so that it is flat on one side and beveled on the other, being used so that the flat side is held against the wood in widening out the saw cut made in the stock. If a round knife of this type is not available, a large-bladed grafting knife may be substituted.

The shallow type of saw-kerf graft is made in essentially the same manner, except that the cut into the stock is qutie shallow. The outer edge of the scion is considerably thicker than the inner edge. It should be cut to just fit into the opening in the stock. Care must be taken to see that the cambium layers match exactly. The scions, in this case, are held in place by nailing with flat-headed wire nails (5/8 in. long. 20 gauge). Finally, of course, the exposed cut surfaces and the end of the scions are thoroughly covered with wax.

Bark Grafting

This method is rapid, simple, readily performed by amateurs, and if properly done, gives a high percentage of "takes." It requires no special equipment and can be performed on branches ranging from 1 in. up to a foot or more in diameter. The latter size is not recommended as it is difficult to heal over such large stubs before decay-producing organisms get started. The bark graft, since it depends on the bark separating readily from the wood, can only be done after active growth of the stock has started in the spring. As dormant scions must be used, it is necessary to gather the scion wood for deciduous species during the dormant season and hold it under refrigeration until the grafting operation is done. For evergreen species, freshly collected scion wood can be used. Scions are not as securely attached to the stock as in some of the other methods and are more susceptible to wind breakage during the first year even though the healing has been satisfactory. Therefore the new shoots arising from the scions probably should be staked during the first year, or cut back to about half their length, especially in windy areas. After a few years' growth, the bark graft union is as strong as the unions formed by other methods.

There are several modifications of the bark graft. Three important types are described.

Bark Graft (Method No. 1)

Several scions are inserted into each stub. For each scion, a vertical knife cut about 2 in. long is made at the top end of the stub through the bark to the wood. The bark is then lifted slightly along both sides of this cut, in preparation for the insertion of the scion. The scion should be of dormant wood, 4 or 5 in. long, containing two or three buds, and be $\frac{1}{4}$ to $\frac{1}{2}$ in. in thickness. One cut about 2 in. long is made along one side at the base of the scion. With large scions, this cut extends about onethird of the way into the scion, leaving a "shoulder" at the top. The purpose of this shoulder is to reduce the thickness of the scion to minimize the separation of bark and wood after insertion in the stock. The scion should not be cut too thin, however, or it will be mechanically weak and break off at the point of attachment to the stock.

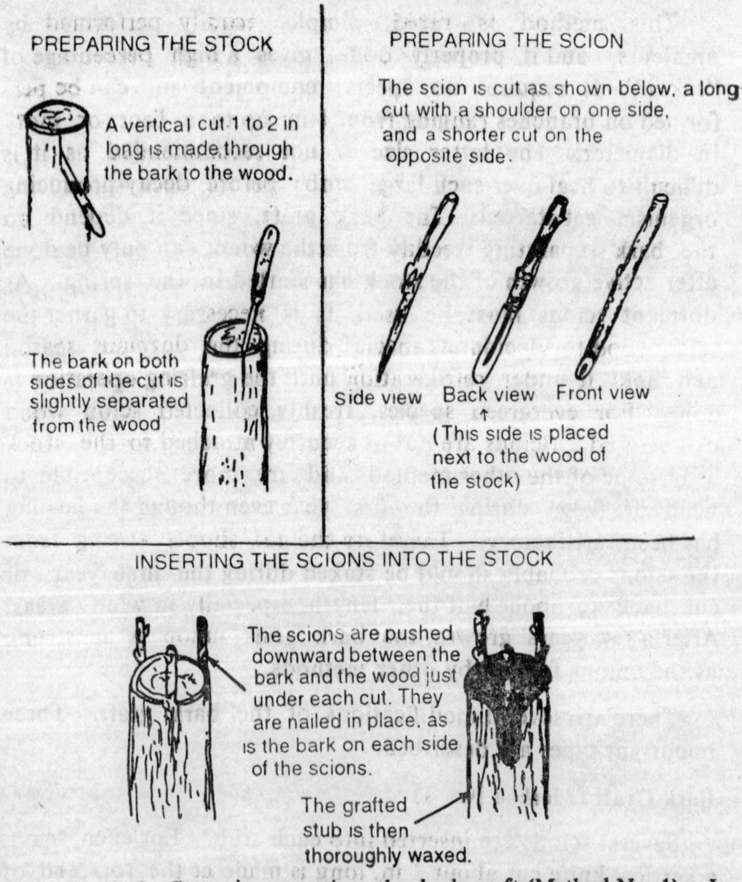

Fig. 8.12. Steps in preparing the bark graft (Method No. 1). In grafting some thick-barked plants the vertical cut in the bark is unnecessary, the scion being inserted between the bark and wood of the stock.

If small scions are used, no shoulder is necessary. On the side of the scion opposite the first long cut, a second shorter cut is made, as shown in Fig. 8.12, thereby bringing the basal end of the scion to a wedge shape. The scion is then inserted between the bark and the wood of the stock, centered directly under the vertical cut through the bark. The longer cut on the scion is placed against the wood, and the shoulder on the scion is brought down until it rests on top of the stub. The scion is then ready to be fastened in place. A satisfactory method is to nail the scion into the wood, using two nails per scion. Flat-headed nails 5/8 to 1 in. long, of 19 or 20 gauge wire, depending on the size of the scions, are satisfactory. The bark on both sides of the scion should also be securely nailed down, or it will tend to peel back from the wood. Another method commonly used with soft-barked trees, such as the avocado, is to insert all the scions in the stub and then hold them in place by wrapping with string, adhesive tape, or waxed cloth around the stub. This is more effective than nailing in preventing the scions from blowing out but probably does not give as tight a fit. Both nailing and wrapping are advisable for maximum strength. If a wrapping material is used, it may be necessary to cut this later to prevent constriction.

In cases in which the bark of the stock is quite thick and small scions are used, it may not be necessary to make the vertical slit in the bark. The scions, cut as described before, can be pushed into place between the bark and the wood. They can then be nailed or tied securely with string or waxed cloth.

After the stub has been grafted and the scions fastened by nailing or tying, all cut surfaces, including the end of the scions, should be thoroughly covered with grafting wax.

Bark Graft (Method No. 2)

This type of graft (Fig. 8.13) is similar to method No. 1, except that the scion is not inserted centered on the vertical cut in the bark of the stock. The bark is lifted only along one

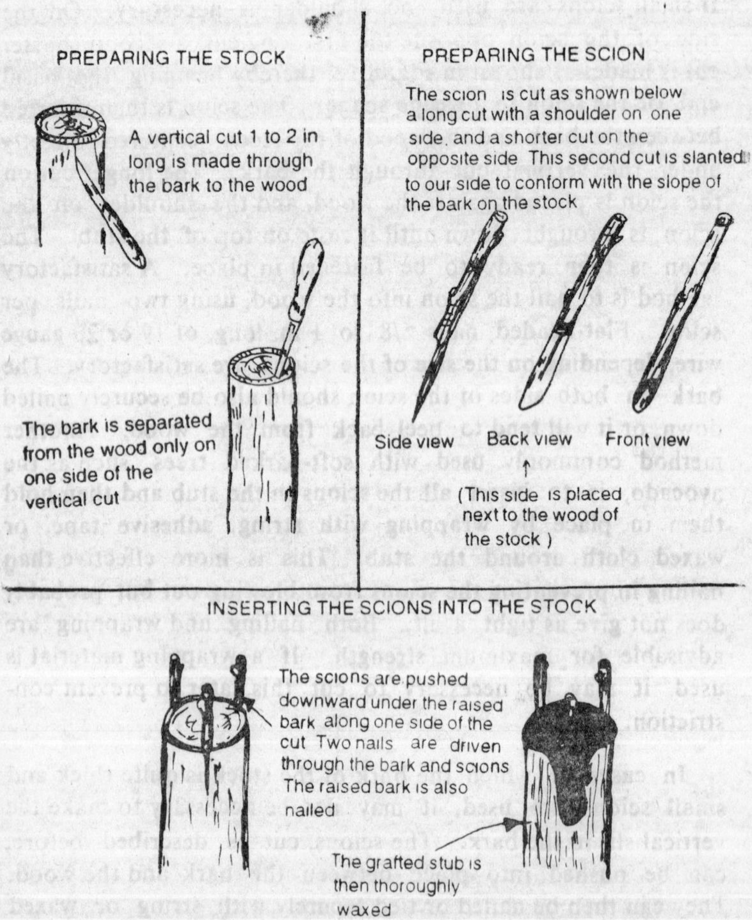

Fig. 8.13. Steps in making the bark graft (Method No. 2).

side of the vertical cut ; the bark on the other side is not disturbed. The scion is inserted under the raised bark and held in place with two nails driven through the bark of the stock, through the scion, and into the wood of the stock. The scion is cut with a shoulder, just as described for method No. 1. However, the short cut on the back of the scion, rather than being parallel to the first longer cut, is slanted to one side slightly to conform to the slope of the bark under which the scion is inserted. As in method No. 1, the raised bark near the scion should be fastened securely in place with two or more nails or wrapped with tape to prevent it from peeling back. This modification of the bark graft has been widely used with good results. It has an advantage over method No. 1, in that one of the edges of the scion is placed against undisturbed bark with its intact cambium cells. This tends to promote more rapid healing of the union. In method No. 1, where the bark is lifted away from the scion on both sides, intact cambium cells are some distance from the scion and healing is usually slower.

Bark Graft (Method No. 3)

In this method, as shown in Fig. 8.14, *two* knife cuts about 2 in. long are made through the bark of the stock down to the wood, rather than just one as in the other two methods. The distance between these two cuts should be exactly the same as the width of the scion. The piece of bark between the cuts should be lifted and the terminal two-thirds cut off. The scion is prepared with a smooth slanting cut along one side at the basal end. This cut should be about 2 in. long but made *without* the shoulder, in contrast to the other two methods. On the opposite side of the scion, a cut about ½ in. long is made, forming a wedge at the base of the scion. The scion should fit snugly into the opening in the bark with the longer cut inward and with the wedge at the base slipped under the flap of remaining bark. Rapid healing can be expected because both sides of the scion are touching undisturbed bark and cambium cells, which is not the case in the other two type of the bark graft.

The scion should be nailed into place with two nails, the lower nail going through the flap of bark covering the short cut on the back of the scion. If the bark along the sides of the scion should accidentally become disturbed, it must be nailed back into place.

Method No. 3 is well adapted for use with thick-barked trees, such as walnuts, on which it is not feasible to insert the scion under the bark.

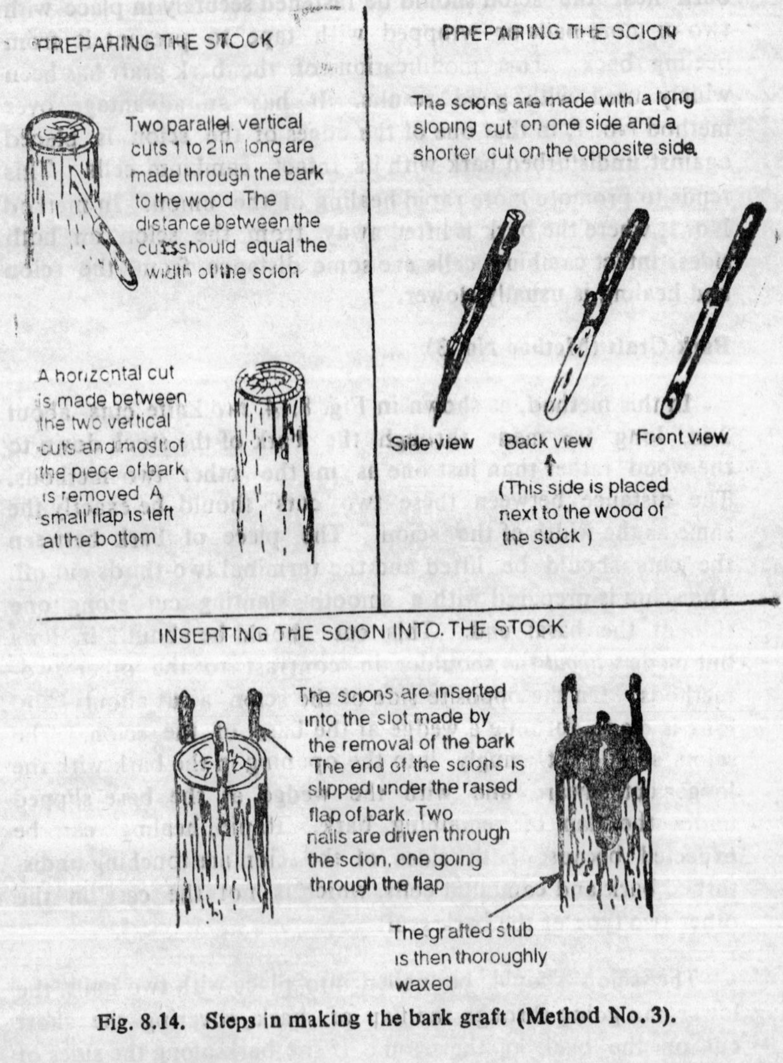

Fig. 8.14. Steps in making the bark graft (Method No. 3).

Approach Grafting

The distinguishing feature of approach grafting is that two independent, self-sustaining plants are grafted together. After a union has occurred, the top of the stock plant is removed above the graft and the base of the scion plant is removed below the graft. Sometimes it is necessary to sever these parts gradually rather than all at once. Approach grafting provides a means of establishing a successful union between certain plants which are difficult to graft together otherwise. It is usually performed with one or both of the plants to be grafted growing in a pot or container. Use is often made of this method by placing seedling plants growing in containers under an established plant which is to furnish the scion part of the new, grafted plant.

This type of grafting can be done at any time of the year, but healing of the union is more rapid if it is performed at a season when growth is active. As in other methods of grafting, the cut surfaces should be securely fastened together, then covered with grafting wax to prevent drying of the tissues.

Three useful methods of making approach grafts are described below and illustrated in Figure 8.15.

Spliced Approach Graft

Preferably the two stems should be approximately the same size. At the point where the union is to occur, a slice of bark and wood 1 to 2 in. long is cut from both stems. This cut should be the same size on each so that identical cambium patterns will be made. The cuts must be perfectly smooth and as nearly flat as possible so that when they are pressed together there will be close contact of the cambium layers. The two cut surfaces are then bound tightly together with string, raffia, or nurseryman's tape. The whole union should then be covered with grafting wax. After the parts are well united, which may require considerable time in some cases, the stock above the union and the scion below the union are cut, and the graft is then completed. It may be necessary to reduce the leaf area of the scion if it is more than the root system of the stock can sustain.

Tongued Approach Graft

This is the same as the spliced approach graft except that after the first cut is made in each stem to be joined, a second cut—downward on the stock and upward on the scion—is made, thus providing a thin tongue on each piece. By inter-

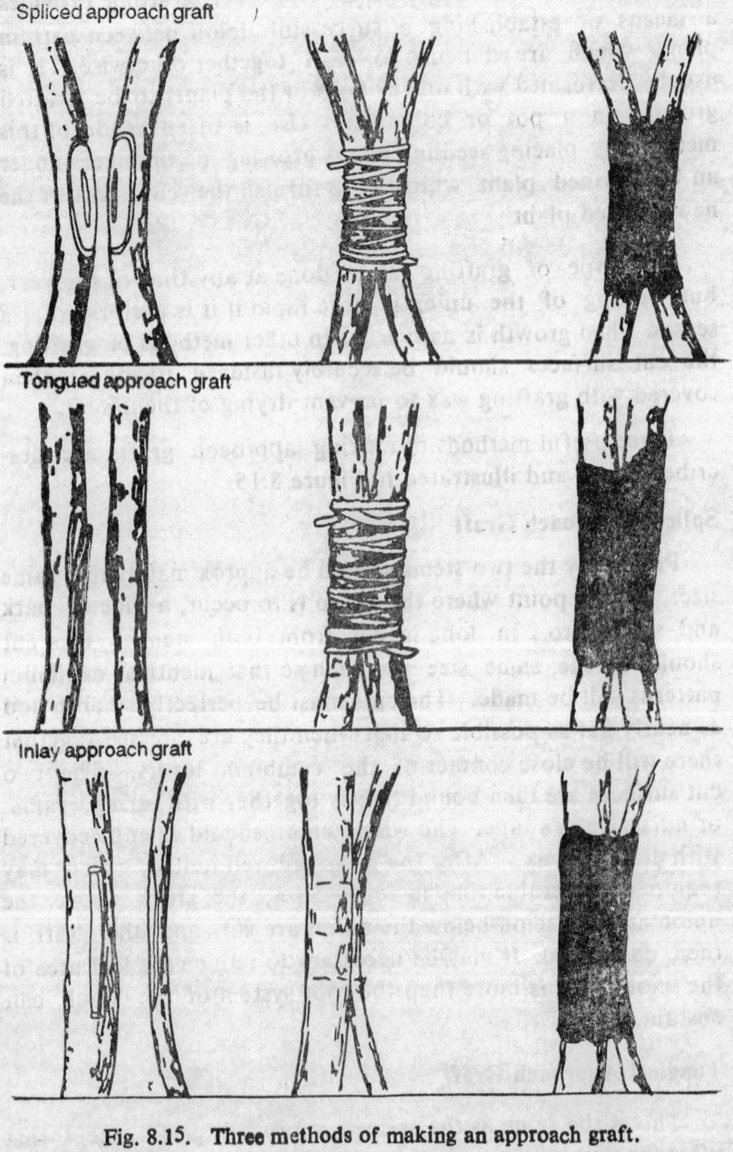

Fig. 8.15. Three methods of making an approach graft.

locking these tongues a very tight, closely fitting graft union can be obtained.

Inlay Approach Graft

This method may be used if the bark of the stock plant is considerably thicker than that of the scion plant. A narrow slot, 3 or 4 in. long, is made in the bark of the stock plant by making two parallel knife cutsand removing the strip of bark between. This can only be done when the stock plant is actively growing and the bark "slipping." The slot should be exactly as wide as the scion to be inserted. The stem of the scion plant, at the point of union, should be given a long, shallow cut along one side, of the same length as the slot in the stock plant and deep enough to go through the bark and slightly into the wood. This cut surface of the scion branch should be laid into the slot cut in the stock plant and held there by nailing with two or more small, flat-headed wire nails. The entire union must then be thoroughly covered with grafting wax. After the union has healed, the stock can be cut off above the graft and the scion below the graft.

Inarching

This method is similar to approach grafting in that both stock and scion plants are on their own roots at the time of grafting ; it differs in that the top of the new rootstock plant usually does not extend above the point of the graft union as it does in approach grafting. Inarching is generally considered to be a form of "repair grafting," being used in cases in which the roots of an established tree have been damaged by such things as cultivation implements, rodents, or disease. It can be used to very good advantage in saving a valuable tree or improving its root system.

Seedlings (or rooted cuttings) planted beside the older, damaged tree, or suckers arising near its base, are grafted into the trunk of the tree for the purpose of providing a new root system to supplant the damaged roots. The seedlings to be inarched into the tree should be spaced about 5 or 6 in. apart around the circumference of the tree if the damage is extensive. The tree will usually stay alive for some time after the injury occurs unless it is very severe. A satisfactory procedure for inarching is to plant seedlings of a compatible species or variety around the tree during the dormant season. Then, as active growth commences in early spring, the grafting operation can be done.

Inarching old, weakly growing trees with strong, vigorous seedling rootstocks has on some occasions proved beneficial in promoting renewed active growth of the old trees.

The seedling plants to provide the new root system are usually considerably smaller than the tree to be repaired. As illustrated in Fig. 8.16, the graft union is made in a manner similar to that described for method No. 3 of the bark graft. The upper end of the seedling, which should be $\frac{1}{4}$ to $\frac{1}{2}$ in. thick, is given a long shallow cut along the side for 4 to 6 in. This cut should be on the side next to the trunk of the tree and made deep enough to remove some of the wood, thus exposing two strips of cambium tissue. At the end of the seedling another shorter cut, about $\frac{1}{2}$ in. long, is made on the side opposite the long cut, this makes a sharp, wedge-shaped end on the seedling stem. A long slot is made in the trunk of the older tree by removing a piece of bark the exact width of the seedling and just as long as the cut surface made on the seedling. A small flap of bark is left at the upper end of the slot, under which the wedge end of the seedling is inserted. Then the seedling is nailed into the slot with four or five small, flat-headed wire nails. The nail at the top of the slot should go through the flap of bark and through the end of the seedling. If any of the bark of the tree along the sides of the seedling should accidentally be pulled loose, it is necessary to nail it back in place. After nailing, the entire area of the graft union should be thoroughly waxed.

In inarching some species, such as the walnuts, better healing of the union takes place if the top of the seedling is retained and allowed to extend above the graft union for a period of time, finally being cut off just above the union.

Owing to the food materials translocated from the larger tree, the seedlings grow very rapidly and soon provide a considerable number of new roots for the grafted tree. If shoots arise from the inarches, they should be suppressed by

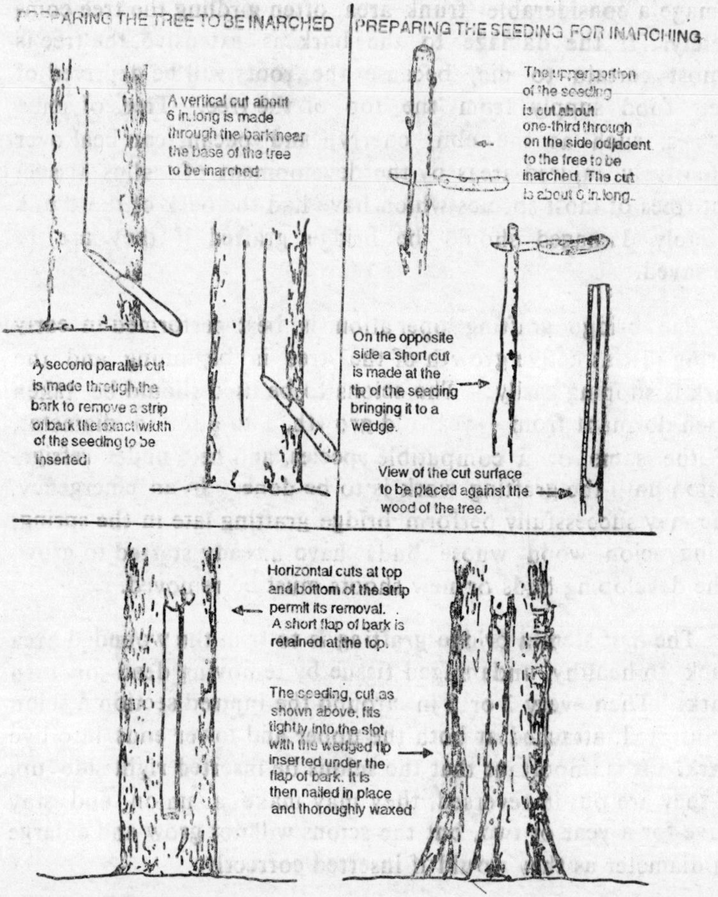

Fig. 8.16. Steps in inarching a large plant with smaller ones planted around its base.

cutting off their tips. The shoots should finally be removed entirely as the union becomes well established.

Bridge Grafting

This is a form of repair grafting, and is used in cases in which the root system of the tree has not been damaged but where there is injury to the bark of the trunk. Sometimes cultivation implements, rodents, disease, the winter injury will

damage a considerable trunk area, often girdling the tree completely. If the damage to the bark is extensive, the tree is almost certain to die, because the roots will be deprived of their food supply from the top of the tree. Trees of some species, such as the elm, cherry, and pecan, can heal over extensively injured areas by the development of callus tissue. But trees of most species which have had the bark of the trunk severely damaged should be bridge grafted if they are to be saved.

The bridge grafting operation is best performed in early spring just as active growth of the tree is beginning and the bark is slipping easily. The scions to be used should be taken when dormant from 1-year-old growth, $\frac{1}{4}$ to $\frac{1}{2}$ in. in diameter, of the same or a compatible species, and held under refrigeration until the grafting work is to be done. In an emergency, one may successfully perform bridge grafting late in the spring, using scion wood whose buds have already started to grow. The developing buds or new shoots must be removed.

The first step in bridge grafting is to trim the wounded area back to healthy, undamaged tissue by removing dead or torn bark. Then every 2 or 3 in. around the injured section a scion is inserted, attached at both the upper and lower ends into live bark. It is important that the scions be inserted right side up. If they are put in reversed, they may make a union and stay alive for a year or two, but the scions will not grow and enlarge in diameter as they would if inserted correctly.

Figure 8.17 shows the details of making a satisfactory type of bridge graft. It is essentially the same as method No. 3 of the bank graft. Just above and below the injured area, a slot 2 to 3 in. long and exactly the width of the scion is cut in the bark of the trunk for each scion. The piece of bark is removed, with the exception of a flap about $\frac{1}{2}$ in. long which is left at the end of the slot. The scions are cut to fit into these slots at each end of the wound, and long enough to bow outward slightly. This bow allows for good contact at each end and permits some swaying of the trunk in the wind without tearing the scions loose. To prepare the scions, a cut is made

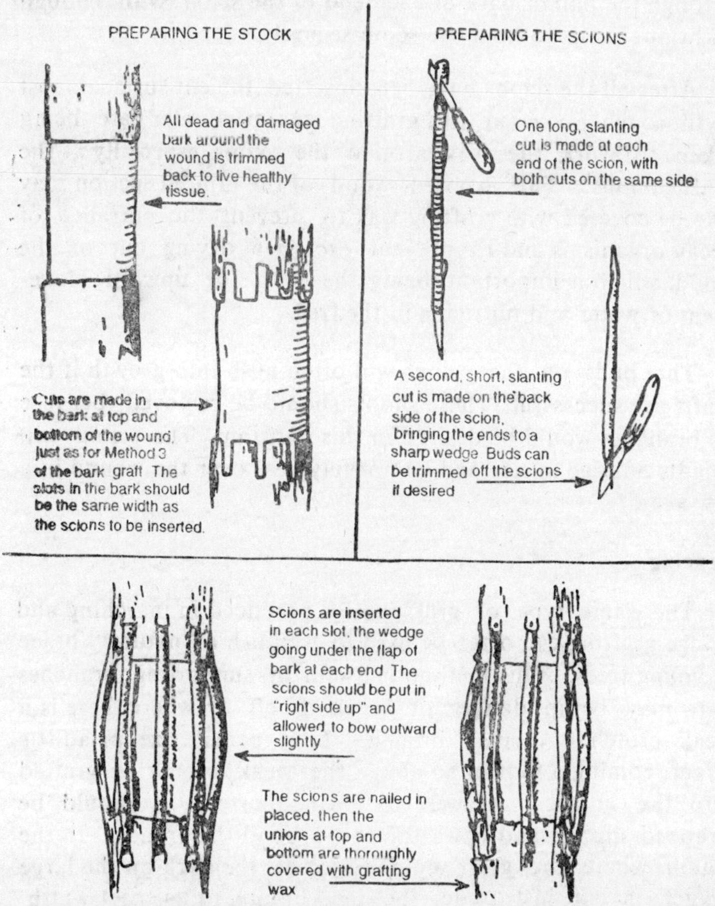

Fig. 8.17. A satisfactory method of making a bridge graft, using a modification of Method No. 3 of the bark graft.

along both ends on the side which is to fit into the slot. The cut should be the same length as the slot and deep enough to remove some bark and wood and expose two strips of cambium tissue. Then, on the opposite side of the scion, at each end, another shorter cut, about ½ in. long, is made to form the end of the scion into a wedge shape. The ends of the scion should be inserted under the flaps of bark and nailed in place, using 3/4 in. 20 gauge flat-headed wire nails. One nail should go through the flap of bark at each end of the scion with enough additional nails to hold the scion securely.

After all the scions have been inserted, the cut surfaces must be thoroughly covered with grafting wax, particular care being taken to work the wax around the scions, especially at the graft unions. The exposed wood of the injured section may also be covered with grafting wax to prevent the entrance of decay organisms and to prevent excessive drying out of the wood, which is important, being the path for upward movement of water and nutrients in the tree.

The buds on the scions will often push into growth if the grafts are successful. These shoots should be removed, because no branches would be desired in this position. The scions will rapidly enlarge in size and completely heal over the wound in a few years.

Bracing

The same type of graft union practiced in inarching and bridge grafting can often be used to establish a "natural" brace in young trees. This method is useful in supporting branches that may be in danger of breaking off or where there is a weak crotch. A small branch, about pencil size or a little larger, coming a foot or so above the weak crotch is grafted into the adjacent branch to be supported. It should be wrapped spirally and upward partly around the branch. In the region where the graft union is to be, the bark on the large branch is cut just under the small branch to its exact width. This should extend for a length of 6 in. or more, and the spiral piece of bark should be removed. Grafting such as this could only be done at a time of year when there is active growth and the bark is slipping easily. Early spring, just as the new growth is starting, is the preferable time.

After the slot in the bark is ready, the "scion"—the small branch—should be smoothly cut on the lower side through about a third of its thickness and for the length of the slot—6 in. or more. If the cuts are well made, the small branch should fit snugly into the slot. It is helpful to trim the end of the small branch to a wedge point by cutting it on the top side so that it can be inserted under the bark at the top end of the slot.

The next step is to nail the branch in place in the slot with small, flat-headed wire nails (5/8 or 3/4 in., 20 gauge) placed at 1- or 2-in. intervals. The graft union should then be thoroughly waxed to prevent drying. It is advisable to tie the two branches together temporarily with a strong cord to prevent whipping by the wind, which might put the graft union loose.

GRAFTING CLASSIFIED ACCORDING TO PLACEMENT

Grafting may be classified according to the part of the plant on which the scion is placed—a root, the crown (the junction of the stem and root at the ground level), or various places in the top of the plant.

Root Grafting

In this class of grafting the rootstock seedling, rooted cutting, or layered plant is dug up, and the roots are used as the stock for the graft. The entire root system may be used (*whole-root graft*—Figure 8.18), or the roots may be cut up into small pieces and each piece used as a stock (*piece-root graft*). Both methods give satisfactory results. As the roots used are relatively small ($\frac{1}{4}$ to $\frac{1}{2}$ in. in diameter), the whip, or tongue, graft is generally used. Root grafting is usually performed indoors during the late winter or early spring. The scion wood collected previously is held in storage, while the rootstock plants are also dug in the late fall and stored under cool (40° to 50° F ; 4° to 10° C) and moist conditions until the grafting is done.

The term *bench grafting* is sometimes given to this process, because it is often performed at benches by skilled grafters as a large-scale operation. A number of plants are propagated commercially by root grafting—apples, pears, grapes, and such ornamentals as the wisteria and rhododendron.

In making root grafts, the root pieces should be 3 to 6 in. long and the scions about the same length, containing two to

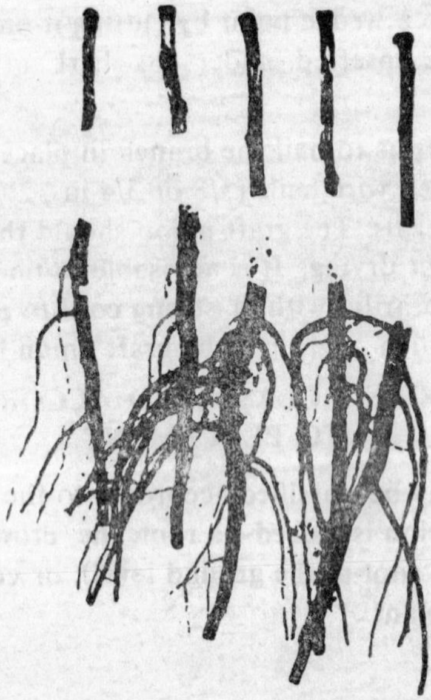

Fig. 8.18. Whole root apple grafts made by the whip or tongue method and wrapped with adhesive nursery tape.

four buds. After the grafts are made and properly tied they are bundled together in groups of 50 to 100 and stored for callusing in damp sand, peat moss, or other packing material. They may be placed in a cool cellar or under refrigeration at approximately 45° F (7° C) for about 2 months. The callusing period for apples can be shortened to around 30 days if the grafts are stored at a temperature of about 70° F (21° C) and at a high humidity. To use this higher callusing temperature the material should be collected in the fall and the grafts made before any cold weather has overcome the rest period of the scion buds. After the unions are well healed, the grafts must be stored at cool temperatures—35° to 40° F (2° to 4° C)—to overcome the "rest period" of the buds and to hold them dormant until planting. The grafts are lined out in early spring to the nursery row directly from the low temperature storage conditions. For general callusing purposes, temperatures below rather than above 70° F (21° C) are the most satisfactory. By the proper regulation of temperature, callusing processes may be accelerated by increased temperature or retarded by decreased temperature so that, within reasonable limits, a desired degree of callus formation may be had within a given length of time. Provision should be made for adequate aeration of the callusing grafts.

As soon as the ground can be prepared in the spring, the grafts are lined out in the nursery row 4 to 6 in. apart. They should be planted before growth, of the buds or roots begins. If this starts before the grafts can be planted, they should be moved to lower temperatures (30° to 35° F ;—1° to 2° C). The grafts are usually planted deep enough so that the graft union is just below the ground level, but if roots are to arise only from the rootstock, the graft should be planted with the union well above the soil level. It is very important to prevent scion rooting where certain definite influences, such as dwarfing or disease resistance, are expected from the rootstock.

After one summer's growth, the grafts should be large enough to transplant to their permanent location. If not, the scion may be cut back to one or two buds, or headed back somewhat to force out scaffold branches, and then allowed to grow a second year. With the older root system a strong, vigorous top is obtained the second year.

Nurse-Root Grafting

Under certain conditions, it is desired to have a stem cutting of a difficult-to-root species on its own roots. One way this can be done is by making a root graft, using the plant to be grown on its own roots as the scion and a root of a compatible species as the stock. The scion may be made longer than usual and the graft planted deeply with the major portion of the scion below ground. In some cases scion-rooting is promoted by rubbing a rooting stimulant, such as indolebutyric acid into several vertical cuts made through the bark at the base of the scion, just above the graft union. This is done just before planting, and the grafts are set deeply so that most of the scion is covered with soil. After one or two seasons of growth many of the scions will have roots. The temporary nurse rootstock is then cut off and the top reduced in proportion to the root system. The rooted scion is replanted to grow on its own roots.

Scion rooting is often better if the nurse-root is buried deeply enough so that a new shoot of the current season's developing from a bud on the scion, will be rooted, rather than trying to root the older, lignified tissue of the original scion. This is essentially a form of mound layering. As the new scion shoot grows, soil is gradually mounded up around it to a height of 5 or 6 in., although the terminal leaves are at no time covered.

Several methods of handling, as illustrated in Figure 8.19, eliminate the necessity of digging up the graft and cutting off the rootstock. The rootstock piece will eventually die if it is grafted onto the scion in an inverted position. The inverted stock piece sustains the scion until scion roots are formed, but the stock fails to receive food from the scion and eventually dies, thus leaving the scion on its own roots. In another method an incompatible rootstock is used. Hence, if the graft is planted deeply, scion roots will gradually become more important in sustaining the plant, and the incompatible rootstock will finally cease to function.

In a third method the base of the scion, just above the graft union, is bound with some type of wrapping material to eventually girdle and cut off the rootstock. Excellent results have been obtained with ordinary budding rubber strips (0.016 gauge). Budding rubbers disintegrate within a month when exposed to sun and air ; when buried in the soil, they will last as long as 2 years, allowing sufficient time for the scion to become rooted. Yet, owing to the slow deterioration of the rubber below ground, the rootstock is finally girdled and cut off.

Crown Grafting

A graft union made at the rootstem transition region—the "crown" of the plant—on an established rootstock is termed a "crown graft." Several methods of grafting are commonly

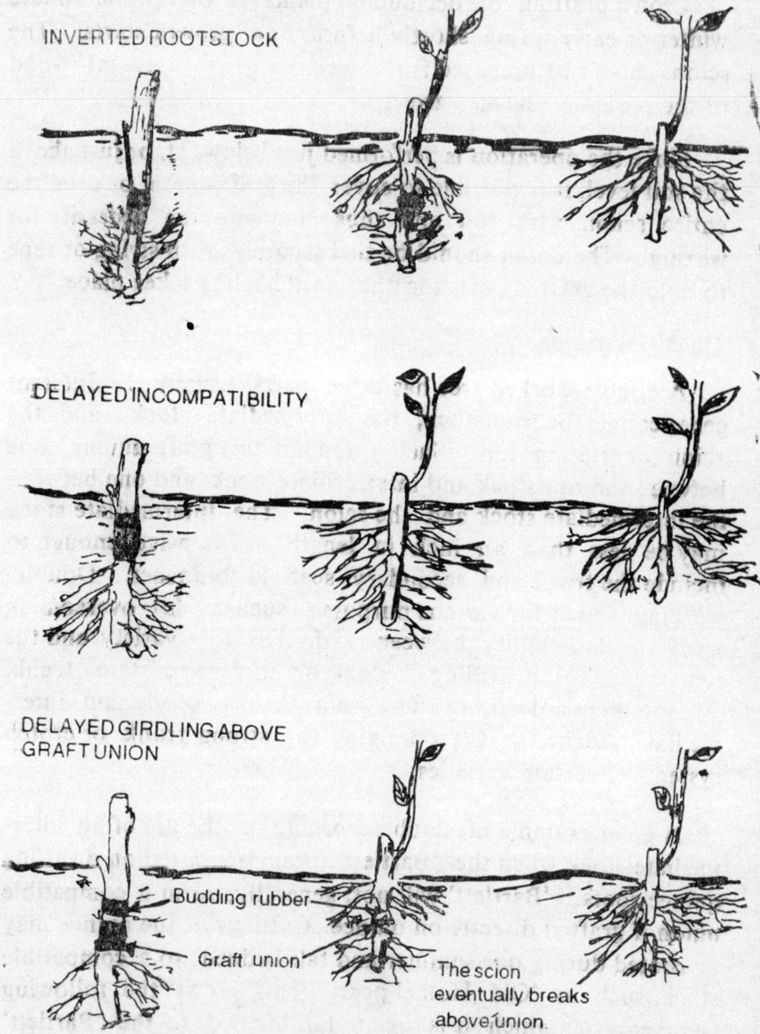

Fig. 8.19. The "nurse root" graft is a temporary graft used to induce the scion to develop its own roots. The nurse root sustains the plant until scion roots form ; then it dies. Three methods are shown here for preparing a nurse root graft.

used in crown grafting, such as the whip, side, cleft, saw-kerf, or bark graft, depending upon the species and size of the rootstock.

Crown grafting of deciduous plants is best done in late winter or early spring, shortly before new growth starts. The scions should be prepared from well-matured, dormant wood of the previous season's growth.

Since the operation is performed just below, at, or just above the soil level, it is possible to cover the graft union, or even the entire scion, with soil and thus eliminate the necessity for waxing. The union should be tied securely with string or tape to hold the grafted parts together until healing takes place.

Double-Working

A double-worked tree has three parts, usually all different genetically—the rootstock, the intermediate stock, and the scion, or fruiting, top. Such a tree has two graft unions, one between the rootstock and intermediate stock, and one between the intermediate stock and the scion. The intermediate stock may be less than an inch in length or extensive enough to include the trunk and secondary scaffold branches. Double-working is used for various purposes, such as (*a*) overcoming graft incompatibility between a desired top variety and the rootstock, (*b*) providing a cold or disease-resistant trunk, (*c*) obtaining a dwarfing effect from the use of certain intermediate stocks, or (*d*) obtaining the strong trunk or crotch systems of certain varieties.

A good example of double-working is the use of an intermediate stock when the 'Bartlett' pear is grafted on dwarfing quince roots. 'Bartlett' will not, generally, form a compatible union if grafted directly on quince. Cuttings of the quince may be rooted during one summer and fall-budded to a compatible stock, such as 'Old Home' pear. This grows the following summer, after which it is again fall-budded to the 'Bartlett' pear, the 'Bartlett' bud being inserted in the intermediate stock. The 'Bartlett' top is developed the next summer, after which the double-worked nursery tree is ready to dig and set in its permanent location.

A quicker method—which also produces a straighter trunk than double-budding— is illustrated ia Figure 8.20. Dormant

'Bartlett' pear scions are bench-grafted onto dormant 'Hardly' pear stem pieces during late winter, using the whip or tongue grafting method. These grafts are then packed in damp material : after several weeks in a cool location, they become well-callused and have united. In early spring such grafts (scion variety on the interstock) are crown-grafted in the nursery row to the rootstock—in this case, rooted quince cuttings. After one year in the nursery row, the double-worked nursery trees are ready to be dug.

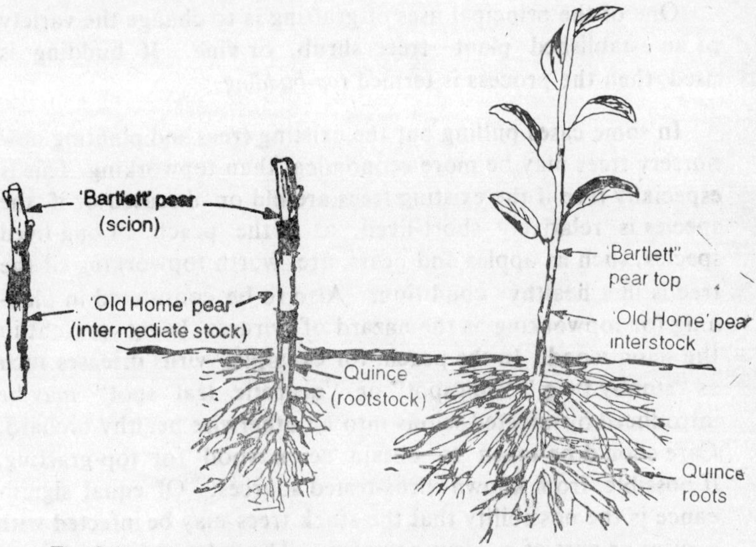

Fig. 8.20. There are three distinct parts and two graft unions in a double-worked plant, illustrated here by a 'Bartlett' pear grafted on an 'Old Home' pear which in turn is grafted on quince as a rootstock. The 'Old Home' pear, in this case, is the intermediate stock, or interstock.

Another method of double-working all at one time is by bench grafting, using the whip or tongue grafting method, but making two graft unions rather than one. The subsequent callusing and planting procedure is the same as for a simple root graft.

A method, termed "double-shield budding" has been developed both in England and Germany for double-working.

Essentially, this method consists of preparing a T-bud but inserting a thin budless shield of the intermediate variety under and below the shield piece containing the bud of the variety desired for the top. However, such short "compatibility bridges" may not always produce the desired effect. It has been shown that the degree of dwarfing is proportional to the length of the interstock.

Top-Grafting (Topworking)

One of the principal uses of grafting is to change the variety of an established plant—tree, shrub, or vine. If budding is used, then the process is termed *top-budding*.

In some cases pulling out the existing trees and planting new nursery trees may be more economical than topworking. This is especially true if the existing trees are old or diseased or if the species is relatively short-lived, as is the peach. Long-lived species, such as apples and pears, are worth topworking if the tree is in a healthy condition. Also to be considered in planning for topworking is the hazard of viruses being present in the scion wood. In the peach, for example, virus diseases such as "stubby twig," "ringspot" or "necrotic leaf spot" may be introduced by infected scions into an otherwise healthy orchard. Care should be taken to obtain scion wood for top-grafting, if possible, from known virus-tested sources. Of equal significance is the possibility that the stock trees may be infected with viruses or part of a virus complex. The value of scion wood from virus-tested sources can be nullified if the stock is virus-infected.

Preparation for Top-Grafting

Any of the methods of grafting described earlier in this chapter—whip, side, cleft, saw-kerf, or bark—can be used for top-grafting, which is usually done in the spring, shortly before new growth starts. The exact time depends upon the method to be used. The cleft, side, whip and saw-kerf graft can be done before the bark is slipping. The bark graft must be done when the bark is slipping, preferably just as the buds of the stock tree are starting growth.

It is usually advisable to obtain an ample amount of good quality scion wood prior to grafting and store it under the proper conditions.

In preparing the stock tree for topworking, one must decide, for each individual tree, which and how many scaffold branches (usually three to five) should be used. If the work is done high in the tree in the smaller, secondary scaffold branches, an earlier return to bearing will result than when fewer and larger limbs are grafted lower in the tree. The branches to be grafted should be well distributed around the tree and up and down the main trunk, avoiding branches with weak, narrow crotches. All others can be removed unless one or more nurse branches are used.

Retaining *nurse branches* is a common practice in topworking deciduous trees in regions where cold winters are experienced. In localities where the winters are mild and damage of succulent growth by winter killing is slight, nurse branches are not used. For broad-leaved evergreens, such as citrus and olives, it is customary to leave nurse branches. The presence of nurse branches, especially if they are on the south and west sides of the tree, is desirable because they protect the grafted branches from sun scald ; otherwise, the branches should be whitewashed to prevent this. Retaining a large scaffold branch as a nurse limb results in less vigorous growth of the scions. They are then less likely to blow out in winds or be winter-killed than the more vigorous, succulent shoots which develop from the scions where no nurse branches are retained. Also, fewer suckers and watersprouts (which must be removed) develop when nurse branches are left. The latter should be so pruned as not to interfere with the growth of the young grafts.

A practice which is often recommended, especially for older trees, is to top work them during a 2-year period, grafting perhaps two main branches on the northeast side of the tree the first year and retaining two branches on the southwest as nurse branches. The second year these are topworked.

Topworking is most successful when done on relatively young trees where the branches to be grafted are not larger than 3 or 4 in. in diameter and are relatively close to the ground. When attempting to topwork large, old trees, it is often necessary to go high up in the trees to find branches with a diameter as small as 4 in. If the grafting is done on such branches, the new top is inconveniently high for the various orchard operations, such as thinning and harvesting. The other alternative is cutting off the branches or main trunk close to the ground and inserting the scions into wood one or more feet in diameter. Although many scions can be inserted around the tree between the bark and the wood by the bark graft method and may grow well for several years, it is quite likely, that wood rot will develop in the center of the stub before the growth of the scions can heal it over. Also, some scions are not mechanically held in place very securely and may be blown out by strong winds after they reach considerable size.

It is important that the branch to be topworked be cut off in such a location that the region just below the cut is smooth and free from knots or small branches, so that there will be a satisfactory place for inserting the scions. The branches are best cut off about 9 to 12 in from the main trunk to keep the tree headed low. The branch should not be cut off more than a few hours before the grafting is to be done.

In Preparation for Topworking (Top-Grafting)

(*a*) Do the work in the spring when the trees are dormant or shortly after growth starts.

(*b*) Select for grafting three to five well-placed scaffold branches which are not larger than about 4 in. in diameter and which are conveniently close to the ground.

(*c*) Retain nurse branches for broad-leaved evergreen trees and for deciduous trees where the winters are severe.

(*d*) Cut off the branches properly so that the bark is not torn down the trunk.

Grafting on a cool, overcast day with no wind blowing offers the most protection from drying of the cut surfaces of the scion and stock until they can be covered with grafting wax. Grafting on hot, sunny, and windy days should be avoided. During grafting, the scion wood must not dry out by being exposed to the sun. It should be kept moist and cool is some container or be wrapped in moist burlap.

The need for prompt and thorough coverage of all cuts, including the tip end of the scions, cannot be stressed too strongly. The wax should be worked into the bark of the stock, sealing all small cuts or cracks where air could penetrate in and around the cut surfaces where healing tissue is expected to develop.

Subsequent Care of Topworked Trees

After the actual top-grafting (or top-budding) operation is finished, much important work needs to be done before the topworking is successfully completed. A good grafting job can be ruined by failure to care for the trees properly.

If the grafting has been done in late spring when growth is active, trees of some species, such as the walnut, will "bleed" to a considerable extent from the grafted stub, even though it has been covered with grafting wax. This flow of sap around the scions can be so heavy as to interfere with the normal healing processes at the graft union. If this condition appears, it can often be corrected by making several slanting cuts with a knife through the bark in the trunk of the tree several feet below the grafted stubs. The bleeding will then take place at these cuts rather than around the graft union. This extensive sap flow is not particularly harmful to the tree and will usually stop within a few days.

In 3 to 5 days after grafting, the trees should be carefully inspected and the graft unions rewaxed if cracks or holes appear in the wax.

A problem needing immediate attention is the prevention of sunburn on the portions of the trunk and large branches which are exposed to the direct rays of the sun by the removal of the protecting top foliage. This is especially important if the grafting has been done late in the season when hot weather can be expected, and when no protecting nurse branches have been retained. The radiant energy from the sun absorbed by the dark-coloured bark can raise the temperature of the living cells below the bark to a lethal level.

It is generally advisable to whitewash the trunk, branches and scions of the grafted trees, unless the grafting is done in late winter or early spring when the days are still cold. Various cold-water paints, some made especially for this purpose, are available. The white colour reflects a considerable portion of the sun's radiant energy, thus keeping the temperature of the living tissues within safe limits. Interior, water-base house paints (both latex and acrylic) prevent sunburn for one season. Exterior paints give longer protection but are more likely to cause injury to the tree.

To Prepare Whitewash for Grafted Trees

Formula I. Quicklime, 5 lb ; salt, $\frac{1}{2}$ lb ; sulphur, $\frac{1}{4}$ lb. Add water to the lime to start it slaking. Then add the salt and sulphur. This mixture should be prepared in a crockery or wood container. Allow to age for several days, then dilute to a consistency just thick enough to apply with a paint brush. Dilute to a thinner consistency for spray application.

Formula II. Hydrated lime, 25 lb ; zinc sulphate, 2 lb ; water, 50 gal. Add water first to a power sprayer with agitator running, then add other components. Spray mixture on trees.

One way to protect the scions from the heat of the sun when the grafting has been done late in the season is to cover the end of the grafted stub and the scions with a large paper bag, the corners of which have been cut to allow for ventilation. The bag is tied securely on the stub with heavy string. However, in a test, of various methods of protecting scions of grafted walnut trees, including the use of paper bags, whitewashing proved the best.

Another help in preventing sunburn is to retain some of the watersprouts which soon start growth along the trunk and branches of the grafted tree. They must be kept under control or they will quickly shade out the developing scions. Rather than removing the watersprouts completely, the grafter can head them back to several inches in length, and they will shade the bark underneath. An additional benefit is that the food manufactured by this leaf area will help sustain the tree until the new scions develop sufficient foliage to do so.

It should be emphasized that, especially during the first summer after topworking the sucker and watersprout growth arising from below the grafted branches must be kept pruned back so as not to interfere with the growth of the scions. If it has been customary to fertilize the trees with nitrogen, this should be withheld for a year or two after grafting, because no stimulation of growth is usually needed. If the trees have been under irrigation, their water requirement will be much less, owing to the removal of a considerable amount of leaf area when the tops were cut back for grafting.

Two to four scions are generally inserted in each stub. If all the scions grow, they should all be retained during the first year, because they will help heal over the stub. However, just one branch, from the best placed and strongest growing scion, should be retained permanently. Growth from the remaining scions should be retarded by rather severe pruning, keeping them alive to help heal the branch, but allowing the permanent scion to become the dominant one. To keep two or more scions for permanent branches at one point will undoubtedly result in a weak crotch, which will eventually break. The first year, then, the best practice is to retain all scions to help heal the stub but, by pruning, to retard the growth of all but the best one. If two shoots arise from the scion to be retained permanently, select the best of these and remove the other. After the stub has healed over, the temporary scions can be removed completely. This may be in the second or third year.

If the permanent scion grows rather vigorously, the danger arises of its becoming topheavy and breaking off during winds.

This may be handled in two ways—either by retarding the growth of the permanent branch by pruning it back or by nailing a lath or other type of stick onto the tree and tying the new branch to this stick. *When tying a cord around a branch, always make a loop so that there is no chance of the branch being girdled as it grows.*

Should only one scion grow at each stub, the problem of securing adequate healing of the stub on the side opposite the living scion may prove to be serious. Healing may be helped by sawing off the stub at an angle away from the surviving scion. The cut surface of the stub can be covered with grafting wax to retard wood rotting. It may take a number of years for the single scion to heal over the stub, and it is possible that wood rotting may start before it can do so.

When none of the scions grow in a stub, there are still some possibilities for getting it topworked. One is by allowing several well-placed watersprouts arising just below the cut surface to grow, then top-budding them during the summer. Or the watersprouts may be allowed to grow so as to keep the branch alive and healthy, then the grafting operation repeated the following year, making a fresh cut a foot or so below the original cut.

If nurse branches have been used, they should be cut back and away from the scions at intervals throughout the summer so that scion shoots are always fully exposed to the sun and have sufficient space to grow. Nurse branches should be removed entirely or grafted by the second or third year.

When the top of the tree has been finally worked over to the new variety, it will grow vigorously for a few years. Good pruning practices are needed to prevent badly placed branches from developing.

Frameworking

In this method of changing the variety of fruit trees, all the main scaffold branches are retained on the tree but most of the small laterals throughout the top of the tree are replaced by a large number of scions of the replacement variety. These are inserted by the side graft method with the cuts made very close to the origin of lateral branches (about 1 in. in diameter) arising from the secondary scaffolds. Long scions having 7 or 8 buds are often used. In frameworking the tree quickly returns to fruiting, but considerable labour is needed for grafting, and much aftercare is required in keeping growth from the original tree below the scions cut away so that only the replacement variety is maintained.

Herbaceous Grafting

Grafting herbaceous types of plants is used for various purposes, such as studying virus transmission, stock-scion physiology, and grafting compatibility, as well as for the commercial greenhouse production of certain cucurbitaceous crops, particularly in Europe. Usually such grafts are made while the plants are quite small, the stock being grafted shortly after seed germination. Such material is generally very soft, succulent, and susceptible to injury. In one technique a simple splice graft is used with a diagonal cut made through the seedling stock, just above the cotyledons. A piece of thin-walled polyethylene tubing of the proper size to give a snug fit is slipped over the cut end of the stock. The basal end of the scion receives a diagonal cut similar in length and angle to that given the stock. The scion is then slipped down into the plastic tubing so that the two cut surfaces make intimate contact. The tubing holds the graft in place until healing occurs—about 12 days after grafting. Then the tubing may be slipped off over the scion if there are no leaves and if the bud has not expanded ; otherwise, it may be cut off with a razor blade.

In another procedure used in grafting older herbaceous stock plants, several leaves are retained below the graft union. The cleft graft is used (but with only one scion), and the graft union is bound with raffia, budding rubbers, or adhesive latex tape. To prevent drying out, the entire plant—following grafting—is covered with a supported polyethylene bag. The grafted plant is then set in the shade until the graft has healed ; then the plastic cover can be removed.

A method of establishing *Vitis vinifera* grapes on phylloxera or nematode-resistant rootstocks is by "greenwood" grafting, in which the grafting is done in the spring, placing the scion—taken from new growth—on a new green shoot which has developed from the rootstock plant. A simple splice graft is made with the sloping cuts 1 to 1½ in. long. The stock and scion pieces must be the same diameter. The scion has only one bud. The cuts are matched as closely as possible, and the graft union is completely covered by wrapping with a budding rubber. The graft union should be healed and the bud growing by two weeks after grafting. Since these grafts are quite fragile, they must be tied to a stake.

Nurse-Seed Grafting

Germination of some large-seeded woody species, such as the chestnut, is hypogeal; that is, the cotyledons remain below ground in the seed coats with the shoot tip appearing above ground. In a grafting procedure using certain species with seeds of this type, when the seeds have just germinated, the petioles of the cotyledons are cut off transversely just at the seed, leaving the cotyledons inside. A knife point is inserted into the seed between the cut petioles, making an opening for the scion. The scion, which is prepared from dormant wood of the previous season's growth, is cut to a wedge shape at the base, as for a cleft graft. The scion is inserted into the cut between the cotyledons so that the exposed cambium surfaces of the scion are in close contact with the cut surfaces of the cotyledons. The "seed-grafts" are then lined out in the rooting medium with the union about 1½ in. below the surface. A graft union takes place, and roots arise from the cut cotyledon petioles. This type of grafting is done in early spring, using properly stratified seed and dormant scion wood. Chestnuts, avocados, and camellias have been grafted successfully by this method; in fact grafted camellia plants of the desired cultivar can be obtained in the same length of time it would take for the seedlings to become large enough to graft by the conventional cleft graft method.

In a modification of the nurse-seed graft, which has worked well in propagating chestnuts, pecans, walnuts, and oaks, the actual graft union is made into the seedling hypocotyl at the point of attachment of the cotyledons. The scion is cut to a thin wedge and inserted into a split made in the hypocotyl. After insertion of the scion, the graft is wrapped firmly with rubber strips or waxed string.

Cutting-Grafts

In this type of graft, a leafy scion is grafted onto a leafy, unrooted stem piece (which is to become the rootstock), and the combination is then placed in a rooting medium under intermittent mist for simultaneous healing of the graft union and rooting of the stock. This procedure was utilized many years ago in studying stock-scion physiology in citrus. More recently it has been used in commercial propagation of various types of citrus on clonal dwarfing rootstocks. It is also of value in propagating certain conifer species that are difficult to start as cuttings.

For citrus a simple splice graft is used. The slope of the cut is at 15-deg, angle, and the union is tied with a rubber band. The base of the stock is dipped into a root-promoting material, such as indolebutyric acid and then the grafts are placed under mist, or in a closed case, in flats of the rooting medium over bottom heat. After healing of the union and rooting of the stock, the grafts are allowed to harden by discontinuing the mist and bottom heat for about 2 weeks. Then the grafts are ready for planting in 1-gal. cans or other containers.

In using this method for conifers the difficult-to-root scion is grafted in early spring to the unrooted (but easily rooted) stock by a side graft. The graft union is made about 2 in. above the base of the stock. The stock and scion are both 10 to 12 in. long. The base of the stock is treated with an indolebutyric acid rooting compound, and then the combination is inserted into a flat of a well-drained rooting medium, such as perlite, so that the medium covers the graft union. The flats containing the grafts are kept in a cool greenhouse under intermittent mist. About 8 weeks later the cuttings should be rooted and the graft union healed. Then the top of the stock is cut off just above the graft union, and the grafted plant is transferred to an appropriate-sized container of soil.

9

Physiological Processes of Budding

In contrast to grafting, in which the scion consists of a short detached piece of stem tissue with several buds, budding utilizes only one bud and a small section of bark, with or without wood. Budding is often termed "bud grafting", since the physiological processes involved are the same as in grafting.

The commonly used budding methods depend upon the bark "slipping". This term indicates the condition when the bark can be easily separated from the wood. It denotes the period of year when the plant is in active growth, when the cambium cells are actively dividing, and newly formed tissues are easily torn as the bark is lifted from the wood. Beginning with new growth in the spring, this period should last until the plant ceases growth in the fall. However, adverse growing conditions, such as lack of water, defoliation, or low temperatures, may lead to a tightening of the bark and can seriously interfere with the budding operation. Getting the stock plants in the proper condition for budding is an important consideration. Of the commonly used methods described here, only one—the chip bud can be done when the bark is not slipping.

The budding operation, particularly T-budding, can be performed more rapidly than the simplest method of grafting, some rose budders inserting as many as 2000 or 3000 or more

T-buds a day if the tying is done by helpers. If performed under the proper conditions, the percentage of successful unions in T budding is very high—90 to 100 percent. Budding is widely used in producing nursery stock of rose and fruit tree cultivars, where hundreds of thousands of individual plants are propagated each year. Therefore, for propagation operations involving large numbers of plants, where speed and low mortality are essential, budding upon selected rootstocks is likely to be chosen.

The use of budding is confined generally to young plants or the smaller branches of large plants where the buds can be inserted into shoots which are from ½ to 1 in. in diameter. Top-working young trees by top-budding is quite successful. Here the buds are inserted in small, vigorously growing branches in the upper portion of the tree.

Budding may result in a stronger union, particularly during the first few years, than is obtained by some of the grafting methods, and thus the shoots are not as likely to blow out in strong winds. Budding makes more economical use of propagating wood than grafting, each bud potentially being capable of producing a new plant of the desired variety. This may be quite important if propagating wood is scarce. In addition, the techniques involved in budding are simple and can be easily performed by the amateur.

Rootstocks for Budding

In propagating nursery stock of the various fruit and ornamental species by budding, a rootstock plant is used. It should have the desired characteristics of vigor, growth habit, and disease resistance, as well as being easily propagated. This rootstock plant may be a rooted cutting, a rooted layer, or more commonly a seedling. Usually, one year's growth in the nursery row before budding is to be done is sufficient to produce a rootstock plant large enough to be budded, but seedling of slow-growing species may require two seasons.

To produce nursery trees free of harmful pathogens (viruses, fungi, or bacteria) it is essential that the rootstock plant, as well as the budwood, be free of such organisms.

Time of Budding—Fall, Spring, or June

The important budding methods are used at seasons of the year when the stock plant is in active growth and the cambial cells actively dividing so that the bark separates readily from the wood. It is also necessary that well-developed buds of the desired variety be available at the same time. These conditions exist for most plant species at three different times during the year. In the Northern Hemisphere, these periods are late July to early September (*fall budding*), March and April (*spring budding*), and late May and early June (*June budding*).

Fall Budding

This is the most important time of budding in the propagation of fruit tree nursery stock, although actually the budding is done mostly in late summer rather than fall. The rootstock plants are usually large enough by late summer to accommodate the bud, and the plants are still actively growing, with the bark slipping easily. Once growth has stopped and the bark adheres tightly to the wood, budding can no longer be done.

In fall budding, the budsticks, consisting of the current season's shoots, are obtained at the time of budding. They should be vigorous and should contain vegetative or leaf buds. Such shoots are sometimes termed *watersprouts*, especially if they are very vigorous. Short, slowly growing shoots on the outer portion of the tree should be avoided, because they may have chiefly flower buds rather than vegetative buds. Flower buds are usually round and plump, whereas leaf buds are smaller and pointed. Some species have mixed buds, the node containing both vegetative and flower buds. These are satisfactory for use in budding.

Every effort should be made to make sure that the trees from which the budsticks are obtained are free of any bacterial, fungus, or virus diseases. Using infected budsticks can infect every budded nursery tree with the disease.

As the budsticks are selected the leaves should be removed immediately, leaving only a short piece of the leaf stalk or petiole attached to the bud ; this will aid in handling the bud later on. The budsticks should be kept from drying by wrapping in some material such as clean, moist burlap and keeping them in a cool, shady location until they are needed. The budsticks should be used promptly after cutting, although they can be stored for a short time if kept cool and moist. It is best, if possible, when a considerable amount of budding is being done, to collect the budsticks as they are being used, a day's supply at a time.

The best buds to use on the stick are usually those in the middle and basal portions. Buds on the succulent terminal portion of the shoot should be discarded. In certain species, such as the sweet cherry, buds on the basal portion of the shoots are flower buds, which of course should not be used.

In fall budding, after the buds have been inserted, there is nothing more to be done until the following spring. *Although eventually the rootstock is to be cut off above the bud, in no case should this be done immediately after the bud has been inserted.* Healing of the bud piece to the stock is greatly facilitated by the normal movement of water and nutrients up and down the stem of the rootstock. This would, of course, be stopped if the top of the rootstock were cut off above the bud.

If the budding operation is done properly, the bud piece should unite with the stock in 2 to 3 weeks, depending upon the growing conditions. If the leaf stalk or petiole drops off cleanly next to the bud, this is a good indication that the bud has united, especially if the bark piece retains its normal light brown or green colour and the bud stays plump.

The bud union has healed, in most deciduous species the bud usually does not grow or "push out" in the fall, since it is in a physiological rest period. It remains just as it is until spring, at which time the chilling winter temperatures have overcome the rest influence and the bud is ready to grow. There are some exceptions to this ; for example, in fall budding of maples, roses, honey locust, and certain other plants, some of the buds may start growth in the fall. In northern areas, if such fall-forced buds do not start early enough for the shoots to mature before cold weather starts, they are likely to be winter-killed.

In the spring, just before new growth begins, the rootstock is cut off, immediately above the bud. It is desirable to make a sloping cut, slanting away from the bud. Although this cut may be waxed, it is usually not essential unless the stock is large in diameter. Cutting back the rootstock forces the inserted bud into growth. In citrus budding, it is a common practice to partially cut the stock above the bud and to lop or bend it over away from the bud. The leaves of the stock still furnish the roots with some nutrients, but the partial cutting forces the bud into growth. After the new shoot from the bud is well established, the top is completely removed.

In northern regions, fall-inserted buds are sometimes covered with soil during the winter until danger of frost has passed and are then uncovered and topped-back in late spring.

Where strong winds occur and in species in which the new shoots grow vigorously, support for the newly developing shoot may be necessary. One practice sometimes followed is to cut off the rootstock several inches above the bud, using this projecting stub as a support on which to tie the tender young shoot arising from the bud. This stub is removed after the shoot has become well established. In another procedure, stakes may be driven into the ground next to the stock to which the developing shoot is tied at intervals during its growth.

Cutting back to force the main bud to grow also forces many latent buds on the rootstock into growth. These must be rubbed off as soon as they appear, or they will soon choke out the desired inserted bud. It may be necessary to go over the budded plants several times before these "sprouts" stop appearing. Nurserymen refer to this procedure as "suckering."

The shoot arising from the inserted bud becomes the top portion of the plant. After one season's growth in the nursery, with favorable conditions of soil, water and nutrients, temperature, and insect and disease control. Such a tree would have a

1-year-old top and a 2-or perhaps 3-year-old root, but it is still considered a "yearling" tree. If the top makes insufficient growth the first year, it can be allowed to grow a second year, and is then known as a 2-year-old tree. However, abnormally slow-growing trees should be discarded.

Spring Budding

This method is similar to fall budding except that the work takes place the following spring as soon as active growth of the rootstock begins and the bark separates easily from the wood. The period for successful spring budding is limited, and budding should be completed before the rootstocks have made much new growth.

Budsticks are chosen from the same type of shoots—in regard to vigor of growth and type of buds—that would be used in fall budding, except that they are not collected until the dormant season the following winter. The leaves would, of course, have fallen by this time, and the buds would have experienced sufficient chilling to overcome their rest period. Budwood must be collected while it is still dormant—before there is any evidence of the buds swelling. Since the buds must be dormant when they are inserted, and since the rootstocks must be in active growth, it is necessary in spring budding that the budsticks be gathered some time in advance of the time of budding and stored at temperatures (32° to 40° F ; 0° to 4° C) to hold the buds dormant. The budsticks should we wrapped in bundles with damp peat moss or some similar material to prevent drying out.

In spring budding, the actual budding operation should be done just as soon as the bark on the rootstock slips easily. Then, about 2 weeks after budding, when the bud unions have healed, the top of the stock must be cut off above the bud to force the inserted bud into active growth. At the same time, latent buds on the rootstock begin to grow and should be removed. Sometimes it is helpful to permit such shoots from the rootstock to develop to some extent to prevent sunburn and help nourish the plant. They must be held in check, however, and eventually removed.

Although the new shoot from the inserted bud gets a later start in spring budding than in fall budding, spring buds will usually develop rapidly enough, if growing conditions are favorable, to make a satisfactory top by fall. Fall budding, however, is to be preferred for several reasons : the higher temperatures at that time promote more certain healing of the union, the budding season is longer, there is no necessity to store the budsticks, the inserted buds start growth earlier in the spring, and the pressure of other work is usually not so great for nurserymen in the late summer as in the spring. Spring budding is used sometimes on rootstocks which were fall-budded but on which the buds failed to take.

June Budding

June budding is used to obtain a "1-year-old" budded tree in a single growing season. Its principal characteristic is that budding is done in the early part of the growing season and the inserted bud forced into growth immediately during the same season. As a method of nursery propagation, June budding is confined to regions which have a relatively long growing season —in the U.S. this includes California and the southern states. In the propagation of fruit trees, June budding is used mostly in producing such stone fruits as peaches, nectarines, apricots, almonds, and plums. Peach seedlings are generally used as the rootstock, but almond seedlings and plum hardwood cuttings can also be used. Budding is done by the T-bud method. If seeds are planted in the fall or stratified seeds as early as possible in the spring, the seedlings usually attain sufficient size (12 in. high and at least 1/8 in. in diameter) to be budded by mid-May or early June (in the Northern Hemisphere). Preferably, June budding should not be done much after mid-June, or a nursery tree of satisfactory size will not be obtained by fall. June-budded trees are not as large by the end of the growing season as those propagated by fall or spring budding but they are of sufficient size (about 3/8 in. to 5/8 in. caliper and 3 to 5 ft tall) to be used commercially.

The budwood used in June budding consists of current season's growth, that is, of new shoots which have developed since growth started in the spring. By late May or early June, these shoots will usually have grown sufficiently to have a well developed bud in the axil of each leaf. At this time of year these buds will not have entered the rest period, so when they are used in budding they continue their growth on through the summer, producing the top portion of the budded seedling.

For June-budded trees handling subsequent to the actual operation of budding is somewhat more exacting than for fall- or spring-budded trees. The rootstocks are smaller and have less stored food than those used in fall or spring budding. The object behind the following procedures—shown in Fig. 9.1—is to keep the rootstock (and later the budded top) actively and continuously growing so as to allow no check in

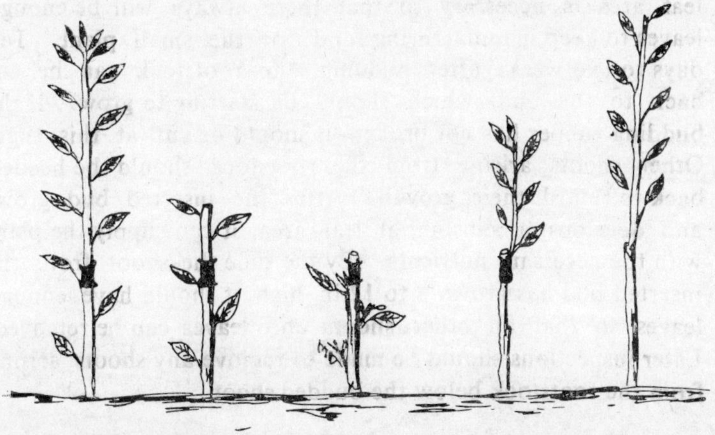

Fig. 9.1. It is important in June budding that the stock be cut back to the bud properly. Far left : The bud is inserted high enough on the stock so that there are several leaves below the bud. Center left : Three or 4 days after budding, the stock is partially cut back-2 to 5 in. above the bud. Center ; Ten days to 2 weeks after budding, the stock is completely removed just above the bud. Center right : This forces the bud and other buds on the stock into growth ; the latter must subsequently be removed. Far right : Appearance of the budded tree after the new shoot has made considerable growth.

growth, while at the same time changing the seedling shoot to a budded top. The bud should be inserted high enough (5 to 6 in.) on the stem so that a number of leaves—at least three or four—can be retained below the bud. The method of T-budding with the "wood out" described later should be used. Healing of the inserted bud should be very rapid at this time of year, since temperatures are relatively high, and rapidly growing, succulent plant parts are used. By 4 days after budding, healing should have started, and the top of the rootstock can be cut back somewhat—2 to 5 in. above the bud—leaving at least one leaf above the bud and several below it. This operation will force the inserted bud into growth and will check terminal growth of the rootstock. It will also stimulate shoot growth from basal buds of the rootstock, which will produce additional leaf area. This continuous leaf area is necessary so that there always will be enough leaves to keep manufacturing food for the small plant. Ten days to two weeks after budding, the rootstock can be cut back to the bud, which should be starting to grow. If the budding rubber has not broken, it should be cut at this time. Other shoots arising from the rootstock should be headed back to retard their growth. After the inserted bud grows and develops a substantial leaf area, it can supply the plant with the necessary nutrients. By the time the shoot from the inserted bud has grown 8 to 12 in. high, it should have enough leaves so that all other shoots and leaves can be removed. Later inspections should be made to remove any shoots arising from the rootstock below the budded shoot.

Another method which works well is to partially cut the stock just above the bud and break it over. Nutrients are still able to pass from the top of the stock to the roots, but this partial blocking forces the inserted bud into growth.

June budding is often of considerable value to nurserymen who find that their regular supply of fall- or spring-budded trees is insufficient to meet their expected demand. By this method they can, as late as the end of June, still propagate trees to be ready by fall, provided that they have a supply of rootstock plants large enough to bud.

The steps in fall, spring, and June budding are compared in Fig. 9.2.

METHODS OF BUDDING

T-Budding (Shield Budding)

This method is known by both names, the "T-bud" designation arising from the T-like appearance of the cut in the stock, whereas the "shield bud" name is derived from the shield-like appearance of the bud piece when it is ready for insertion in the stock.

T-budding is by far the most common method of budding and is widely used by nurserymen in propagating nursery stock of most fruit tree species, roses, and many ornamental shrubs. Its use is generally limited to stocks which are about ½ to 1 in. in diameter, with fairly thin bark, and which are actively growing so that the bark will separate readily from the wood. If the bark is so tight on the wood that it has to be forcibly pried loose, the chances of the bud healing successfully are rather remote. The operation should then be delayed until the bark is slipping easily.

The bud is inserted into the stock 2 to 10 in. above the soil level in a smooth bark surface. There are different opinions as to which is the proper side of the stock in which to insert the bud. If extreme weather conditions are likely to occur during the critical healing period just following budding, it may be desirable to place the bud on the side of the stock on which as much protection as possible may be obtained. Some believe that if the bud is placed on the windward side, there is less chance of the young shoot breaking off. Otherwise, it probably makes little difference where the bud is inserted, the convenience of the operator and the location of the smoothest bark being the controlling factors. When rows of closely planted rootstocks are budded, it is more convenient to have all the buds on the same side for later inspection and manipulations.

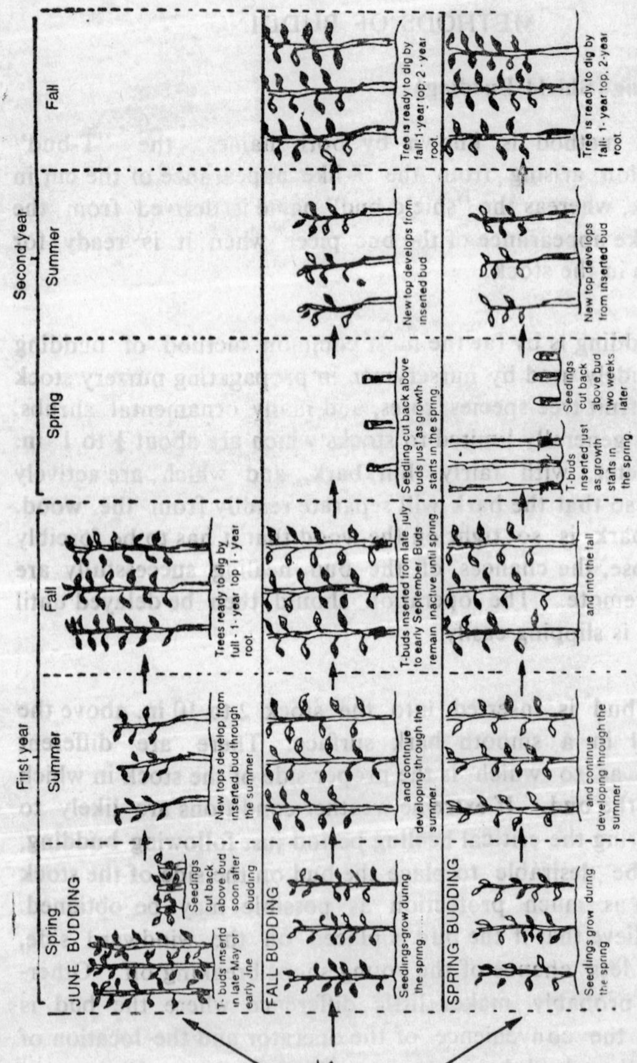

Fig. 9.2. Comparison of the steps in June, fall, and spring budding. The actual techniques in budding are not difficult, but it is very important that the various operations be done at the proper time.

The cuts to be made in the stock plant are illustrated in Fig. 9.3. There are various modifications of this technique; most budders prefer to make the vertical cut first, then the horizontal crosscut at the top of the T. As the horizontal cut is made, the knife is given a twist to throw open the flaps of bark for insertion of the bud. It is important that neither the vertical nor horizontal cut be made longer than necessary, because this requires additional tying later to close the cuts.

After the proper cuts are made in the stock and the incision is ready to receive the bud, the shield piece is cut out of the budstick.

To remove the shield of bark containing the bud, a slicing cut is started at a point on the stem about ½ in. below the bud, continuing under and about an inch above the bud. The shield piece should be as thin as possible but still thick enough to have some rigidity. A second horizontal cut is then made ½ to ¾ in. above the bud, thus permitting the removal of the shield piece.

There are two methods of preparing the shield—with the "wood in" or with the "wood out." This refers to the little sliver of wood just under the bark of the shield piece and which will remain attached to it if the second horizontal cut is deep and goes through the bark and wood, joining the first slicing cut. Some professional budders believe it is best to remove this sliver of wood, but others retain it. In budding certain species, however, such as maples and walnuts, much better success is usually obtained with "de-wooded" buds. If it is desired to prepare the shield with the wood out, the second horizontal cut should be just deep enough to go through the bark and not through the wood. Then if the bark is slipping easily, the bark shield can be snapped loose from the wood (which still remains attached to the budstick) by pressing it against the budstick and sliding it sideways. A small core of wood comprising the vascular tissues supplying the bud is present, and this should remain in the bud, rather than adhering to the wood and leaving a hole in the bud. If the shield is pulled outward rather than being slid sideways

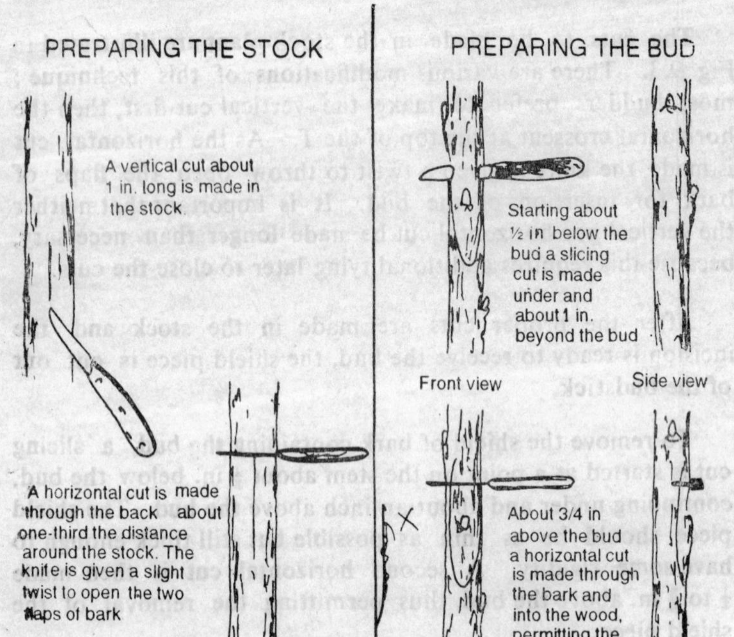

Fig. 9.3. Steps in making the T-bud (shield bud) There are several variations of the procedure shown here.

from the wood, this core usually pulls out of the bud, eliminating the chances of success. In June budding of fruit trees, the shield piece is usually prepared with the wood out. In most

other instances, however, the wood is left in. In spring budding, using dormant budwood, this sliver of wood is tightly attached to the bark and cannot be removed.

The next step is the insertion of the shield piece containing the bud into the incision in the stock plant. The shield is pushed under the two raised flaps of bark until its upper horizontal cut matches the same cut on the stock. The shield should fit snugly in place, will covered by the two flaps of bark, but with the bud itself exposed.

No waxing is necessary, but the bud union must be wrapped, using materials to hold the two components firmly together until healing is completed. Rubber budding strips, especially made for wrapping, are widely used for this purpose. Their elasticity provides sufficient pressure to hold the bud securely in place. The rubber, being exposed to the sun and air, usually deteriorates, breaks, and drops off after several weeks, at which time the bud should be healed in place. If the budding rubber is covered with soil, the rate of deterioration will be much slower. This material has the advantage of eliminating cutting the wrapping ties, which can be a costly operation if thousands of plants have been budded. The rubber will expand as the rootstock grows, and thus there is little danger of constriction.

In tying the bud, the ends of the budding rubbers are held in place by inserting them under the adjacent turn. The bud itself should not be covered. The amount of tension given the budding rubber is quite important. It should not be too loose, or there will be too little pressure holding the bud in place. On the other hand, if the rubber is stretched extremely tight, it may be so thin that it will deteriorate rapidly and break too soon—before the bud union has taken place. Often the tying is done from the top down to avoid forcing the bud out through the horizontal cut.

Raffia (fiber-like leaf segments of certain *Raphia* species) has been widely used for wrapping buds. This material is soaked in water over-night before it is used so that it will be flexible. Raffia must be cut later—about 10 days after budding—to prevent constriction at the bud union as the plant grows. If such nonelastic ties are not promptly cut, the resultant constriction can have a very adverse affect on subsequent growth of the bud.

Plastic ties of polyvinyl chloride (PVC) film, 3/8 in. wide, are quite useful for budding. Such material is moisture-proof and elastic, and since it is transparent, it permits inspection of the buds after covering.

Inverted T-Budding

In localities where a great deal of rain occurs, water running down the stem of the rootstock will often enter the T-cut, soak under the bark, and cause decay of the shield piece. Under such conditions an inverted T-bud may give better results, since it is more likely to shed excess water. In citrus budding, the inverted T method is widely used, even though the conventional method also gives excellent results. In species which bleed badly during budding, such as chestnuts, the inverted T-bud allows better drainage and better healing.

The techniques of the inverted T-bud method are the same as those already described, except that the incision in the stock has the transverse cut at the bottom rather than at the top of the vertical cut, and in removing the shield piece from the budstick the knife starts above the bud and cuts downward below it. The shield is removed by making the transverse cut ½ to ¾ in. below the bud. The shield piece containing the bud is inserted into the lower part of the incision and pushed upward until the transverse cut of the shield meets that made in the stock.

It is important in using this inverted T-bud method that a normally oriented shield bud piece should not be inserted into an inverted incision in the stock. The bud would then have a reversed polarity. Although such upside down buds do live and grow, at least in some species, their use may not result in normal shoot development.

Patch Budding

The distinguishing feature of patch budding and related methods is that a rectangular patch of bark is removed completely from the stock and replaced with a patch of bark of the sams size containing a bud of the variety to be propagated.

Patch budding is somewhat slower and more difficult to perform than T-budding, but it is widely and successfully used on thick-barked species, such as walnuts and pecans, in which T-budding gives poor results, presumably owing to the poor fit around the margins of the bud. Patch budding, or one of its modifications, is also extensively used in propagating various tropical species, such as the rubber tree (*Hevea brasiliensis*).

Patch budding requires that the bark of both the stock and budstick be slipping easily. It is usually done in late summer or early fall, but can also be done in the spring. In propagating nursery stock, the diameter of the rootstock and the budstick should preferably be about the same, from ⅝ to 1 in. Although the budstick should not be much larger than about 1 in. in diameter, the patch can be inserted successfully into stocks as large as 4 in. in diameter, although adequate healing of such large stubs may be a problem.

Special knives have been devised to remove the bark pieces from the stock and the budstick. Some type of double-bladed knife that will make two transverse parallel cuts 1 to 1-3/8 in. apart is necessary. These cuts, about an inch in length, are made through the bark to the wood in a smooth area of the rootstock several inches above the ground. Then the two transverse cuts are connected at each side by vertical cuts made with a single-bladed knife.

The patch of bark containing the bud is cut from the budstick in the same manner in which the bark patch is removed from the stock. Using the same two-bladed knife, the budder makes two transverse cuts through the bark, one above and one below the bud. Then two vertical cuts are made on each side of the bud so that the bark piece will be about an inch wide. The bark piece containing the bud is now ready to be removed. It is important that it be slid off sideways rather than being lifted or pulled off. There is a small core of wood, the bud trace, which must remain inside the bud if a successful "take" is to be obtained. By sliding the bark patch to one side, this core is broken off, and it stays in the bud. If the bud patch is lifted off, this core of wood is likely to remain attached to the wood of the budstick, leaving a hole in the bud.

After the bud patch is removed from the budstick, it must be inserted immediately on the stock, which should already be prepared, needing only to have the bark piece removed. The patch from the budstick should fit snugly at the top and bottom into the opening in the stock, since both transverse cuts were made with the same knife. It is more important that the bark piece fit tightly at top and bottom than that it fit along the sides. These procedures are illustrated in Figures 9.4.

The inserted patch is now ready to be wrapped. Often the bark of the stock will be thicker than the bark of the inserted bud patch so that, upon wrapping, it is impossible for the wrapping material to hold the bud patch tightly against the stock. In this case it is necessary that the bark of the stock be pared down around the bud patch so that it will be of the same thickness, or preferably slightly thinner, than the bark of the bud patch. Then the wrapping material will hold the bud patch tightly in place.

One type of tool, has a pair of vertical cutting blades in addition to the horizontal blades. With this, all four cuts of the rectangle are made at once.

In cases in which difficulty is experienced in obtaining successful unions with the patch bud method, it may help to make the four cuts of the rectangle in the stock 1 to 3 weeks ahead of the time when the actual budding is to be done. This bark patch is not removed from the stock, however, until the patch containing the bud is ready to be inserted. When the

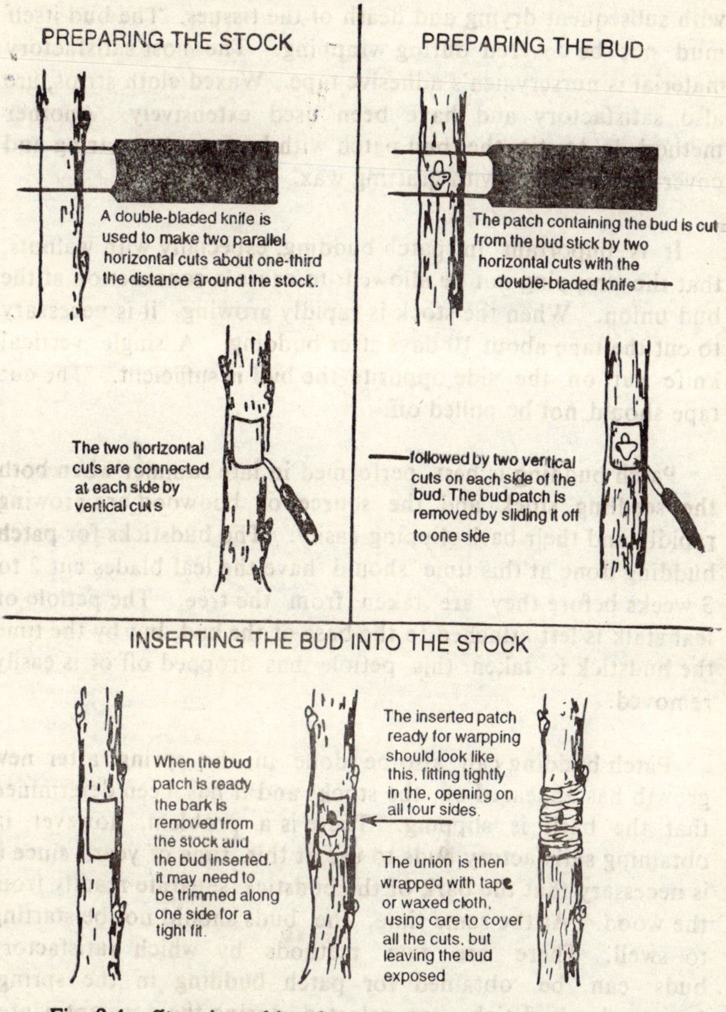

Fig. 9.4. Steps in making the patch bud. This method is widely used for propagating thick-barked plants.

cuts are made ahead of time, the wounding causes the callusing process to start, so that when the new bark patch is inserted, it heals very rapidly.

In wrapping the patch bud, a material should be used which not only will hold the bark tightly in place but will cover all

the cut surfaces to prevent the entrance of air under the patch, with subsequent drying and death of the tissues. The bud itself mud not be covered during wrapping. The most satisfactory material is nurserymen's adhesive tape. Waxed cloth strips, are also satisfactory and have been used extensively. Another method is to tie the bud patch with heavy cotton string and cover the cut edges with grafting wax.

It is important in patch budding, especially with walnuts, that the wrapping not be allowed to cause a constriction at the bud union. When the stock is rapidly growing, it is necessary to cut the tape about 10 days after budding. A single vertical knife cut on the side opposite the bud is sufficient. The cut tape should not be pulled off.

Patch budding is best performed in late summer when both the seedling stock and the source of budwood are growing rapidly and their bark slipping easily. The budsticks for patch budding done at this time should have the leaf blades cut 2 to 3 weeks before they are taken from the tree. The petiole or leaf stalk is left attached to the base of the bud, but by the time the budstick is taken this petiole has dropped off or is easily removed.

Patch budding can also be done in the spring after new growth has appeared on the stocks and it has been determined that the bark is slipping. There is a problem, however, in obtaining satisfactory buds to use at this time of year, since it is necessary that the bark of the budstick separate readily from the wood. At the same time, the buds should not be starting to swell. There are two methods by which satisfactory buds can be obtained for patch budding in the spring. In one, the budsticks are selected during the dormant winter period and stored at low temperatures (about 36° F ; 2° C) and wrapped in moist sphagnum or peat moss to prevent their drying out. Then, about 3 weeks before the time the spring budding is to be done, they are brought out into a warm room and set with their bases in a container of water, or they may be left packed in damp peat or sphagnum moss. The increased temperature will cause the cambium layer to become active, and soon the bark will slip sufficiently for the buds to be used. Although a few of the more terminal buds on each stick may start swelling in this time and cannot be used, there should be a number of buds in a satisfactory condition.

The second method of obtaining buds for spring patch budding is to take them directly from the tree which is the source of the budwood, but at the time the budding is to be done. If the trees are inspected carefully, it will be seen that not all of the buds start pushing at once. The terminal ones are usually more advanced than the basal ones. There is a period when the bark is slipping easily throughout the shoot containing the desired buds, but when only a few of the buds have developed so far that they cannot be used. The remaining buds, which are still dormant but upon bark that can be removed readily, may be taken and used immediately for budding. It is easier to obtain suitable buds from young trees which made vigorous shoot growth the previous year than it is from old trees. When the budding is done will be governed then by the stage of development of the buds. This will vary considerably with the species and variety being used. Stage of development of the rootstock is not as critical. Its bark must be slipping well, and the budding should be done before the stock plant has made much new growth.

Flute Budding

Flute budding is similar to patch budding, except that the patch of bark removed from the rootstock almost completely encircles it, leaving a narrow connection (about one-eighth of the circumference) between the upper and lower parts of the stock. In taking the bud patch from the budstick, a two-bladed knife is used, with the two transverse cuts completely encircling the budstick. A single vertical cut connects the two horizontal cuts, permitting the bud patch to be removed. In fitting this bud patch to the stock, it may be necessary to shorten its circumference by a vertical cut to remove the surplus amount of bark. If the bud patch fails to unite, the narrow connecting strip of bark on the stock keeps the top of the stock alive (Fig. 9.5).

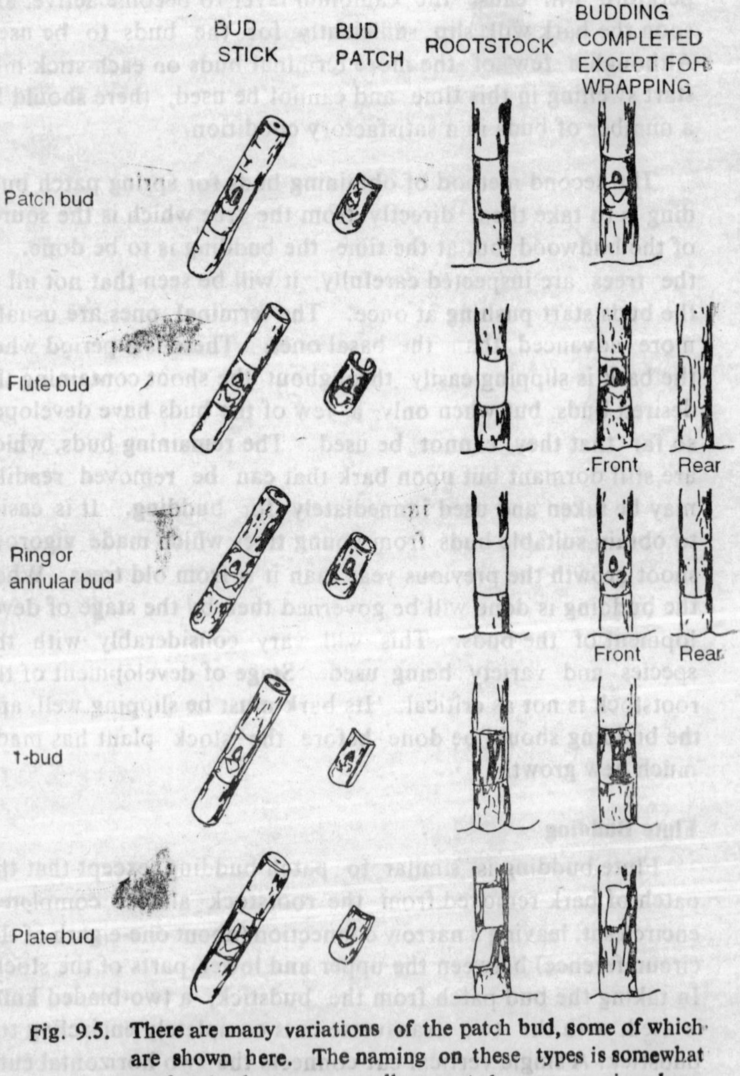

Fig. 9.5. There are many variations of the patch bud, some of which are shown here. The naming on these types is somewhat confused ; the most generally accepted names are given here.

Ring or Annular Budding

By this method, a complete ring of bark is removed from the stock and a complete ring from the budstick. In order for the two to match, the size of the stock and that of the budstick should be about the same. Since the stock is completely girdled, if the bud patch fails to heal in, the stock above the ring may eventually die. This method is not as widely used as the ordinary patch bud, since it has no particular advantages and is rather cumbersome to perform (Fig. 9.5).

I-Budding

In this type of budding, the bud patch is cut just as for patch budding, that is, in the form of a rectangle or square. Then, with the same parallel-bladed knife, two transverse cuts are made through the bark of the stock. These are joined at their centers by a single vertical cut to produce the figure I. The two flaps of bark can then be raised for insertion of the bud patch beneath them. It may make a better fit to slant the side edges of the bud patch. In tying the I-bud, care should be taken to see that the bud patch does not buckle upward and fail to touch the stock.

This method should be considered for use when the bark of the stock is much thicker than that of the budstick. In such cases, if the patch bud were used, considerable paring down of the bark of the stock around the patch would be necessary. This operation is not necessary in the I-bud method (Fig. 9.5).

Chip Budding

This is a budding method that can be done at times when the bark is not slipping *e.g.*, early in the spring before growth starts, or during the summer when active growth stops prematurely owing to lack of water or some other cause. It is generally used with fairly small material, ½ to 1 in. in diameter. Although chip budding is quite successful, it is not as fast or simple as T-budding and is not likely to be used if conditions are favorable for the use of T-budding. Chip budding in the fall has given consistently excellent results, however, in budding grape varieties or phylloxera or nematode-resistant rootstocks. It is not commonly used on deciduous tree fruit species.

As illustrated in Fig. 9.6, a chip of bark is removed from a smooth place between nodes near the base of the stock and re-

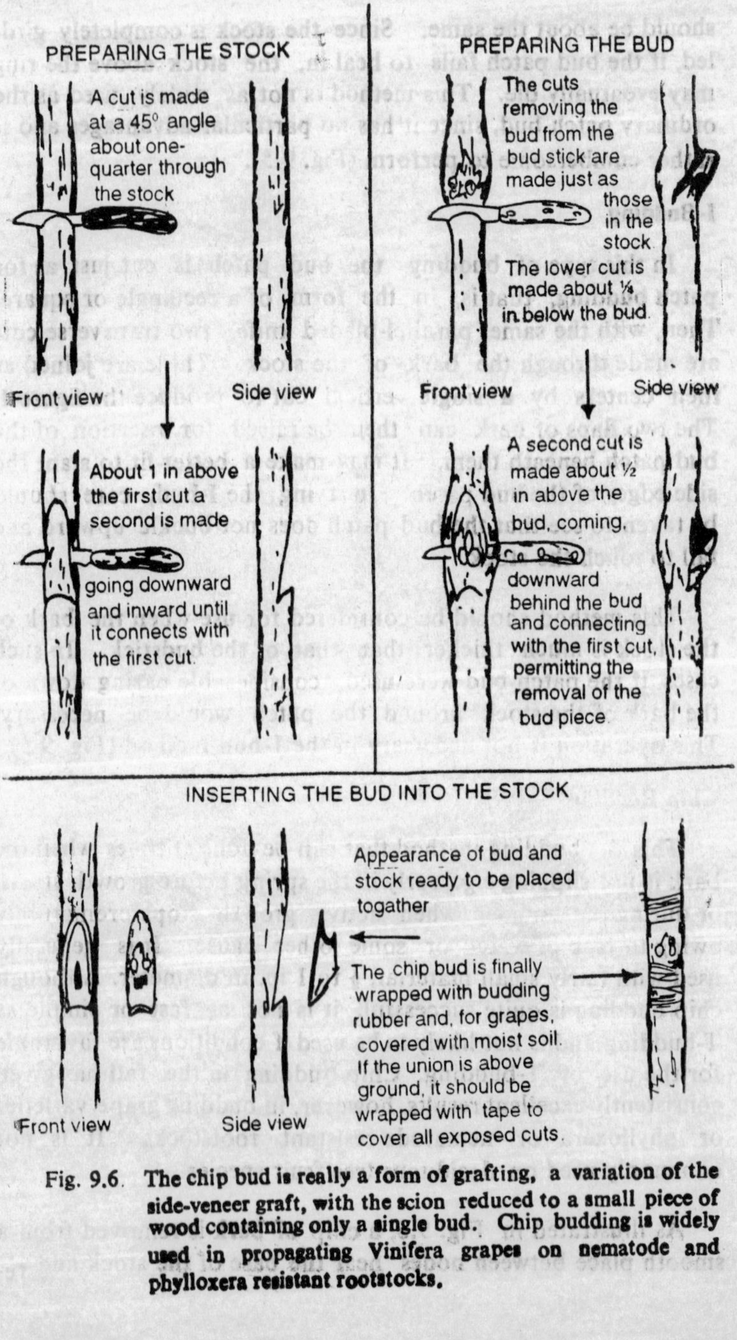

Fig. 9.6. **The chip bud is really a form of grafting, a variation of the side-veneer graft, with the scion reduced to a small piece of wood containing only a single bud. Chip budding is widely used in propagating Vinifera grapes on nematode and phylloxera resistant rootstocks.**

placed by another chip of the same size and shape from the budstick which contains a bud of the desired variety. The chips in both stock and budstick are cut out in the same manner. The first cut is made just below the bud and down into the wood at an angle of about 45 deg. The second cut is started about ½ in. above the bud and goes inward and downward behind the bud until it intersects the first cut. The order of making these two cuts may be reversed. The chip is removed from the stock and replaced by the one from the budstick. If they have both been cut to the same size and shape—as they should be—a good fit is obtained. It is important that the cambium layer of the bud piece be placed so as to coincide with that of the stock, preferably on both sides of the stem, but at least on one side.

There are no protective flaps of bark to prevent the bud piece from drying out, as there are in T-budding. It is very important, then, that the chip bud be wrapped to seal the cut edges as well as to hold the bud piece tightly into the stock. Nurseryman's adhesive tape works very well for this, although string can be used if all the cut edges are covered with grafting wax. If the bud is inserted into the stock close to the ground level, as it is in grape propagation, it is sufficient to wrap the bud with budding rubber and cover the whole bud union immediately with several inches of finely pulverized moist soil, which can be removed after the bud has united. Budding rubbers under the soil level arc very slow to disintegrate, so they may need to be cut or removed before constriction occurs.

In chip budding, as in the other methods, the stock is not cut back above the bud until the union is completed. If the chip bud is inserted in the fall, the stock is cut back just as growth starts the next spring. If the budding is done in the spring, the stock is cut back about 10 days after the bud has been inserted.

Top-Budding

In young trees when there is an ample supply of vigorous shoots at a height of 4 to 6 ft, top-budding provides a fast and certain method of topworking. It can be used in older trees, too, if they are cut back rather severely the year before to provide a quantity of vigorous watersprout shoots fairly close to the ground.

Depending upon the size of the tree, 10 to 15 buds are placed in vigorously growing branches ½ to ¾ in. in diameter in the upper portion of the tree—about shoulder height. Although a number of buds could be placed in a single branch, usually only one will be saved to develop into secondary branches, which will then form the permanent new top of the tree. The T-bud method is used on thin-barked species, and the patch bud on those with thick bark.

Top-budding is usually done in midsummer, as soon as well-matured budwood can be obtained and while the stock tree is still in active growth with the bark slipping easily. Orchard trees generally stop growth earlier in the season than young nursery trees; therefore the budding must be done earlier. When top-budding is done at this time of year, the buds will remain inactive until the following spring. At that time, just as vegetative growth starts, the stock branches are cut back just above the buds. This forces the buds into active growth, and they should develop into good-sized branches by the end of the summer. At the time the shoots are cut back to the buds, all other unbudded branches should be removed at the trunk. It is important that the trees be inspected carefully through the summer and any shoots removed that are arising from any but the inserted buds.

Top-budding can also be done in the spring just as the tree to be top-worked is starting active growth and the bark is slipping easily.

Double-Working by Budding

In propagating nursery trees, some of the budding methods can be used in developing double-worked trees. The intermediate stock can be budded on the rootstock; then the following year the desired variety is budded on the interstock. Although quite effective, this is a rather lengthy process, taking

3 years. As illustrated in Fig. 9.7, it is, possible, to develop a double-worked tree in one operation in 1 year by the double-shield bud method. A T-bud is used, but just under and below i a budless shield piece of the desired interstock is inserted.

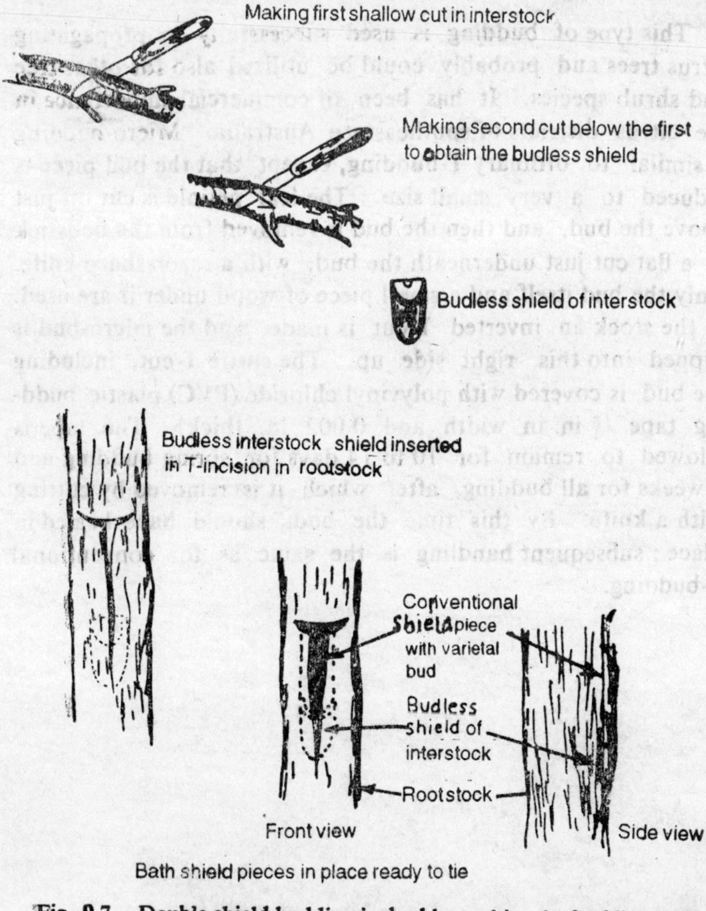

Fig. 9.7. Double-shield budding is double-working by budding in one operation. In Garner's method (2), as illustrated here, the intermediate stock is reduced to a small, budless shield piece inserted under and just below the regular shield piece used in T-budding.

In studies comparing the two methods of "double-shield" budding—that of Garner and that of Nicolin—it appears that the Nicolin method, with a full-length intermediate piece, is simpler and more easily done, but the results by either method are satisfactory, at least for some combinations.

Micro-Budding

This type of budding is used successfully in propagating citrus trees and probably could be utilized also for other tree and shrub species. It has been of commercial importance in the citrus districts of southeastern Australia. Micro-budding is similar to ordinary T-budding, except that the bud piece is reduced to a very small size. The leaf petiole is cut off just above the bud, and then the bud is removed from the budstick by a flat cut just underneath the bud, with a razor-sharp knife. Only the bud itself and a small piece of wood under it are used. In the stock an inverted T-cut is made, and the micro-bud is slipped into this, right side up. The entire T-cut, including the bud, is covered with polyvinyl chloride (PVC) plastic budding tape ($\frac{3}{4}$ in. in width and 0.002 in. thick). The tape is allowed to remain for 10 to 14 days for spring budding and 3 weeks for all budding, after which it is removed by cutting with a knife. By this time the buds should have healed in place ; subsequent handling is the same as for conventional T-budding.

10

Propagation by Layering

Layering is the development of roots on a stem while it is still attached to the parent plant. The rooted stem is then detached to become a new plant growing on its own roots. A layered stem is known as a *layer*. This may be a natural means of reproduction, as in black raspberries and trailing blackberries, or it may be induced by the "artificial" methods described in this chapter.

Factors Affecting the Propagation of Plants by Layering

Root formation during layering is stimulated by various stem treatments (as shown in Figure 10.1) which cause an interruption in the downward translocation of organic materials—carbohydrates, auxin, and other growth factors—from the leaves and growing shoot tips. These materials accumulate near the point of treatment, and rooting occurs in this general area even though the stem is still attached to the parent plant.

Water and minerals are supplied to the layered shoot, because the stem is not severed and the xylem remains intact. Thus layering does not depend upon the length of time during which a severed shoot (cutting) can be maintained before rooting occurs. This is an important reason for layering being more successful with many plants than propagation by cuttings.

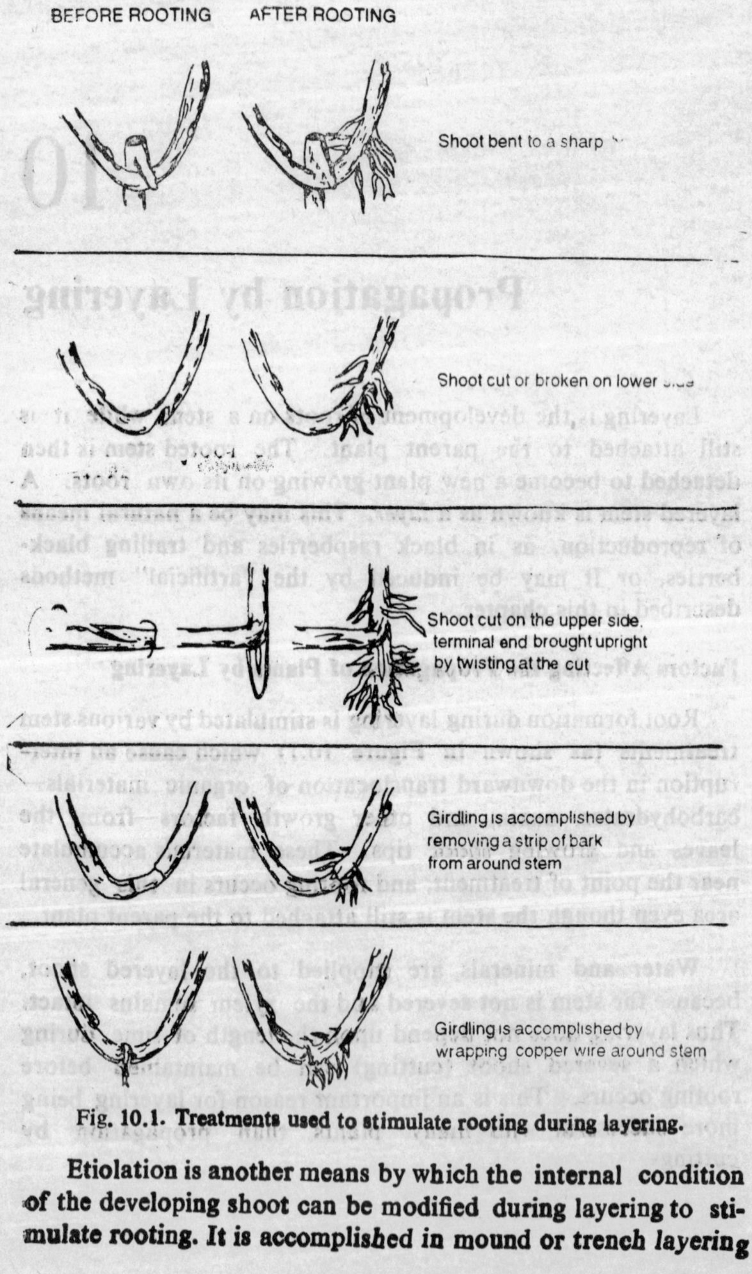

Fig. 10.1. Treatments used to stimulate rooting during layering.

Etiolation is another means by which the internal condition of the developing shoot can be modified during layering to stimulate rooting. It is accomplished in mound or trench *layering*

by covering the newly developing shoot as it grows so that the basal part of the layered shoot is not exposed to the light. This apparently accounts in large measure for the success with which shy-rooting plants are rooted by these two methods.

Applying root-promoting substances, such as indolebutyric acid, during layering is sometimes beneficial, as it is with cuttings, although the method of application may be somewhat different. Applying the material to girndling cuts as a powder, in lanolin, or as a solution in 50 per cent alcohol can be utilized effectively.

Root formation on layers depends upon continuous moisture, good aeration, and moderate temperatures in the rooting zone. These conditions are best provided by a rooting medium such as sawdust. Prolonged dry spells and compact, heavy soils hinder root development, particularly during the initial stages of rooting. The addition of granulated peat moss to the soil mounded around apple and quince stock plants has promoted rooting. Excessively high temperatures in the upper layers of soil during the spring and summer may reduce the moisture content and cause compaction, not only inhibiting but rooting injuring the shoots as well.

Characteristics and Uses of Layering

The principal advantage of layering is the success with which plants can be rooted by this method. Many clones which will not root easily by cuttings can be propagated by layering, enabling the plant to be established on its own roots. Most methods of layering are relatively simple to perform and can be practiced out-of-doors in the nursery or garden. When small numbers of plants are involved, layering can give a high degree of success with somewhat less skill, effort, and equipment than is necessary with cuttings. With those few kinds of plants in which layering occurs naturally, it is a simple and economical method of propagation.

In some cases, a larger plant can be produced in a shorter time than it would if started as a cutting. However, since transplanting becomes increasingly difficult as the size of the layer increases, special precautions are necessary to establish successfully the larger plant on its own roots.

On the other hand, layering is an expensive method of propagation and does not lend itself to the large-scale techniques of mechanization used in modern nurseries. Part of the increased cost of propagation is due to the additional hand labour required. A layered plant requires a certain amount of individual attention, depending upon the particular method in use, even though the operations involved are in themselves simple. Also, the number of salable plants from a given number of stock plants is smaller than with cuttings, buds, or scions. The methods tend to be cumbersome, and the stock plants take up a considerable area which is difficult to cultivate and maintain free of weeds.

Layering is usually limited by American nurserymen to those plants which propagate naturally in this manner—*e.g.*, black raspberry, trailing blackberries, and those plants too difficult or impossible to propagate by other methods, yet of sufficient value to justify the cost. For instance, the filbert (*Corylus* sp), Muscadine grape (*Vitis rotundifolia*), and litchi (*Litchi chinensis*) are propagated commercially in this manner. Layering is used in propagating certain clonal rootstocks, such as the Malling apple stocks which are not easily rooted as cuttings or which require special equipment, such as mist installations.

Some ornamental trees and shrubs are propagated by layering particularly in Europe. Established blocks of plants to be used for layering (known as *stool beds*) have been in production for many years in some cases, and nurserymen are skilled in their management

Layering is perhaps best utilized by the amateur horticulturist who wishes to propagate a relatively small number of plants, or by specialists involved with reproducing certain kinds of plants. In these ceses, expense per plant and the individual attention necessary would not be factors in the choice of method.

PROCEDURES IN LAYERING

Tip Layering

In tip layering, rooting takes place near the tip of the current season's shoot which is bent to the ground. The shoot tip begins to grow downward into the soil but recurves to produce a sharp bend in the stem, from which roots develop. This natural method of reproduction is characteristic of trailing blackberries, dewberries, and black and purple raspberries.

Stems of these plants are biennial in that the canes are vegetative during the first year, fruitful the second, and pruned out after fruiting. "Summer topping" new canes by pinching off 3 to 4 in. of the tip after growth of 18 to 30 in. encourages

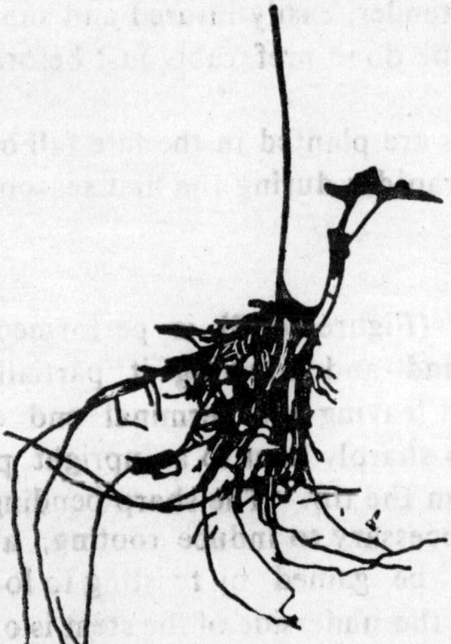

Fig. 10.2. Tip layer of boysenberry. In this method the entire tip of the cane should be covered with soil.

lateral shoot production. This will increase the number of potential tip layers, and also next year's fruit crop. By late summer, the canes begin to arch over, and their tips assume a characteristic appearance in that the terminal ends become elongated and the leaves small and curled to give a "rat-tail" appearance. The best time for layering is when only part of the lateral tips have attained this appearance. If the operation is done too soon, the shoots may continue to grow instead of forming a terminal bud. If it is done too late, the root system will be small.

The tips are preferably layered by hand, a spade or trowel being used to make a hole 3 to 4 in. deep. The end of the shoot is inserted into the hole and covered with soil. For large-scale operations, a shallow furrow may be plowed along the row into which the tips are placed. Rooting takes place rapidly, and the plants are ready for digging by the end of the same season. The rooted tip consists of a terminal bud, a large mass of roots, and 6 to 8 in. of the old cane to serve as a "handle" and to mark the location of the new plant (Figure 10.2). Since the tip layers are tender, easily injured and subject to drying out, digging should be done preferably just before replanting.

Rooted tip layers are planted in the late fall or early spring. New canes develop rapidly during the first season.

Simple Layering

Simple layering (Figure 10.3) is performed by bending a branch to the ground and covering it partially with soil or rooting medium, but leaving the terminal end exposed. The end of the branch is sharply bent to an upright position about 6 to 12 in. back from the tip. The sharp bending of the shoot may be all that is necessary to induce rooting, although additional benefit may be gained by twisting to loosen the bark. Cutting or notching the underside of the stem is often practiced. Other methods are illustrated in Figure 10.1. The bent part of the shoot is next inserted into the soil so that it can be covered to a depth of 3 to 6 in. A wooden peg, bent wire, or stone may be used to hold the layer in place, and a vertical wooden stake should be inserted beside the layer to support the exposed shoot and hold it upright. If the branch is relatively inflexible and hard to bend, the tension may be lessened by notching the upper side of the shoot on the highest part of the bend back from the layer itself.

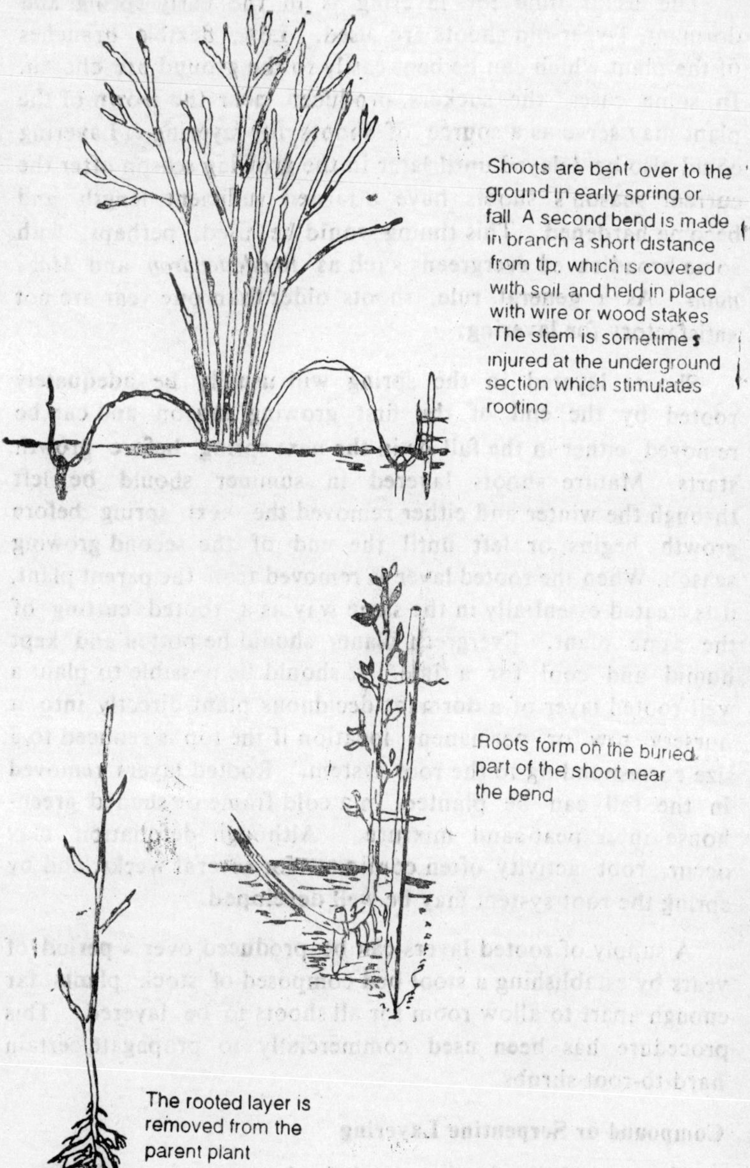

Fig. 10.3. Steps in propagation by simple layering.

The usual time for layering is in the early spring, and dormant, 1-year-old shoots are used. Low, flexible branches of the plant which can be bent easily to the ground are chosen. In some cases, the suckers produced near the crown of the plant may serve as a source of shoots for layering. Layering could also be delayed until later in the growing season after the current season's shoots have attained sufficient length and become hardened. This timing would be used, perhaps, with some broadleaved evergreens such as *Rhododendron* and *Magnolia*. As a general rule, shoots older than one year are not satisfactory for layering.

Shoots layered in the spring will usually be adequately rooted by the end of the first growing season and can be removed either in the fall or in the next spring before growth starts. Mature shoots layered in summer should be left through the winter and either removed the next spring before growth begins or left until the end of the second growing season. When the rooted layer is removed from the parent plant, it is treated essentially in the same way as a rooted cutting of the same plant. Evergreen plants should be potted and kept humid and cool for a time. It should be possible to plant a well-rooted layer of a dormant deciduous plant directly into a nursery row or permanent location if the top is reduced to a size corresponding to the root system. Rooted layers removed in the fall can be planted in a cold frame or shaded greenhouse in a peat-sand mixture. Although defoliation may occur, root activity often continues for several weeks, and by spring the root system may be well developed.

A supply of rooted layers can be produced over a period of years by establishing a stool bed composed of stock plants far enough apart to allow room for all shoots to be layered. This procedure has been used commercially to propagate certain hard-to-root shrubs.

Compound or Serpentine Layering

Compound layering is essentially the same as simple layering, except that the branch is alternately covered and exposed along its length (Figure 10.4). Generally, the stem is injured

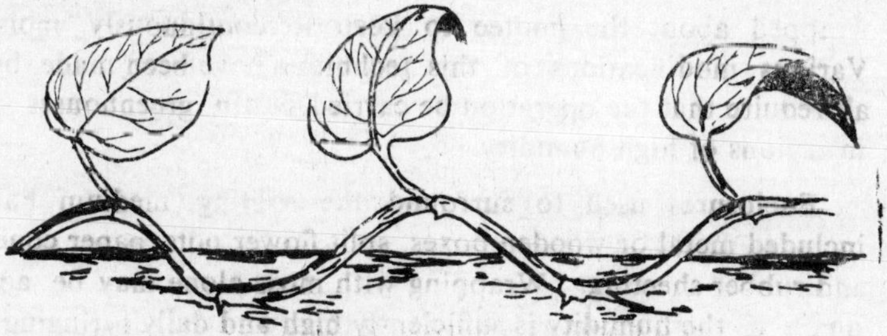

Fig. 10.4. Compound layering of a *Philodendron* vine.

or girdled at the lower part of the stem and covered in the same manner as for simple layering. Roots develop at each of these buried sections. The exposed part of the stem should have at least one bud to develop a new shoot. After rooting takes place or at the end of the growing season, the branch is cut in sections made up of the new shoot and the portion containing roots. Several new plants are thus possible from a single branch.

This method is used for propagating plants which have long, flexible shoots, such as the Muscadine grape. Ornamental vines, such as *Wisteria* and *Clematis*, can also be rooted this way, although such efforts are generally limited to use by amateurs rather than by commercial propagators.

Air Layering

In air layering, roots form on the aerial part of a plant after the stem is girdled or slit at an angle and enclosed in a moist rooting medium at the point of injury. This method has been used for more than a thousand years, being known in various parts of the world as *Chinese layerage*, *pot layerage*, *circumposition*, *marcottage*, and *gootee*.

The principal limiting factor in air layering has been the difficulty in keeping the rooting medium properly moistened. The ancient Chinese gootee or marcottage method consists of plastering a ball of clay or other soil mixture about the ring or girdle, which is then covered with moss or fiber to hold it together. Above the gootee is a receptacle of water (perhaps a bamboo joint) from which extends a string leading to and wrapped about the gootee to keep it continuously moist. Various modifications of this technique have been made, but all require that the operation be carried out in greenhouses or in regions of high humidity.

Enclosures used to surround the rooting medium have included metal or wooden boxes, split flower pots, paper cones, and rubber sheeting. Wrapping with moss alone may be adequate if the humidity is sufficiently high and daily syringing is practiced. These devices have been replaced to a large extent by sheets of polyethylene film, which have the properties of high permeability to gases (carbon dioxide and oxygen), low transmission of water vapour, and sufficient durability to withstand weathering for long periods. By this means it is possible to keep the enclosed rooting medium damp from the plant's natural moisture for long periods of time without continuous attention. Not all plastic films have the same degree of the necessary characteristics ; only *polyethylene* film should be used.

Air layering is used to propagate a number of tropical and subtropical trees and shrubs. In Florida, the litchi and the Persian lime (*Citrus aurantifolia*) are propagated commercially by this method. With polyethylene film for wrapping the layers, it is possible to extend the method of air layering out-of-door plants from the tropics to the temperate zones. However, its greatest application is for the amateur horticulturist or the propagator of selected plants for cultural or for scientific purposes rather than for the commercial nurseryman.

Air layers are made in the spring on wood of the previous season's growth or, in some cases, in the late summer with partially hardened shoots. Wood older than one year can be used in some cases, but rooting is less satisfactory and the larger plants produced are somewhat more difficult to handle after rooting. With tropical greenhouse plants, layering should be done after several leaves have developed during a period of growth.

The first step in air layering is to girdle or cut the bark of the stem at a point 6 to 12 in. or more from the tip end. A strip of bark ½ to 1 in. wide, depending upon the kind of plant, is completely removed from around the stem. Scraping the exposed surface to insure complete removal of the phloem and cambium is desirable to retard healing. Another procedure is to make a slanting cut about 2 in. long up and to the center of the stem, keeping the two surfaces apart by sphagnum or a piece of wood. Application of a root promoting material, such as indolebutyric acid to the exposed stem, especially to the upper edges or between the two exposed surfaces of a cut, may be beneficial. About two handfuls of only *slightly* moistened sphagnum moss are placed around the stem to enclose the cut surfaces. If the moisture content of the sphagnum moss is too high, decay of the stem tissues will occur.

A piece of polyethylene film 8 to 10 in. square is wrapped carefully about the branch so that the sphagnum moss is completely covered. The ends of the sheet should be folded (as in wrapping meat) with the fold placed on the lower side. The two ends must be twisted to make sure that no water can seep inside. Adhesive tape, such as electricians' waterproof tape, serves well to wrap the ends ; the winding should be started well above the end of the plastic film to enclose the ends, particularly the upper one, securely. Budding rubbers and florist's ties are other materials which can be used for this purpose.

The time for removal of the layer from the parent plant is best determined by observing root formation through the transparent film. In some plants, rooting occurs in 2 to 3 months or less. Layers made in spring or early summer are best left until the shoots become dormant in the fall, and are removed at that time. Holly, lilac, *Azalea*, and *Magnolia* should be left for two seasons. In general, it is desirable to remove the layer at a time when it is not actively growing. The period between the removal of the rooted layer from the plant and establishment on its own roots is very critical, and the plant may be lost at that time even though it rooted well. The root system at this time is small in proportion to the top and is incapable of supporting the plant without special precautions. The difficulties of transplanting increase in proportion to the size of the top. Severe pruning usually is advisable to bring the top more in line with the size of the roots, but it may not be imperative if the following precautions are followed. The rooted layer should be potted into a suitable container and placed under cool, humid conditions, such as an enclosed frame, where the plants can be frequently syringed. If this operation is done in the fall, a sufficiently large root system may develop by spring to permit successful growth in the open. Placing the rooted layers under mist for several weeks, followed by gradual hardening-off, is probably the most satisfactory procedure.

Mound (Stool) Layering

Mound or stool layering involves cutting a plant back to the ground during the dormant season and mounding soil or other media around the base of the newly developing shoots in the spring to encourage roots to form on them. Covering with soil keeps the shoots etiolated and encourages root formation. Plants with stiff branches that do not bend easily and which are capable of producing an abundance of shoots from the crown year after year are particularly adapted to this method. Plants commonly propagated in this manner include the clonal apple stocks, certain Mazzard cherry stocks, currants, and gooseberries.

A stool bed can remain in use for 15 to 20 years ; therefore it should be on loose, fertile, well-drained soil and established 1 year before the operations begin. The mother plants should be set 12 to 15 in. apart in the row, but the spacing between rows varies under different conditions and types of nursery equipment. Width between rows should be sufficient to allow for cultivation and hilling operations during spring and summer. In England 3½-ft rows are recommended, but in New York a minimum of 8 ft was found desirable. Planting in a shallow trench, as shown in Fig. 10.5 will permit the shoots to arise from the crown at a low level. Plants are cut back to 15 to 18 in. from the ground and left to grow unchecked for a season, the space between the rows being kept cultivated.

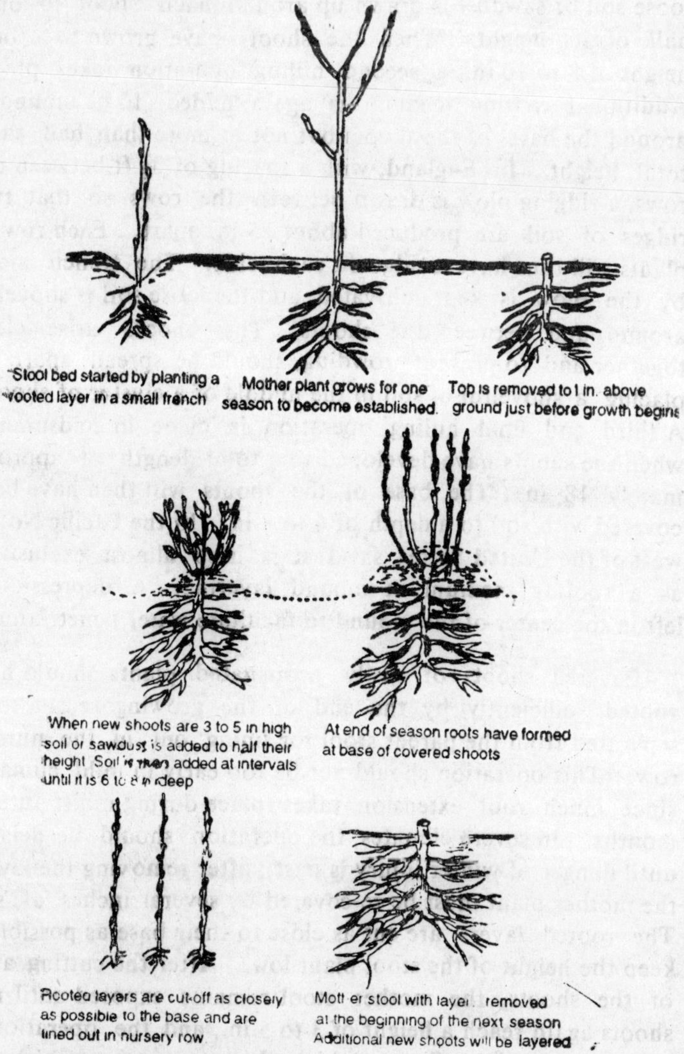

Fig. 10.5. Steps in propagation by mound (stool) layering.

Before new growth starts the next spring, all plants are cut back to an inch above the ground level. Two to five new shoots usually develop from the crown the second year, more in later years. When these shoots have grown 3 to 5 in.,

loose soil or sawdust is drawn up around each shoot to one-half of its height. When the shoots have grown to a total height of 8 to 10 in., a second hilling operation takes place. Additional rooting medium is again added, to be mounded around the bases of the shoots but not to more than half their total height. In England, with a spacing of 3½ ft between the rows, a ridging plow is drawn between the rows so that two ridges of soil are produced about 15 in. apart. Each row of plants will thus have a ridge on each side. The trench made by the plow is kept cultivated, and the loose soil is shoveled around and between the shoots. The shoots arise close together and, to prevent crowding, should be spread apart by placing a shovelful of soil in the middle of a cluster of shoots. A third and final hilling operation is made in midsummer when the shoots have developed to a total length of approximately 18 in. The base of the shoots will then have been covered with soil to a depth of 6 to 8 in. In the Pacific Northwest of the United States, sawdust is used almost exclusively as a rooting medium in mound layering. A depression is left in the center of the mound to facilitate water penetration.

Layered shoots of easily propagated plants should have rooted sufficiently by the end of the growing season to be separated from the parent stool for lining out in the nursery row. This operation should not be too early in mild climates, since much root extension takes place during moist autumn months. In severe climates, the operation should be delayed until danger of winter injury is past ; after removing the layers, the mother plant must be re-covered by several inches of soil. The rooted layers are cut as close to their base as possible to keep the height of the stool plant low. After the cutting away of the shoots, the mother stool remains exposed until new shoots again reach a height of 3 to 5 in., and the operation is underway again. Shoots which do not root or which root poorly often can be cut off at the same time as the well-rooted ones and treated as hardwood cuttings ready for planting in the nursery row.

Several points which are important to the proper utilization of a stool bed should be kept in mind. To prolong the life of the stool bed and to maintain the plants in a vigorous condition, good fertility should be maintained by annual fertilizer application. Disease, insects, and weeds must be controlled. Supplementary overhead irrigation is essential in some areas to maintain proper moisture conditions in the rooting zone. Soil must be added around the base of the shoots while they are still soft and succulent, not delayed until the shoots have hardened. With more difficult plants, such as certain plum varieties, it is best to add soil before the shoots begin to grow so as to induce etiolation at their bases. Girdling the bases of the shoots by wiring about 6 weeks after they begin to grow will stimulate rooting in many plants. To gradually rejuvenate old stool beds, new layers may be planted in spaces between old ones.

Trench Layering

Trench layering consists of growing a plant or a branch of a plant in a horizontal position in the base of a trench and filling in soil around the new shoots as they develop. Roots develop from the bases of these new shoots. (Fig. 10.6).

The first step in this procedure involves the establishment of the mother bed which, as in mound layering, can be used over a period of years. Rooted layers or 1-year-old nursery-budded or grafted trees are planted 18 to 30 in. apart at an angle of 30 to 45 deg. down a row. The rows should be 4 to 5 ft apart—wide enough to allow for cultivation and to draw soil up around the plants to a height of 6 in. The plants are then cut back to a uniform length—18 to 24 in.—and left to grow one season. In some cases—in layering walnuts, for instance—the plants are planted horizontally in the trench and the developing shoots layered the first year.

Before the beginning of growth in the spring, the parent layers are bent over and laid flat on the bottom of a trench dug along the row, about 2 in. deep and wide enough to receive the entire layer. Weak lateral branches are cut to ½ in. in length and strong laterals merely tipped back. Wooden pegs may be used to hold the mother layer in place. Short lengths of wire bent to form a U, or a longer single wire rolled on

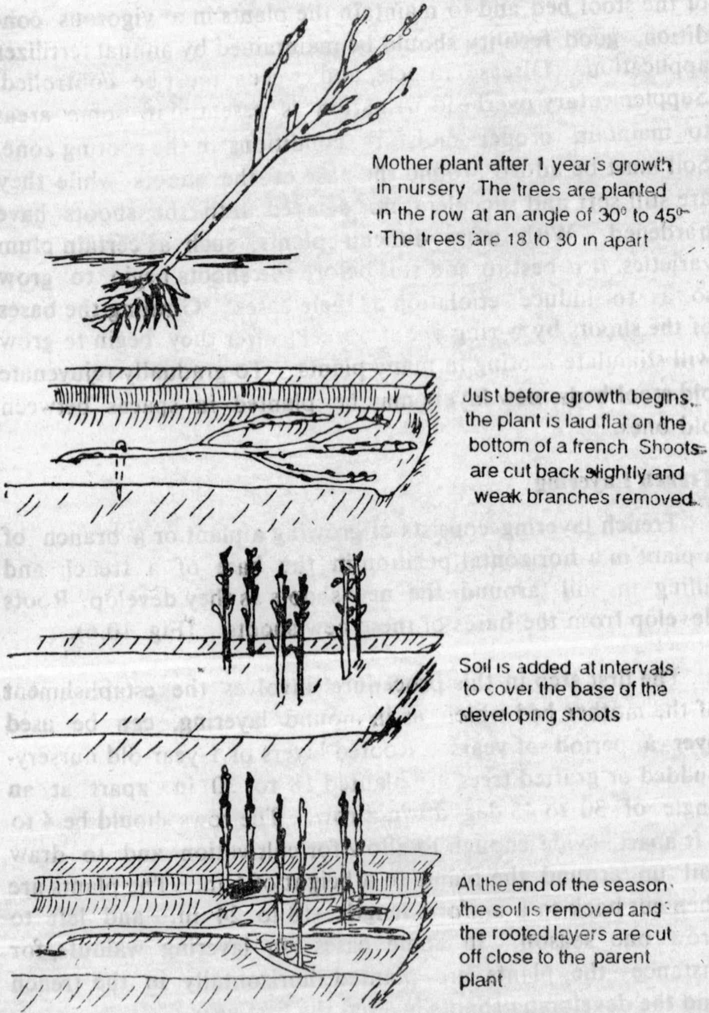

Fig. 10.6. Steps in propagation by trench layering.

the end and set at an angle, will be useful to hold down the small shoots. It is important that all parts of the shoot be completely flat on the floor of the trench.

Before the buds swell, the entire layer is covered with 1 to 2 in. of fine soil or other rooting medium such as peat moss,

sawdust, or wood shavings. Another inch of medium is added when the developing shoots have pushed through the first soil layer but before their leaves have expanded. Several more additions of soil are made the first 2 or 3 weeks of the growing season to ensure the etiolation of the basal 2 to 3 in. of the shoot. When the shoots have grown an additional 3 to 4 in., soil is again added to half the height of the exposed shoot. Further additions are made at intervals until the bases of the shoots are finally covered to a depth of 6 to 8 in. by midsummer. Roots form on the base of the current season's shoot.

At the end of the growing season after the plants have become dormant—or the next spring—the soil is removed from around the layered shoots, and rooted layers are cut off from the original layered stock as close to the base as possible. Unrooted shoots can be left to be pegged down the nexy year to fill in gaps or to rejuvenate an old bed. New shoots may arise from adventitious buds along the old original layer underground. However, some of the vigorous shoots growing from the mother plant can be pegged down each year to produce a new supply of rooted layers.

Trench layering is primarily a nursery method for propagating particular fruit tree rootstocks difficult to propagate by other methods. It could also be practiced on established shrubs or trees by bending long, flexible shoots or vines to the ground under the same conditions as is done in simple layering. The shoot is covered along its entire length, but the tip is left exposed. New shoots which develop from the buds along the stem grow upward through the soil, with roots forming at their base. This latter procedure is sometimes known as *continuous layering.*

Plant Modifications Suitable for Natural Layering

Some plants exhibit modifications of their vegetative structure or method of growth which lead to their natural vegetative increase. Those listed below could be considered natural forms of layering and often can be utilized for propagation.

Runners

A *runner* is a specialized stem which develops from the axil of a leaf at the crown of a plant, grows horizontally along the ground, and forms a new plant at one of the nodes. The strawberry is a typical plant propagated in this way. Other plants propagated by runners include bugle (*Ajuga*) and the strawberry geranium (*Saxifraga sarmentose*). Plants of these species grow as a typical rosette or crown.

In most strawberry varieties, runner formation is related to the length of day, runners being produced under daylengths of 12 to 14 hours or more. New plants are produced at alternate nodes. These take root but remain attached to the mother plant for some time. New runners may in turn be produced by the daughter plants. The connecting stems die in the late fall and winter, and each daughter plant becomes separate from the others.

In propagating by runners, the rooted daughter plants are dug when they have become well rooted, and then transplanted to the desired locations.

Stolons

Stolon is a term used to describe various types of horizontally growing stems that produce adventitious roots when in contact with the soil. Specifically these are prostrate or sprawling stems, as found in *Cornus stolonifera*, for instance. The underground stem of the potato that terminates in the tuber is a stolon. The term is sometimes used to describe the tip layers of black raspberry and blackberry or the runners of strawberry. In such plants, the stolon could be treated as a naturally occurring rooted layer, cut from the parent plant and transplanted.

Offsets

An *offset* is a characteristic type of lateral shoot or branch which develops from the base of the main stem in certain plants. This term is applied generally to a shortened, thickened stem of rosette-like appearance. Many bulbs reproduce by producing typical offset bulblets from their base. The term offset (or *offshoot*, as is sometimes used) also applies to lateral branches arising on stems of monocotyledons. The date palm produces lateral shoots from the base of the plant by which it is propagated. The pineapple is also propagated by offsets, although in commercial culture these are termed ratoons, suckers, or slips, depending upon the location on the plant where they are produced. Lateral shoots arising from rhizomes, as in the banana or orchid, are also offsets or offshoots.

Offsets are removed by cutting them close to the main stem with a sharp knife. If it is well rooted, the offset can be potted as is done with any rooted cutting. If insufficient roots are present, the shoot is placed in a favorable rooting medium and treated as a leafy stem cutting.

In cases in which offset development is meager, cutting off the main rosette may stimulate the development of offsets from the old stem just as removing the terminal bud stimulates lateral shoots in any other type of plant. For instance, in *Echeveria* the main stem may elongate so that the plant becomes a fleshy rosette borne on top of a fleshy, bare stem. The rosette, with a short piece of stem, may be removed and rooted, and new offsets will develop from buds at the base. It is desirable that this operation be carried out while the stem is somewhat soft and succulent, rather than allowing it to become hard and woody.

Offshoots of the date palm do not root readily if separated from the parent plant. They are usually layered for a year prior to removing

Suckers

A *sucker* is a shoot which arises on a plant from below ground. The most precise use of this term is to designate a shoot which arises from an adventitious bud on a root. However, in practice, shoots which arise from the vicinity of the crown are also referred to as suckers even though originating from stem tissue. Nurserymen generally designate any shoot produced from the rootstock below the bud union of a budded tree as a sucker and refer to the operation of removing them as "suckering." In contrast, a shoot arising from a latent bud of a stem several years old, as, for instance, on the trunk or main branches, should be termed a *watersprout*.

The tendency to "sucker" is a characteristic possessed by some plants and not by others. The ability of a plant to sucker and the ability of a plant to grow from root cuttings are closely related. In such cases, however, nurserymen prefer to use root cuttings rather than depend on naturally produced suckers. Since suckers arise from adventitious buds, they may show juvenile characteristics.

Suckers are dug out and cut from the parent plant. In some cases part of the old root may be retained, although most new roots arise from the base of the sucker. It is important to dig the sucker out rather than pull it, to avoid injury to its base. Suckers are treated essentially as a rooted layer or as a cutting, in case few or no roots have formed. They are usually dug during the dormant season.

Crowns

The term *crown* as generally used in horticulture designates that part of a plant at the surface of the ground from which new shoots are produced. In trees or shrubs with a single trunk, the crown is principally a point of location near the ground surface marking the general transition zone between stem and root. In herbaceous perennials, the crown is the part of the plant from which new shoots arise annually. The crown of many herbaceous perennials consists of many branches, each being the base of the current season's stem, which originated from the base of the preceding year's branch. These new shoots are stimulated to grow from the base of the old stem as it dies back after blooming. Adventitious roots develop along the base of the new shoots. These new shoots eventually flower either the same year they are produced or the following year. As a result of the annual production of new shoots and the dying back of old shoots, the crown may become extensive within a period of a relatively few years.

In certain plants for example, the strawberry or the African violet (*Saintpaulia*)—the stem is a short and thickened structure from which the leaves are produced in a rosette-like arrangement. The entire body of the plant is often referred to as the crown. Lateral shoots or offsets are produced from the base of the crown. An old plant may be composed of a number of "crowns" or "crown divisions" which have been produced in this manner.

Multibranched woody shrubs may develop extensive crowns. Although an individual woody stem may persist for a number of years, new, vigorous shoots are continuously produced from the crown, and eventually crowd out the older shoots. If left undisturbed, such shrubs develop into extensive thickets. Under normal handling, the older shoots are regularly removed by pruning to give way to the younger, more vigorous shoots.

Division of the crown is an important method of propagation for herbaceous perennials, and to some extent for woody shrubs, because of its simplicity and reliability. Such characteristics make this method particularly useful to the amateur or professional gardener who is generally interested in only a modest increase of a particular plant. Many herbaceous perennials must be divided every 2 to 3 years to prevent the plants from becoming overcrowded.

Crowns of outdoor herbaceous perennials are usually divided in the spring just before growth begins or in late summer or autumn at the end of the growing season. As a general rule, those plants which bloom in the spring and summer and produce new growth after blooming should be divided in the fall. Those which bloom in summer and fall and make little or no new growth until spring should be divided in early spring. Potted plants are divided when they become too large for the particular container in which they are growing.

In crown division, plants are dug and cut into sections with a knife. In herbaceous perennials such as the Shasta daisy or aster, where an abundance of new rooted offshoots are produced from the crown, each may be broken from the old crown and planted separately, the older part of the plant clump being discarded. If a large clump is desired, then a section of the old crown bearing a number of new shoots from its base may be used. In the case of plants in which the crown consists of a number of rosettes or offsets, as occurs in the African violet, division should be made between each of the rosettes, but roots should be present on each section.

Shrubs may be divided in the same manner with a shovel or hatchet. Such an operation should be carried out at a time when the plant is dormant. The top should be cut back and the roots trimmed at the time of division, and each section planted as a new shrub.

11

Propagation by Specialized Stems and Roots

This chapter deals with propagation by specialized vegetative structures—*bulbs, corms, tubers, tuberous roots, rhizomes*, and *pseudobulbs*. These organs are primarily modified plant parts specialized for food storage. Plants possessing them are invariably herbaceous perennials in which the shoots die down at the end of a growing season, and the plant lives over in the ground as a dormant, fleshy organ which bears buds to produce new shoots the next season. Such plants are admirably suited to withstanding periods of adverse growing conditions in their yearly growth cycle. The two principal climatic cycles for which such performance is geared are the warm cold cycle of the temperate zones and the wet-dry cycle of tropical and subtropical regions.

The second function of these specialized organs is that vegetative reproduction. Table 11.1 shows the list of flowering plants grown from specialized stems or roots.

BULBS

Definition and Structure

Bulbs are produced by monocotyledonous plants in which the usual plant structure is modified for storage and reproduction. A *bulb* is a specialized underground organ consisting of a short, fleshy, usually vertical stem axis (*basal plate*) bearing at its apex a growing point or a flower primordium enclosed by thick, fleshy scales. (See Figure 11.1).

Table 11 1. Flowering Plants Grown from Specialized Stems or Roots.

Bulbs	*Corms*	*Tubers*	*Tuberous Roots*	*Rhizomes*
		I		
		Hardy ; spring flowering ; planted in the fall		
Camassia	*Crocus*		*Anemone* (Wind flower)	*Convallaria* (Lily-of-the-valley)
Chionodoxa (Glory-of-the-snow)			*Arisaema* (Jack-in-the-pulpit)	*Iris* (bearded types)
Erythronium (Trout lily)			*Claytonia* (Spring beauty)	
Fritillaria (Guinea hen flower)			*Dicentra* (Bleeding heart)	
Galanthus (Snowdrop)			*Eranthis* (Winter aconite)	
Hyacinthus (Hyacinth)			*Eremurus* (Desert candle)	
Iris (bulbous)				
Leucojum (Snowflake)				
Muscari (Grape hyacinth)				
Narcissus (Daffodil)				
Scilla (Squill)				
Tulipa (Tulip)				
		II		
		Semi-hardy to hardy ; summer and fall-flowering ; planted in the fall		
Lilium (Lily)	*Colchicum* (Autumn crocus)	*Begonia evansiana*		
Lycoris (Hardy amaryllis)	*Crocus* (some species)			
	Sternbergia			

Table 11.1 (*Continued*)

Bulbs	*Corms*	*Tubers*	*Tuberous Roots*	*Rhizomes*
		III		
		Tender ; summer-and fall-flowering ; dug in the fall and stored over winter ; planted in the spring when danger of freezing is over ; these can be kept outdoors over winter in mild climates		
Amaryllis belladonna	***Freesia***	***Caladium***	***Begonia*** **(Tuberous)**	***Canna***
Galtonia **(Summer hyacinth)**	***Gladiolus***		***Dalilia***	
Hymenocallis	***Ixia***		***Gloriosa***	
Polianthes **(Tuberose)**	***Tritonia***			
Tigridia **(Tiger flower)**				
Watsonia				
Zephyranthes **(Zephyr lily)**				
		IV		
		Tender ; mostly grown in greenhouses or as house plants ; grown outdoors in mild climates		
Clivia **(Kafir lily)**		***Caladium***	***Agapanthus*** **(Lily-of-the-Nile)**	***Achimenes***
Crinum			***Arum*** **(Black calla)**	***Zantedeschia*** **(Calla lily)**
Eucharis grandiflora **(Amazon lily)**			***Colorusia*** **(Elephant's car)**	
Haemanthus **(Blood lily)**			***Cyclanien***	
Hippeastrum **(Amaryllis)**			***Ranuculus*** **(florist's types)**	
Nerine			***Sinningia*** **(Gloxinia)**	
Oxalis (bulbous types)				
Vallota **(Scarborough lily)**				
Velthernia				

Most of the bulb consists of *bulb scales*, which morphologically are the continuous, sheathing leaf bases (see Figure 11.2). The outer bulb scales are generally fleshy and contain reserve food materials, whereas the bulb scales toward the center function less as storage organs and are more leaf-like. In the center of the bulb, there will be either a vegetative

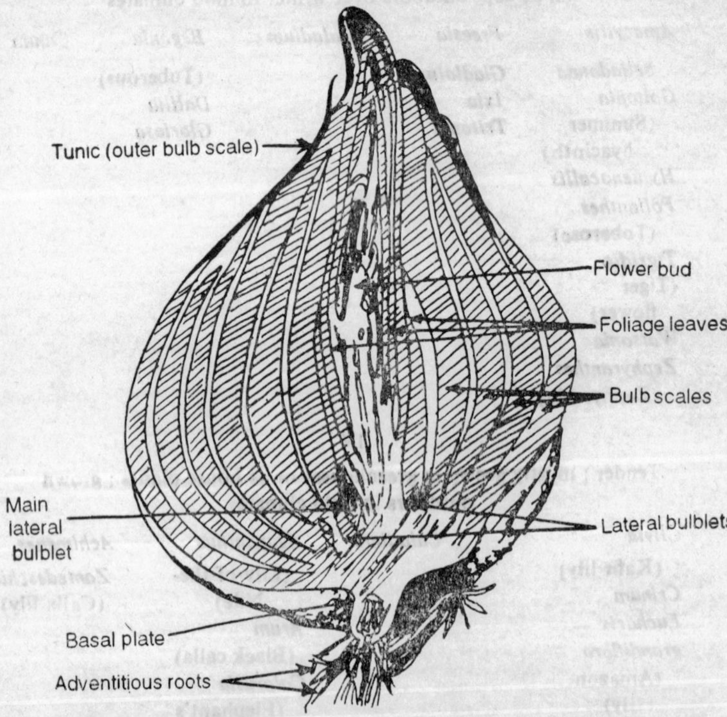

Fig. 11.1. The structure of a tulip bulb—an example of a typical laminate bulb. Longitudinal section representing stage of development shortly after the bulb is planted in the fall and as new bulb begins to appear. Redrawn from Mulder and Luyten.

growing point or an unexpanded flowering shoot. Growing points develop in the axil of these scales to produce miniature bulbs, known as *bulblets*, which when they grow to full size are known as *offsets*. In various species of lilies, bulblets may form in the leaf axils either on the underground portion or on

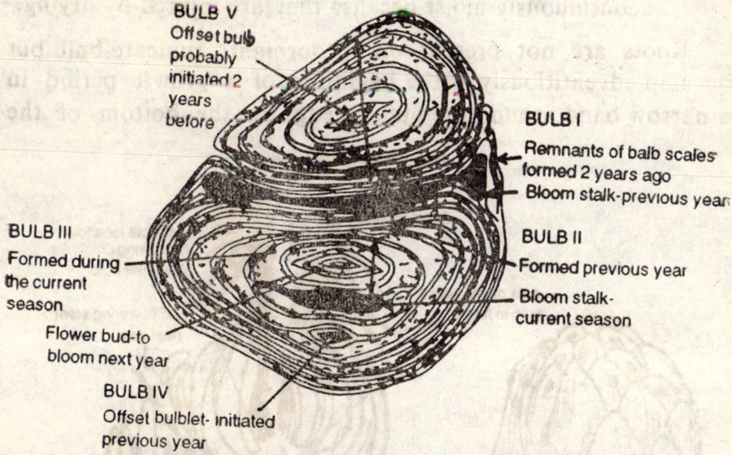

Fig. 11.2. Cross section of a daffodil bulb. The continuous, concentric leaf scales found in the laminate bulb are shown. Also shown is the perennial nature of the daffodil bulb, which continues to grow by producing a new bulb annually at the main growing point. Lateral offset bulbs are also produced ; parts of five individual, differently aged bulbs are shown here. Redrawn from Huisman and Hartsema.

the aerial portion of the stem. Aerial bulblets are called *bulbils.*

There are two types of bulbs ;

(*a*) *Tunicate* (*laminate*) bulbs, represented by the onion and tulip. These bulb have outer bulb scales which are dry and membranous. This covering, or *tunic*, provides protection from drying and mechanical injury to the bulb. The fleshy scales are in continuous, concentric layers, or *lamina,* so that the structure is more or less solid.

(*b*) *Non-tunicate* (*scaly*) bulbs, represented by the lilies (Figure 11.3). These bulbs do not possess the enveloping dry covering. The scales are separate and give the bulb a scaly appearance. In general, the non-tunicate bulbs are easily damaged and must be handled more carefully than the tunicate bulbs ; they must be kept continuously moist because they are injured by drying.

Roots are not present on a dormant, tunicate bulb but develop adventitiously at the beginning of a growth period in a narrow band around the outside edge on the bottom of the

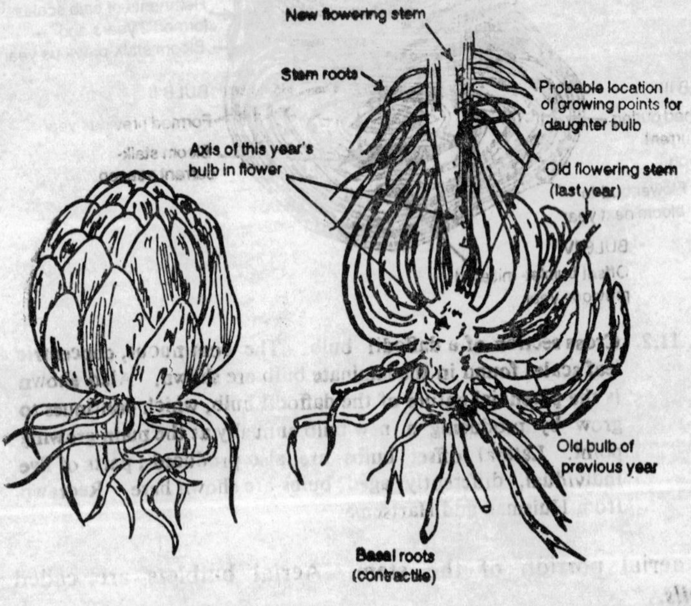

Fig. 11.3. Left : Outer appearance of a scaly bulb of lily (*Lilium hollandicum*). Right : Cross section of bulb of *L. speciosum* showing how a new axis and bulb scales form in the center with outer scales and old stem disintegrating.

basal plate. In the non-tunicate lily bulb, roots are produced in midsummer or later, and persist through the following year. In most lily species, roots also form on the stem above the bulb.

Contractile roots are produced in many species that appear to "pull" the bulb or corm to a given level in the ground.

Growth Behaviour

An individual bulb goes through a characteristic cycle of development, beginning with its initiation as a growing point and terminating in flowering and seed production. This general developmental cycle is composed of two stages : (*a*) vegetative and (*b*) reproductive. In the vegetative stage the bulblet grows to flowering size and attains its maximum weight. The subsequent reproductive stage includes the induction of flowering, differentiation of the floral parts, elongation of the flowering shoot, and finally flowering and (sometimes) seed production. Various bulb species have specific environmental requirements for the individual phases of this cycle which determine their seasonal behaviour, environmental adaptations, and methods of handling. They can be grouped into classes according to their time of bloom and method of handling.

Spring-Flowering Bulbs

Important commercial crops included in this group are the tulip, daffodil, hyacinth, and bulbous iris, although other kinds are grown in gardens.

(*a*) *Bulb formation.* The vegetative stage begins with the initiation of the bulblet on the basal plate in the axil of a bulb scale. In this initial period, which usually occupies a single growing season, the bulblet is insignificant in size, since it is present within another growing bulb and can be observed only if the bulb is dissected. Its subsequent pattern of development and the time required for the bulblet to attain flowering size are somewhat different for different species. The bulbs of the tulip and the bulbous iris, for instance, disintegrate upon flowering but leave a cluster of new bulbs and bulblets which were initiated the previous season. The largest of these may have attained flowering size at this time, but smaller ones require several additional years of growth (see Figures 11.1—11.3). The flowering bulb of the daffodil, on the other hand, continues to grow from the center year by year, producing new offsets which may remain attached for several years (see Figures 11.2). The hyacinth bulb also continues to grow year by year, but because the number of offsets produced is limited, artificial methods of propagation are usually used.

The size and quality of bloom is directly related to the size of the bulb and the amount of stored food. Only bulbs greater than a certain minimum size are capable of initiating flower buds. Commercial value is largely based on bulb size although the condition of the bulb and freedom from disease are also important quality factors.

The greatest increase in size and weight of the developing bulb take place in the period during and (mostly) after flowering, as long as the foliage remains in good condition. Cultural operations that encourage vegetative growth include irrigation ; weed, disease, and insect control ; and fertilization. Their greatest benefit, however, is to the next year's flower, because larger bulbs are produced. Conversely, adverse situations, such as poor growing conditions, removal of foliage, and premature digging of the bulb, result in smaller bulbs and inferior flower production.

Moderately low temperatures tend to prolong the vegetative period, whereas higher temperatures may cause the vegetative stage to cease and the reproductive stage to begin. Thus a shift from cool to warm conditions early in the spring, as occurs in mild climates, will shorten the vegetative period, result in smaller bulbs, and consequently produce inferior blooms the following year. Commercial bulb-producing areas for hardy spring-flowering bulbs are largely in regions of cool springs and summers, such as Holland or the Pacific Northwest of the United States.

The relative length of photoperiod apparently is not an important factor affecting bulb formation in most species. It has been shown, however, to be significant in some *Allium* species, such as onion and garlic.

(*b*) *Flower bud formation and flowering*. The beginning of the reproductive stage and the end of the vegetative stage is indicated by the drying of the foliage and the maturation of the bulb. From then on, no additional increase in size or weight of the bulb takes place. The roots disintegrate, and the bulb enters a seemingly "dormant" period. However, important internal changes take place, and in some species the vegetative growing point undergoes transition to a flowering shoot. In nature, all bulb activity takes place underground during this period ; in horticultural practice, the bulbs are dug, stored, and distributed during this 3-month period.

Temperature plays a key role in the progression from vegetative growth to flower bud formation. The optimum temperature—determined as the shortest time in which the bulb would flower—has been established for various developmental phases. In general, a deviation from these temperatures or an increase in the period of time at any one temperature results in an increase in the time required to bloom. Too high a temperature may prevent flower development entirely.

In the tulip, the differentiation of the flower parts takes place after the bulb becomes dormant in the summer. The optimum temperature for this process is 68° F (20° C). After 3 weeks of storage at this temperature, the optimum drops abruptly to 48° F (9° C), and remains there for 10 to 12 weeks, the lower temperatures being necessary to overcome the rest period and to stimulate the flower shoot to elongate. The optimum temperature then gradually increases as the flower shoot elongates and a flower is produced.

According to the studies from which the above information is quoted, the hyacinth has somewhat more exacting requirements. The optimum temperature immediately upon digging was quite high (90° to 95° F ; 32° to 35° C) for a week or two. When the first flower primordium on the raceme had appeared, this shifted to 78° F (26° C). When the last flower primordium had appeared, it had shifted to 63° F (17° C), and 3 weeks later it had shifted to 55° F (13° C), where it remained for about 12 weeks. Maintaining the bulbs at a temperature above 86° F (30° C) or below 55° F (13° C) during this period inhibits flower bud development entirely. For the hyacinth, a storage temperature of 78° F (26° C) is generally recommended for the flower forming period.

The daffodil ('King Alfred') has an optimum temperature of 62° F (17° C) for flower formation followed by a period of about 8 weeks at 48° F (19° C).

Holding the bulbs continuously at high temperatures (86° to 90° F ; 30° to 32° C ; or more) or at temperature near freezing will stop bulb development and can be used to increase the period required for flowering. With a shift to more optimum temperatures, flower bud development will continue unimpaired. Such storage conditions (below or above optimum) can be used when shipping bulbs from the Northern to the Southern Hemisphere.

In bulbous iris, a flower bud is not initiated within the bulb. It begins to appear when the shoot is a few inches out of the ground.

Lilies

A non-tunicate bulb, *e.g.*, the lily exhibits certain important differences from a tunicate bulb. Different species of lily have different methods of bulb reproduction, some examples being given in the following section on propagation. After flowering, the old bulb may divide or form offsets which can be used for propagation. In the subsequent vegetative period during summer the new bulb for next year grows in size and accumulates reserve materials. Lily bulbs do not go dormant in the same manner as other bulbs, and the roots do not disintegrate. The bulb should be out of the ground as little as pessible and, when it is for the purpose of propagation, should be handled carefully and kept from drying. The commercial value of the bulb depends not only on its size and weight but also on the condition of the fleshy roots and the freedom of the basal plate from decay.

In the Easter lily (*Lilium longiflorum*), transition of the growing point to a flowering shoot does not take place in the bulb, occurring after the shoot protrudes through the "nose" of the bulb. Storing the mature bulbs at 45° to 50° F (7° to 10° C) for 6 weeks and then growing them at temperatures of 60° F (16° C) or more (night) will produce blooms within 80 to 100 days. Storing the mature bulbs at room temperatures or at low (31° F ; —1° C) temperatures will keep the bulbs dormant and delay flowering. In the latter case, the method of bulb storage is also important. If they dry out, they will deteriorate. If they absorb moisture from the packing material, they may decay.

Tender, Winter-Flowering Bulbs

There are a number of flowering bulbs from tropical areas whose growth cycle is geared to a wet-dry climatic cycle rather than a cycle related to cold-warm conditions.

The amaryllis (*Hippeastrum vittata*) is an example of this kind. Its bulb is perennial, growing continuously from the center, the outer scales disintegrating. New leaves are produced continuously from the center during the vegetative period extending from late winter to the following summer. In the axil of every fourth leaf (or scale) that develops, a growing point is initiated. Thus throughout the vegetative period a series of vegetative offsets are produced. By fall the leaves mature and the bulb goes dormant, during which time the bulb should be dry. In this period, the *fourth* growing point from the center and any external to it differentiate into flower buds, and the shoot begins to elongate slowly. After 2 to 3 months of dry storage, the bulbs are watered, the flowering shoots elongate rapidly, and flowering takes place in midwinter. Maximum foliage development and bulb growth are essential to produce a bulb large enough to form a flowering shoot.

PROPAGATION

Offsets

Offsets provide a simple and reliable method for propagating many kinds of bulbs. This method is sufficiently rapid for the commercial production of tulip, dallodil, balbous iris, and grape hyaciath bulbs, but in general it is too slow for the lily, hyacinth, and amaryllis.

If undisturbed, the offsets may remain attached to the mother bulb for several years. They can also be removed at the time the bulbs are dug and replanted into beds or nursery rows to grow into flowering sized bulbs. This may require several growing seasons, depending upon the kinds of bulb and size of the offset.

(*a*) *Tulip*. Bulb planting takes place in the fall. Two systems of planting are used : the bed system, used mostly in Europe, and the row or field system, nsed mostly in the United States. Beds are usually 3 ft wide and separated by 12- to 18-in. paths. The soil is removed to a depth of 4 in., the bulbs set in rows 6 in. apart, and the soil replaced. In the other system, single or double rows are placed wide enough apart to permit the use of machines. To improve drainage, two or three adjoining rows may be planted on a ridge. Bulbs are spaced one to two diameters apart, with small bulbs scattered along the row. A mulch may be applied after planting but removed the following spring before growth.

Planting stock consists of those bulbs of the minimum size for flowering (9 to 10 cm in circumference or smaller). Since the time required to produce flowering sizes varies with the size of the bulb, the planting stock is graded so that those of one size can be planted together. For instance, an 8-cm or larger bulb normally requires a single season to become of flowering size ; a 5- to 7-cm bulb, two seasons ; and those 5 cm or less, 3 years.

During the flowering and subsequent bulb growing period of the next spring, good growing conditions should be provided so that the size and weight of the new bulbs will be at a maximum. Foliage should not be removed until it dries or matures. Important cultural operations include removal of competing weed growth, irrigation, fungicidal sprays to control *Botrytis blight* and fertilization. It is desirable to remove the flower heads at blooming time, because they may serve as a source of *Botrytis* infection and can lower bulb weight.

Bulbs are dug in midsummer when the leaves have turned yellow or the outer coats of the bulb have become dark brown in colour. In the Pacific Northwest of the United States, where temperatures are cool and the leaves remain green for a longer period, digging may take place before the leaves dry.

If the bulbs are dug too early or if warm weather causes early maturation, the bulbs may be small in size. The bulbs are dug by machine or by hand with a short-handled spade. After the loose soil is shaken from the bulbs, they are placed in trays in well-ventilated storage houses for drying, cleaning, sorting, and grading. General storage temperatures are 65° to 68° F (18° to 20° C). For early flowering, the bulbs should be held at 68° F for 3 weeks and then placed at 48° F (9° C) for 8 weeks. Later flowering will be produced by holding bulbs at 72° F (22° C) for 10 weeks. For shipment to the Southern Hemisphere, the bulbs can be held at 31° F (—1° C) until late December, when they are shifted to a higher temperature (78° F ; 25.5° C).

(*b*) *Daffodil* : Daffodil bulbs are perennial and produce a new growing point at the center every year. Offsets are produced which grow in size for several years until they break away from the original bulb, although they are still attached at the basal plate. An offset bulb, when it first separates from the mother bulb, is known as a "split," "spoon," or "slab," and can be separated from the mother bulb and planted. Within a year it become a "round," or "single-nose," bulb containing a single flower bud. One year later a new offset should be visible, enclosed within the scales of the original bulb, indicating the presence of two flower buds. At this stage the bulb is known as a "double-nose.". By the next year the offsets split away, and the bulb is known as a "mother bulb." Grading of daffodil bulbs is principally by age, that is, as splits, round, double-nose, and mother bulbs. The grades marketed commercially are the round and the double-nose. The mother bulbs are used as planting stock to produce additional offsets, and only the surplus is marketed. Offsets, or splits, are replanted for additional growth.

Storage should be at 55° to 60° F (13° to 16° C) with a relative humidity of 75 per cent. For earlier flowering, they can be stored at 48° F (9° C) for 8 weeks. To delay flowering, store at 72° F (22° C) for 10 weeks. For shipment from the Northern to the Southern Hemisphere, bulbs can be held at

86° F (30° C) until October, then stored at 31° F (—1° C) until late December, and then at 77° F (25° C).

Hot water treatment plus a fungicide, as formalin, for stem and bulb eelworm control is important. A three- or four-hour treatment at 110° F (43° C) is used, but that temperature must be carefully maintained or the bulbs may be damaged.

(*c*) *Lilies*. Lilies increase naturally, but except for a few species this increase is slow and of limited propagation value except in home gardens. Several methods of bulb increase are found among the different species. For instance, *Lilium concolor*, *L. hansoni*, *L. henryi*, and *L. regale* increase by bulb splitting. Two to four lateral bulblets are initiated about the base of the mother bulb, which disintegrates during the process, leaving a tight cluster of new bulbs. *Lilium bulbiferum*, *L. canadense*, *L. paradalinum*, *L. parryi*, *L. superbum*, and *L. tigninum* multiply from lateral bulblets produced from the rhizome-like bulb. This is sometimes called "budding-off."

Bulblet Formation on Stems

The production of *underground stem bulblets* is the usual commercial method of propagating the Easter lily (*Lilium longiflorum*) and some other lily species. Flowering of the Easter lily occurs in early summer. Bulblets form and increase in size from spring through summer. Between mid-August and mid-September in the Northern Hemisphere the stems are pulled from the bulbs and stacked upright in the field. Periodic sprinkling keeps the stems and bulblets from drying out. Similarly, the base of the stem can be "heeled in" the ground at an angle of 30 to 45 deg. or laid horizontally in trays at high humidity.

About mid-October the bulblets are planted in the field 4 in. deep and an inch apart in double rows which are spaced 36 in. apart. Here they remain for the following season. They are dug in September as yearling bulbs and again replanted, this time 6 in. deep and 4 to 6 in. apart in single rows. At the end of the second year, they are dug and sold as commercial bulbs.

Digging is done in September after the stem is pulled. The bulbs are graded, packed in peat moss, and shipped. Commercial bulbs have a minimum size of 7 to 10 in. in circumference. Lily bulbs must be handled carefully so they will not be injured and must be kept from drying out. The fleshy roots should also be kept in good condition. For long-term storage that will prevent flowering, the bulbs should be packed in polyethylene-lined cases containing peat moss at 30 to 50 per cent moisture and stored at 31° F (—1° C). To produce flowering in the shortest period of time, they should be stored at 45° to 50° F (7° to 10° C) for 6 weeks and then grown at temperatures of 60° F (16° C) (night) or more.

Lilies are attacked by certain viruses, fungus diseases, and nematodes which can be carried in or on the bulbs. Their control during propagation is an important consideration in bulb production. Methods of control include using disease-free stocks for propagation, growing plants in disease-free locations with good sanitary procedures, and treating the bulbs with fungicides. A pre-planting dip in 2 lb of pentachloronitrobenzene (PCNB) plus 2 lb of ferbam in 100 gal. of water has been found useful.

To obtain commercial planting stock of lilies free of *Rhizoctonia*, stem and bulb nematode, and root-lesion nematode, cure bulbs at 05° F (95° C) and 95 per cent relative humidity for 1 to 2 weeks, presoak for 2 days in cool water, soak for 2 hours at 115° F (46° C) in water plus 37 per cent formaldehyde diluted 1 : 200 ; aftersoak the bulbs in a fungicide such as Puratized Agricultural Spray at 1 : 1000. Scale the bulbs, dust the dulblets with ferbam, and place in vermiculite for bulblets to form. Plant new bulblets in treated soil.

Aerial stem bulblets, commonly known as *bulbils*, are formed in the axil of the leaves of some lily species, such as *Lilium bulbiferum*, *L. sargentiae*, *L. sulphureum*, and *L. tigrinum*. Bulbils develop in the early part of the season and fall to the ground several weeks after the plant flowers. They are harvested shortly before they fall naturally and are then handled in essentially the same manner as underground stem bulblets. Increased bulbil production can be induced by disbudding as soon as the flower buds have formed. Likewise, some lily species which do not form bulbils naturally can be induced to do so by pinching out the flower buds and a week later cutting off the upper half of the stems. Species which respond to the latter procedure include *Lilium candidum*, *L. chalcedonicum*, *L. hollandicum*, *L. maculatum* and *L. testaceum*.

Stem Cuttings

Lilies may be propagated as stem cuttings. The cutting is made shortly after flowering. Instead of roots and shoots forming on the cutting, as would occur in other plants, bulblets form at the axils of the leaves and then produce roots and small shoots while on the cutting.

Leaf-bud cuttings, made with a single leaf and a small heel of the old stem, may be used to propagate a number of lily species. A small bulblet will develop in the axil of the leaf. It is handled in the same manner as for the other methods described here.

Bulblet Formation on Scales (Scaling)

An efficient method for producing lily bulbs is the technique known as *scaling*, in which individual bulb scales are separated from the mother bulb and placed in growing conditions so that adventitious bulblets form at the base of each scale. Three to five bulblets will usually develop from each scale. This method is particularly useful for rapidly building up stocks of a new variety or to establish disease-free stocks. Almost any lily species can be propagated by scaling.

Scaling is usually done soon after flowering in midsummer, although it might be done in late fall or even in midwinter. The bulbs are dug, the outer two layers of scales are removed and the mother bulb is replanted for continued growth. It is possible to remove the scales down to the core, but this will reduce subsequent growth of the mother bulb. The scales should be kept from drying and handled so as to avoid injury. Scales with evidence of rot should be discarded and the remaining ones dusted with a fungicide. Naphthaleneacetic acid will stimulate bulblet formation. Mix 1000 parts of thiram or ferbium with 1 part of naphthalene acetic acid and apply as a dust. The PCNB-thiram mixture is also an effective fungicidal treatment.

Scales can be planted directly in the field in open beds or frames, planted no more than 2½ in. deep. Somewhat better results are obtained if they are placed in trays or flats of moist sand, peat moss, sphagnum moss, or vermiculite for 6 weeks at 65° to 70° F (18° to 21° C). The scales are inserted vertically to about half their length. Small bulblets and roots should form at the base within 3 to 6 weeks. The scales are transplanted either into the open ground or into pots or flats of soil, and then planted in the field the following spring. Subsequent treatment is the same as described for underground bulblets.

Basal Cuttage

The hyacinth is the principal plant propagated by this method, although others such as *Scilla* can be handled in this way. Specific methods include "scooping," "scoring," and "coring". Mature bulbs which have been dug after the foliage has died down and are 17 to 18 cm or more in circumference are used. In *scooping*, the entire basal plate is scooped out with a special curve-bladed scalpel, a round-bowled spoon, or a small-bladed knife. Adventitious bulblets develop from the base of the exposed bulb scales. Depth of cutting should be enough to destroy the main shoot. In *scoring*, three straight knife cuts are made across the base of the bulb, each deep enough to go through the basal plate and the growing point. Growing points in the axils of the bulb scales grow into bulblets. In *coring*, the growing point in the center of the bulb is removed entirely with an apple corer or a cork borer 3/8 to ½ in. in diameter. All of the growth potential is concentrated in the development of the bulblets which grow from the basal plate.

To combat decay that may develop during the later incubation period, any infected bulbs should be discarded ; the tools disinfected frequently with an alcohol, formalin, or mild carbolic acid solution, and the cut bulbs dusted with a fungicide. Most important is to callus the bulbs at about 70° F (21° C) for a few days to a few weeks in dry sand or soil or in open trays, cut side up. After callusing, the bulbs are incubated in trays or flats, in dark or diffuse light, at 70° to 90° F (21° to 32° C) and high humidity for 2½ to 3 months.

The mother bulbs are planted about 4 in. deep in nursery beds in the fall. The next spring bulblets produce leaves profusely. Normally the mother bulb disintegrates during the first summer. Annual digging and replanting of the graded bulblets is required until they reach flowering sizes. Bulbs for greenhouse forcing should be 17 cm (6¾ in.) or more in circumference ; bulbs for bedding should be 14 to 17 cm (5½ to 6¾ in.) in circumference. On the average, a scooped bulb will produce 60 bulblets, but 4 to 5 years will be required to produce flowering sizes ; a scored bulb will produce 24, requiring 3 to 4 years ; and a cored bulb will produce 10, requiring 2 to 3 years.

Leaf Cuttings

This method has been reported successful for blood lily (*Haemanthus*), grape hyacinth (*Muscari*), hyacinth, and *Lachenalia* although the range of species is probably wider.

Leaves are taken at a time when they are well developed and green, generally near full bloom. An entire leaf is cut from the top of the bulb and may in turn be cut into two or three pieces. Each section is placed in a rooting medium with the basal end several inches below the surface, as described for rooting cuttings. The leaves should not be allowed to dry out, and bottom heat is desirable. Within 2 to 4 weeks small bulblets form on the base of the leaf and roots develop. At this stage the bulblets are planted in soil.

Bulb Cuttings

Among those plants which have been reported to respond to this method of propagation are the *Albuca*, *Chasmanthe*,

Cooperia, *Haemanthus*, *Hippeastrum*, *Hymenocallis*, *Lycoris*, *Narcissus*, *Nerine*, *Pancratium*, *Scilla*, *Sprekelia*, and *Urceolina*.

A mature bulb is cut into a series of 8 to 10 vertical sections, each containing a part of the basal plate. These sections are further divided by sliding a knife down between each third or fourth pair of concentric scale rings and cutting through the basal plate. Each of these fractions makes up a bulb cutting consisting of a piece or basal plate and segments of three or four scales.

The bulb cuttings are planted vertically in a rooting medium, such as peat moss and sand, with just their tips showing above the surface. The subsequent technique of handling is the same as for ordinary leaf cuttings. A moderately warm temperature, slightly higher than for mature bulbs of that kind, is required. New bulblets develop from the basal plate between the bulb scales within a matter of a few weeks, along with new roots. At this time they are transferred to flats of soil to continue development.

CORMS

Definition and Structure

A *corm* is the swollen base of a stem axis enclosed by the dry, scale-like leaves. In contrast to the bulb, which is predominantly leaf scales, a corm is a solid stem structure with distinct nodes and internodes. The bulk of the corm consists of storage tissue composed of parenchyma cells. In the mature corm, the dry leaf bases persist at each of these nodes and enclose the corm. This covering, known as the *tunic*, protects it against injury and water loss. At the apex of the corm is a terminal shoot bud which will develop into the leaves and the flowering shoot. Axillary buds are produced at each of the nodes. In a large corm, several of the upper buds may develop into flowering shoots, but those nearer the base of the corm are generally inhibited from growing. However, should something prevent the main buds from growing, these lateral buds would be capable of producing a shoot. (Fig. 11.4).

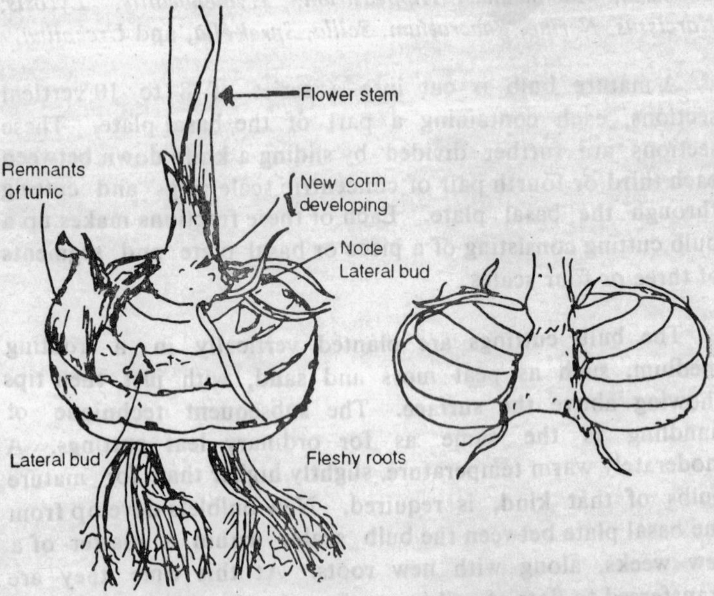

Fig. 11.4. Gladiolus corm. Left : External appearance, Right : Longitudinai section showing solid stem structure.

Two types of roots are produced from the corm : a fibrous root system developing from the base of the old corm and enlarged, fleshy contractile roots developing from the base of the new corm.

GROWTH BEHAVIOUR

Gladious and crocus are typical cormous plants. The gladiolus is a semi-hardy to tender plant which, in areas with severe winters, must be stored over winter and replanted in the spring. At the time of planting, the corm is a vegetative structure. New roots develop from the base of the corm, and one or more of the buds begin to develop leaves. Differentiation of the inflorescence takes place within a few weeks after the bud begins to grow. At the same time the base of the shoot axis thickens, and a new corm for the succeeding year begins to form above the old corm. Root-like structures bearing miniature corms or *cormels* on their tip develop from the base of the new corm. The new corm continues to enlarge, and the old corm begins to shrivel and disintegrate as its contents are utilized in flower production. After flowering, the foliage continues to manufacture food materials, which are stored in the new corm. At the end of the summer, when the foliage dries, there are one or more new corms and perhaps a great number of little cormels. The corms are dug and stored over winter until they are planted the following spring.

PROPAGATION

New Corms

Propagation of cormous plants is principally by the natural increase of new corms. Flower production in corms, as in bulbs, depends upon food materials stored in the corm the previous season, particularly during the period following bloom. In gladiolus, cool nights and long growing periods are favourable for production of very large corms. Fertilization and other good management practices during bloom have their greatest effect on the next year's flowers. Plants are left in the ground for 2 months following blooming, or until frost kills the tops. After digging, the plants are placed in trays with a screen or slat bottom arranged to allow air to circulate between them, and cured at 95°F (35°C) at 80 to 85 per cent relative humidity. Then the new corms, old corms, cormels, and tops can be easily separated. The corms are graded according to size, sorted to remove the diseased ones, treated with a fungicide (Spergon wettable powder), and returned to a 95°F temperature for an additional week. This process suberizes the wounds and helps combat *Fusarium* infection. The corms are then stored at 40°F (5°C) with a relative humidity of 70 to 80 per cent in well-aerated rooms to prevent excessive drying. It is also desirable to treat them immediately before planting with ½ per cent Lysol solution for 2 hours or with ⅛ per cent solution of New Improved Ceresan for 15 minutes.

Cormels

These are miniature corms which develop between the old and the new corms. One or 2 years growth is required for them to reach flowering size. Shallow planting of the corms, only a few inches deep, results in greater production of cormels ; increasing the depth of planting reduces cormel production.

Cormels are separated from the mother corms and stored over winter for planting in the spring. Dry cormels become very hard and may be slow to start growth the following spring, but if they are stored at about 40°F (5°C) in slightly moist peat moss, they will stay plump and in good condition. Soaking dry cormels in warm water for 2 to 5 days just prior to planting will hasten the onset of growth.

Disease-free cormels can be obtained by hot-water treatment. Treatment should be done between 2 and 4 months after digging. The cormels are soaked in water at air temperature for 2 days, then placed in a 1 : 200 dilution of commercial 37 per cent formaldehyde for 4 hours, and then immersed in a water bath at 135°F (57°C) for 30 minutes. The temperature should be maintained within 1°F, plus or minus, of this temperature. At the end of the treatment, the cormels are cooled quickly, dried immediately, and stored at 40°F (5°C) in a clean area with good air circulation. Dusting the dry cormels with a fungicide at this time is also desirable.

The cormels are planted in the field in furrows about 2 in. deep in the manner of planting large seeds. Only grass-like foliage is produced the first season. The cormel does not increase in size but produces a new corm from the base of the stem axis, in the manner described for full-sized corms. At the end of the first growing season, the beds are dug and the corms separated by size. A few of the corms may attain flowering size, but most require an additional year of growth.

Size grades in gladiolus are determined by diameter. There are seven grades, the smallest 3/8 to ½ in. in diameter, the largest 2 in. or more.

Division of the Corm

Large corms may be cut into sections, retaining a bud with each section. Each of these should then develop a new corm. Segments should be dusted with a fungicide because of the great likelihood of decay of the exposed surfaces.

TUBERS

Definition and Structure

A *tuber* is a modified stem structure which develops below ground as a consequence of the swelling of the subapical portion of a stolon and subsequent accumulation of reserve materials.

The most notable example of a plant which produces tubers by which it is propagated is the Irish potato (*Solanum tuberosum*). (See Figure 11.5). The *Caladium*, grown for its striking foliage, is also propagated by tubers, as is the Jerusalem artichoke (*Helianthus tuberosus*).

A tuber has all the parts of a typical stem. The "eyes," present in regular order over the surface, represent nodes, each consisting of one or more small buds subtended by a leaf scar. The arrangement of the nodes is a spiral, beginning with the terminal bud on the end opposite the scar resulting from the attachment to the stolon. The terminal bud is at the apical end of the tuber, oriented farthest (distally) from the

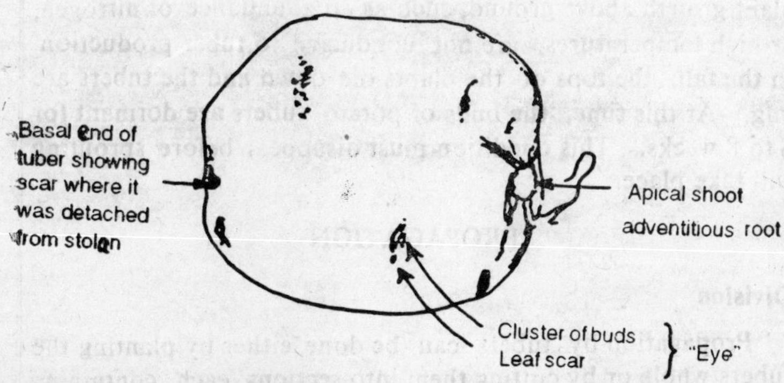

Fig. 11.5. Stem tuber of potato. Note nodes ("eyes") and internodes.

crown of the plant. Consequently it shows the same apical dominance as any stem.

Growth Behaviour

A tuber is a storage and propagative organ that is produced in one growing season, remains dormant during the winter, and then produces new shoots the following spring to start a new cycle. Most information on growth cycles in tubers is from studies of the potato. At the base of the main shoot adventitious roots are initiated, and lateral buds grow out horizontally to become the stolons. Continued growth of the stolon takes place during long photoperiods and is associated with the presence of auxin and a high gibberellin level. Tuberization begins with inhibition of terminal growth and the initiation of cell enlargement and division in the subapical region of the stolons. This stage is associated with short or intermediate day lengths, reduced temperatures (particularly at night), good light, low mineral content, and a reduction in gibberellin levels in the plant.

Tuberization has been explained as being caused by the production of a tuber-inducing substance (perhaps a gibberellin inhibitor) that is produced in the leaves and the mother tuber. It seems to be necessary for the stolon tip to have attained a particular physiological age. Continued tuber enlargement is dependent on a continuing adequate supply of photosynthate. Conditions which favour rapid and luxurious plant growth above ground, such as an abundance of nitrogen, or high temperatures, are not conducive to tuber production. In the fall, the tops of the plants die down and the tubers are dug. At this time, the buds of potato tubers are dormant for 6 to 8 weeks. This condition must disappear before sprouting will take place.

PROPAGATION

Division

Propagation by tubers can be done either by planting the tubers whole or by cutting them into sections, each containing a bud or eye. These small pieces of tuber to be used for propagation of the potato are commonly referred to as "seed." The weight of the tuber piece should be 1 to 2 oz to provide sufficient stored food for the new plant to become well established.

Division of tubers is done with a sharp knife shortly before planting. The cut pieces should be stored at warm (68°F; 20°C) temperatures and relatively high humidities (90 per cent) for 2 to 3 days prior to planting. During this time, the cut surfaces heal (*suberization*) and the "seed" piece is effectively protected against drying and decay. Treatment of potato tubers prior to cutting for the control of *Rhizoctonia* and scab may be desirable. Caladium tubers are produced commercially in Florida. The tubers are cut into sections, usually two buds per piece. These are planted 3 to 4 in. deep, 4 to 6 in. apart in rows 18 to 24 in. apart. Harvest begins in November. After harvest, the tubers are dried in open sheds for 6 weeks or artificially dried for 48 hours. Further storage should be at a temperature not below 60°F (16°C).

TUBERCLES

Begonia evansiana and the cinnamon vine (*Dioscorea batatas*) produce small aerial tubers, known as *tubercles*, in the axils of the leaves. These tubercles may be removed in the fall, stored over winter, and planted in the spring.

TUBEROUS ROOTS

Definition and Structure

Certain herbaceous perennials produce thickened tuberous roots which contain large amounts of stored food. Although the appearance of such roots may vary considerably from one species to another, they have the internal and external features of a typical root. Thus they differ from the true (stem) tuber in that they lack nodes and internodes, buds are present only at the crown or stem (proximal) end, fibrous roots are commonly produced toward the opposite (distal) end. The polarity of the tuberous root is consequently the reverse of that of the true tuber.

Growth Behaviour

Three typical examples of these structures are shown in Figure 11.6. The sweet potato and dahlia produce swollen sections on lateral roots. In the latter case, these are borne in a cluster, each tuberous root attached to the crown of the

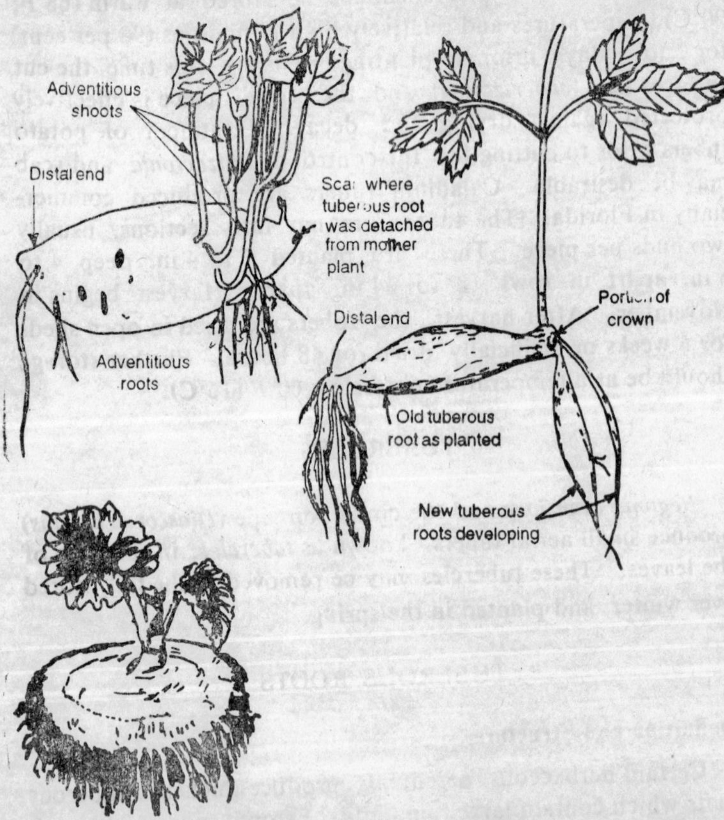

Fig. 11.6. Types of tuberous roots. *Top left* : Sweet potato showing adventitious shoots. *Top right* : Dahlia during early stages of growth. The old root piece will disintegrate in the production of the new plant; the new roots can be used for propagation. *Below left* : Tuberous begonia, showing its vertical orientation. This type continues to enlarge each year.

plant. These roots are biennial. They are produced in one season, after which they go dormant as the herbaceous shoots die. The following spring, buds from the crown produce new shoots which utilize the food materials from the old root during their initial growth. The old root then disintegrates, and new roots are produced, which in turn maintain the plant through the following dormant period.

In the tuberous begonia, on the other hand, the primary taproot becomes a single enlarged tuberous root. Buds are produced at the proximal end (the crown). Fibrous roots are produced from the distal portion of the swollen root. This tuberous root is perennial and lives for a number of years, continuing to increase in size and produce additional buds from the crown. In the tuberous begonia reduced day length results in the enlargement of the tuberous root and the cessation of flowering.

PROPAGATION

Adventitious Shoots

The fleshy roots of a few species of plants such as sweet potato have the capacity to produce adventitious shoots if subjected to the proper conditions of temperature and moisture. The roots are laid in sand so that they do not touch one another and covered to a depth of about 2 in. The bed is kept moist. The temperature should be about 80°F (27°C) at the beginning and about 70° to 75°F (21° to 24°C) after sprouting has started. As the new shoots, or *slips*, come through the covering, more sand is added so that eventually the stems will be covered for 4 to 5 in. Adventitious roots develop from the base of these adventitious shoots. After the slips are well rooted, they are pulled from the parent plant and transplanted into the field. An increase in slip production takes place if sweet potato roots are cut in half and the pieces are subjected to 110°F (43°C) for about 26 hours. This treatment overcomes the apical dominance and also controls nematodes and fungus diseases.

Division

Most plants with fleshy roots must be propagated by dividing the crown so that each section bears a shoot bud. This is necessary in the dahlia, for example. The plant is dug with its cluster of roots intact, dried for a few days, and stored at 40° to 50°F (4° to 10°C) in dry sawdust, dry peat, or dry shavings. Storage in the open may result in shriveling. The cluster of roots is divided in the late winter or spring shortly before planting. In warm, moist conditions the buds begin to grow, and division can be carried out with better assurance that each section will have a bud.

The perennial tuberous root of the tuberous begonia can also be divided as long as each section has a bud. To combat decay, the cut surface should be dusted with a fungicide and each section dried for several days after cutting and before placing in a moist medium.

Vegetative propagation in such plants as these can often be better carried out with stem, leaf, or leaf-bud cuttings. The cuttings will develop tuberous roots at their base. This process can be stimulated if the stem cutting initially includes a small piece of the fleshy root.

RHIZOMES

Structure

A *rhizome* is a specialized stem structure in which the main axis of the plant grows horizontally just below or on the surface of the ground. A number of economically important plants, such as bamboo, sugar cane, banana and many grasses, as well as a number of ornamentals, such as *Iris* and lily-of-the-valley, have rhizome structures. Most are monocotyledons, although a few dicotyledons—for example, low bush blueberry (*Vaccinium angustifolium*)—have analogous underground stems classed as rhizomes. Many ferns and lower plant groups have rhizomes or rhizome-like structures.

Figure 11.7 shows structural features of a rhizome. The stem appears segmented because it is composed of nodes and

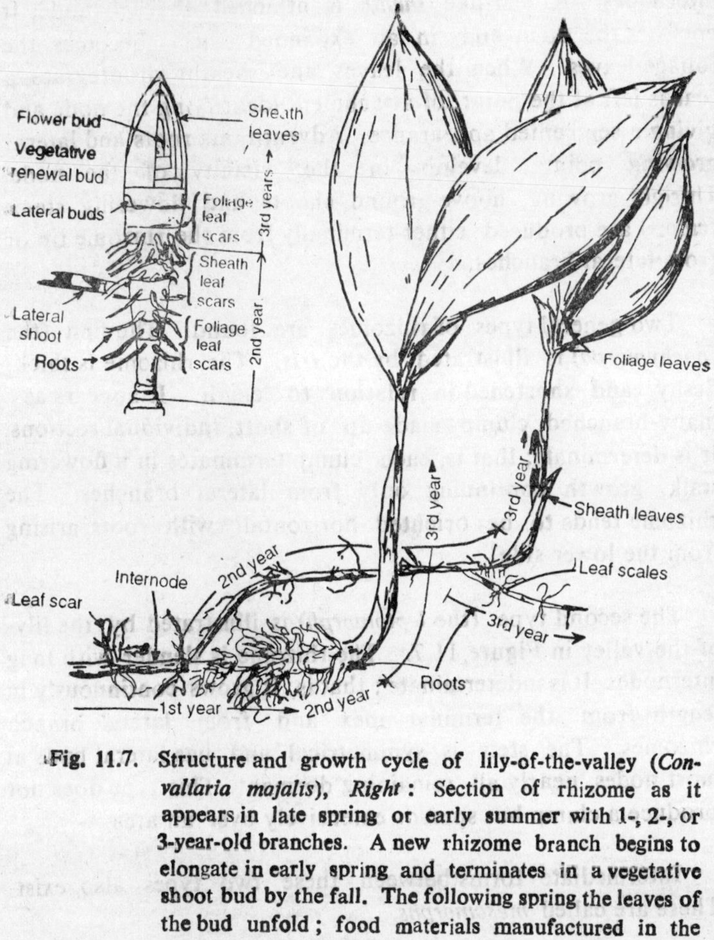

Fig. 11.7. **Structure and growth cycle of lily-of-the-valley (*Convallaria majalis*), *Right* : Section of rhizome as it appears in late spring or early summer with 1-, 2-, or 3-year-old branches. A new rhizome branch begins to elongate in early spring and terminates in a vegetative shoot bud by the fall. The following spring the leaves of the bud unfold ; food materials manufactured in the leaves by photo-synthesis are accumulated in the rhizome. Growth the second season is again vegetative. Early in the third season a flower bud begins to form, and at the same time a vegetative growing point forms in the axil of the last leaf. *Top left* : Section of the 3-year-old branch showing terminal flower bud and lateral shoot bud enclosed in leaf sheaths. Such a section is sometimes known as a *pip* or *crown* and is forced for spring bloom. In the early spring the flowering shoot expands, blooms, and then dies down, the shoot bud beginning a new cycle of development Redrawn from Zwee de.**

internodes. A leaf-like *sheath* is attached at each node; it encloses the stem and, in an expanded form, becomes the foliage leaves. When the leaves and sheaths disintegrate, a scar is left at the point of attachment identifying the node and giving a segmented appearance. Adventitious roots and lateral growing points develop in the vicinity of the node. Upright-growing, above-ground shoots and flowering stems (*culms*) are produced either terminally from the rhizome tip or from lateral branches.

Two general types of rhizomes are found. The first (the (*pachymorph*) is illustrated by the *Iris*. The rhizome is thick, fleshy, and shortened in relation to length. It appears as a many-branched clump made up of short, individual sections. It is determinate; that is, each clump terminates in a flowering stalk, growth continuing only from lateral branches. The rhizome tends to be oriented horizontally with roots arising from the lower side.

The second type (the *leptomorph*) is illustrated by the lily-of-the-valley in Figure 11 7. The rhizome is slender with long internodes. It is indeterminate ; that is, it grows continuously in length from the terminal apex and from lateral branch rhizomes. The stem is symmetrical and has lateral buds at most nodes, nearly all remaining dormant. This type does not produce a clump but spreads extensively over an area.

Intermediate forms between these two types also exist. These are called *mesomorphs.*

Growth Behaviour

Rhizomes grow by elongation of the growing points produced at the terminal end and on lateral branches. Length also increases by growth in the intercalary meristems in the lower part of the internodes. As the plant continues to grow and the older part dies, the several branches arising from one plant may eventually become separated to form individual plants of a single clone.

Rhizomes exhibit consecutive vegetative and reproductive stages, but the growth cycle differs somewhat in the two types described. In the pachymorph rhizome of *Iris*, a growth cycle begins with the initiation and growth of a lateral branch on a flowering section. The flowering stalk dies, but these new lateral branches produce leaves and grow vegetatively during the remainder of that season. Continued growth of the underground stem, storage of food, and the production of a good flower bud at the conclusion of the vegetative period depend upon photosynthesis in the leaves. Consequently foliage should not be removed during this period. A flowering stalk is produced the following spring and no further terminal growth can take place. In general, plants with this structure flower in the spring and grow vegetatively during the summer and fall.

Plants with a leptomorph habit as a general rule (with exceptions) grow vegetatively during the beginning of the growth period and flower later in the same period. The length of time during which an individual rhizome remains vegetative varies with different kinds of plants. An individual branch in the lily-of-the-valley in Figure 11.7 for instance, is vegetative three years before a flower bud forms. Some bamboo species remain vegetative for many years, but then they change abruptly and the entire plant produces flowers.

PROPAGATION

Division of Clumps

Clump division is the usual procedure for propagating many plants from rhizomes, but the procedure may vary somewhat with the two types. In pachymorph rhizomes, individual sections (or culms) are cut off at the point of attachment to the rhizome, the top is cut back, and the piece is transplanted to the new location. Leptomorph rhizomes can be handled in essentially the same way by removing a single lateral "offshoot" from the rhizome and transplanting it. The tip of the lily-of-the-valley rhizome bearing a flower bud (see Fig. 11.7) called a "pip" is removed along with the rooted section below and transplanted.

Division is usually carried out at the beginning of a growth period (as in early spring) or at or near the end of a growth period (*i.e.*, in late summer or fall).

Division of Rhizomes

Dividing a rhizome is done by cutting the rhizome into sections, being sure that each piece has at least one lateral bud, or "eye," it is essentially a stem cutting. Bananas, for instance, are propagated in this way. This general method works well for the leptomorph rhizomes, in which a dormant lateral growing point is present at most nodes. The rhizomes are cut or broken into pieces, and adventitious roots and new shoots develop from the nodes. Rhizome producing turf grasses, for instance, are cut up into sections and the individual "sprigs" transplanted, New plants can be established readily by this method.

Culm Cuttings

In large rhizome-bearing plants, such as bamboos, the aerial shoot, or culm, may be used as a cutting. These may be whole culm cuttings, in which the entire aerial shoot is laid horizontally in a trench. New branches arise at the nodes. A more common alternative procedure is to cut the stem into three- or four-node sections and place them vertically in the ground as for an ordinary stem cutting.

PSEUDOBULBS

Definition and Structure

A *pseudobulb* (literally "false bulb") is a specialized storage structure produced by many orchid species, consisting of an enlarged, fleshy section of the stem made up of one to several nodes. In general appearance the psendobulb varies with different species of orchids, the differences being sufficiently characteristic to aid in identification.

Growth Behaviour

These pseudobulbs arise during the growing season on upright growths which develop laterally or terminally from the horizontal rhizome. Leaves and flowers from either at the terminal end or at the base of the pseudobulb, depending upon the species. During the growth period, they accumulate stored food materials and water and assist the plants in surviving the subsequent dormant period.

PROPAGATION

Offshoots

In a few orchids, such as the *Dentrobium* species, the pseudobulb is long and jointed, being made up of many nodes. Offshoots develop at these nodes. From the base of these offshoots roots develop. The rooted offshoots are then cut from the parent plant and potted.

Division

Most important commercial species of orchids, including the *Cattleya*, *Laelia*, *Miltonia*, and *Odontoglossum*, may be propagated by dividing the rhizome into sections, the exact procedure used depending upon the particular kind of orchid. Division is done during the dormant season, preferably just before the beginning of a new period of growth. The rhizome is cut with a sharp knife back far enough from the terminal end to include four to five pseudobulbs in the new section, leaving the old rhizome section with a number of old pseudobulbs, or "back bulbs," from which the leaves have dehisced. The section is then potted, whereupon growth begins from the bases of the pseubobulbs and at the nodes. The removal of the new section of the rhizome from the old part stimulates new growth, or "back breaks," to occur from the old parts of the rhizome. These new growths grow for a season and can be removed the following year.

An alternate procedure is to cut partly through the rhizome and leave it for 1 year. New back breaks will develop which can later be removed and potted.

Back Bulbs and Green Bulbs

"Back bulbs" (*i.e.*, those without foliage) are commonly used to propagate clones of *Cymbidium*. These are removed from the plant, the cut surface is painted with a grafting compound, and they are placed in a rooting medium for new shoots to develop. The back bulb can be repropagated and a second shoot developed from it.

"Green bulbs" (*i.e.*, those with leaves) can also be used in *Cymbidium* propagation. Treatment with indolebutyric acid, either by soaking or by painting with a paste, has been shown to be beneficial.

●●